CISM Courses and Lectures

The series presents lecture notes, monographs, edited works and proceedings in the field of Mechanics, Engineering, Computer Science and Applied Mathematics.
Purpose of the series is to make known in the international scientific and technical community results obtained in some of the activities organized by CISM, the International Centre for Mechanical Sciences.

International Centre for Mechanical Sciences

Courses and Lectures Vol. 559

For further volumes:
www.springer.com/series/76

Luis Dorfmann · Raymond W. Ogden
Editors

Nonlinear Mechanics of Soft Fibrous Materials

Springer

Editors

Luis Dorfmann
Tufts University, Medford, MA, USA

Raymond W. Ogden
University of Glasgow, UK

ISSN 0254-1971
ISBN 978-3-7091-1837-5 ISBN 978-3-7091-1838-2 (eBook)
DOI 10.1007/978-3-7091-1838-2
Springer Wien Heidelberg New York Dordrecht London

All contributions have been typeset by the authors
Printed in Italy

Printed on acid-free paper

Springer is part of Springer Science+Business Media (www.springer.com)

PREFACE

This volume consists of Lecture Notes based on lectures delivered at the Advanced School on the "Nonlinear Mechanics of Soft Fibrous Materials" held at the International Centre for Mechanical Sciences (CISM) in Udine, Italy, during July 1-5, 2013. The course was presented by 6 lecturers, from Canada, Ireland, Italy, the UK and the USA (2).

The objective of the six chapters herein is to provide a state-of-the-art overview of the nonlinear continuum theory of soft fibrous materials and its applications, ranging from the microscopic to the macroscopic level and including reference to rubberlike solids and soft biological tissues. The chapters include mathematical and computational aspects of the modelling of these materials with particular reference to the constitutive laws that describe the macroscopic properties of the materials.

The first chapter sets the scene by providing the necessary theoretical background for analyzing the properties of fibre-reinforced materials based on fibre directions (so-called preferred directions) and the associated structure tensors, from which are constructed deformation invariants that form the basis of the constitutive description of the elastic behaviour of the materials via a strain-energy function. The notion of the structure tensor is then used to develop the constitutive law for a material with residual stress, which is closely allied to that for fibre-reinforced materials. The theory is used to solve the problem of a residually-stressed circular cylindrical tube under extension and inflation with and without fibre reinforcement.

The second chapter, following some mathematical preliminaries, analyzes the effect of statistically oriented fibres, with particular application to fibre-reinforced porous materials and their finite element implementation. The statistical mechanics of polymeric chains and their networks is the subject of the next chapter, together with a discussion of single fibre mechanics within a continuum setting and the problems of fitting theory to experimental data.

The fourth chapter focuses on an application to a representative example of a soft tissue, that of caterpillar muscle, which is a fibre-reinforced materials with the (muscle) fibres essentially aligned in a single direction. Experimental data are presented therein and, based on these, a pseudoelastic constitutive model is developed that accounts for the observed material response. The model is then adapted for implementation within a finite element framework and applied to the simulation of a bio-actuated micro-pump.

The theory of nonlinear anisotropic elasticity is developed in the fifth chapter and used to describe small-amplitude motions and static deformations superimposed on large pre-strains of fibre-reinforced solids. The equations are solved and illustrate the distinct effects of anisotropy. Particular attention is paid to the propagation of homogeneous plane waves and to the appearance of wrinkling deformations.

The final chapter is concerned with developing a model of nonlinearly elastic solids reinforced by continuously distributed embedded fibres with the elastic resistance of the fibres to extension, bending and twist accounted for.

The combination of these chapters provides a unique perspective on the subject of soft fibre-reinforced materials that is not available elsewhere in a single work.

We have pleasure in thanking our colleagues, Professors Michel Destrade, Salvatore Federico, Giuseppe Saccomandi and David Steigmann, for presenting their lectures and for preparing chapters for this volume and the students for attending the lectures and contributing to the discussions.

We particularly thank the Rector and officers at CISM for their encouragement, enthusiasm, assistance and hospitality, which helped towards the smooth running of the School. We are also grateful to Professor Paolo Serafini, Executive Editor of CISM, for his encouragement to publish these lecture notes.

Luis Dorfmann
Ray Ogden

CONTENTS

Nonlinear Elasticity with Applications to Soft Fibre-reinforced Materials
by R.W Ogden 1

Porous Materials with Statistically Oriented Reinforcing Fibres
by S. Federico 49

Nonlinear Elasticity for Soft Fibrous Materials
by G. Saccomandi 121

Modeling of Bioactive Materials
by L. Dorfmann and C. Paetsch 175

Incremental Equations for Soft Fibrous Materials
by M. Destrade 233

Effects of Fibre Bending and Twisting Resistance on the Mechanics of Fibre-reinforced Elastomers
by D.J. Steigmann 269

Nonlinear Elasticity with Application to Soft Fibre-reinforced Materials

Ray W. Ogden
School of Mathematics and Statistics, University of Glasgow, UK
E-mail: raymond.ogden@glasgow.ac.uk

Abstract. There are many fibrous materials, both natural and man made, ranging from hard material such as wood and carbon fibre composites to relatively soft materials such as rubber and softer materials such as biological tissues. With particular reference to fibre-reinforced soft materials, this chapter provides the theoretical background for analyzing the elastic properties of such materials based on nonlinear continuum mechanics through so-called structure tensors from which are constructed the deformation invariants that form the independent variables in the elastic constitutive laws. The theory is applied to some simple deformations by way of illustration of the material response. The notion of structure tensor is then used to incorporate residual stresses into the constitutive law and the important effects of residual stresses are illustrated in a prototype boundary-value problem related to the elastic behaviour of arteries.

1 Introduction

Fibrous materials are important both in nature and in manufacturing. Examples of fibrous materials include hard materials such as wood, both naturally occurring and manufactured wood board, and carbon fibre composites, which are used for an enormous range of products, including major components of the new generation of commercial airliners, examples being the Airbus 350 and the Boeing Dreamliner. Then there are relatively softer materials such as vulcanized rubber as used in the composite structure of car tyres, for example. Softer still is natural rubber and its elastomeric offshoots including magneto-sensitive and electro-sensitive materials used, for example, as activators and sensors. Much softer are biological tissues such as artery walls, the heart and skin, in which the protein collagen is an important structural component, contributing to the fibrous structure

L. Dorfmann, R. W. Ogden (Eds.), *Nonlinear Mechanics of Soft Fibrous Materials*,
CISM International Centre for Mechanical Sciences
DOI 10.1007/978-3-7091-1838-2_1 © CISM Udine 2015

of tissues, along with, depending on the tissue in question, elastin fibres, smooth muscle cells and myocytes. At the cellular level, the cytoskeleton (the skeleton of a cell) contains fibrous material such as a network of actin fibres, fibrous actin bundles, intermediate filaments and microtubules which are all important for the mechanical functioning of cells.

Thus, fibrous materials are abundant both in nature and in man-made products and it is important to be able to understand the influence of the fibre structure on the overall mechanical response of the composite materials of which the fibres are constituents. The mechanical behaviour of all such materials can be treated within the same framework from the macroscopic perspective, that being the framework of continuum mechanics. In particular, we are concerned here with the nonlinear elasticity theory that underpins the basic elastic response of fibrous materials and is a starting point for more general continuum theories. It is important to understand the theory in order to first of all describe the elastic behaviour of these materials through constitutive laws informed by data from experimental tests and secondly to predict their behaviour for situations in which experimental testing is not feasible. Fibres endow the material locally with so-called preferred directions which lead to the anisotropic response of the material, typically transversely isotropic or orthotropic for one or two families of fibres, respectively. We shall refer to these as *fibre directions* and *preferred directions* interchangeably.

In this chapter we focus on the behaviour of the soft materials whose elastic response is highly nonlinear, having in mind particularly the behaviour of soft biological tissues. And we are concerned only with the passive, time-independent description of their behaviour. For background information on the structure and mechanical behaviour of arteries, in particular, we refer to the work of Humphrey (1995, 2002) and for details of the developments of the continuum description of arterial wall mechanics to Holzapfel et al. (2000), Gasser et al. (2006) and Holzapfel and Ogden (2010) and references therein, for example, and for the myocardium to Holzapfel and Ogden (2009a). See also the collections of papers in the edited volumes by Holzapfel and Ogden (2003, 2006, 2009b).

First, in Section 2, we summarize the main ingredients of the continuum theory of nonlinear elasticity that are required for the subsequent analysis, including a description of the kinematics of deformation, stress tensors, equilibrium equations (we do not treat time-dependent behaviour here), and elastic constitutive laws based on a strain-energy function, with particular reference to the role of structure tensors and invariants associated with preferred directions in the development of the strain-energy function for different material symmetries. For more detailed background on the theory of nonlinear elasticity we refer to standard works such as Ogden (1997), Holzapfel (2000) and Fu and Ogden (2001).

The theory is then illustrated for simple homogeneous deformations, and we examine the range of experiments involving homogeneous deformations that are used in helping to characterize the properties of fibrous materials. Next, a prototype boundary-value problem involving a non-homogeneous deformation relevant to the behaviour of arteries is examined, that of the extension and inflation of a thick-walled elastic tube reinforced symmetrically with two families of fibres, and its thin-walled approximation. The difference between the response with and without fibres and the effect of different fibre orientations is highlighted by using specific material models.

In Section 3, adopting the notion of structure tensors, we include residual stress in the material constitutive law as a structure tensor leading to additional invariants whose number depends on the material symmetry. The effect of residual stress is then illustrated by application to the problem of extension and inflation of a thick-walled elastic tube, and a particular form of the residual stress is chosen along with simple constitutive laws, enabling explicit expressions to be obtained for the pressure and axial load applied to achieve a prescribed axial extension and inflation of the tube. For purposes of numerical calculation and graphical illustration, a range of particular values of the parameters of the problem is selected, leading to comparative plots of the pressure and axial load for a residually-stressed tube, with and without fibre reinforcement.

2 The Elements of Nonlinear Elasticity

2.1 Description of the Deformation

Let $\mathcal{B}_\mathrm{r}$, with boundary $\partial\mathcal{B}_\mathrm{r}$, denote the *reference configuration* of a material body, i.e. the location in space of the body when undeformed and not subject to any external forces, and suppose that a generic material point be labelled by its position vector $\mathbf{X}$ in $\mathcal{B}_\mathrm{r}$. Let the body be deformed quasi-statically into a new configuration, the *deformed configuration*, denoted $\mathcal{B}$, with boundary $\partial\mathcal{B}$ so that the material point $\mathbf{X}$ takes up the position $\mathbf{x}$ (the origins of these position vectors need not be specified at this point).

The deformation from $\mathcal{B}_\mathrm{r}$ to $\mathcal{B}$ is described by the *deformation* $\boldsymbol{\chi}$, as depicted in Figure 1, so that

$$\mathbf{x} = \boldsymbol{\chi}(\mathbf{X}) \quad \text{for all} \quad \mathbf{X} \in \mathcal{B}_\mathrm{r}. \tag{1}$$

The function $\boldsymbol{\chi}$ is one-to-one and onto and possesses appropriate regularity properties for the ensuing analysis.

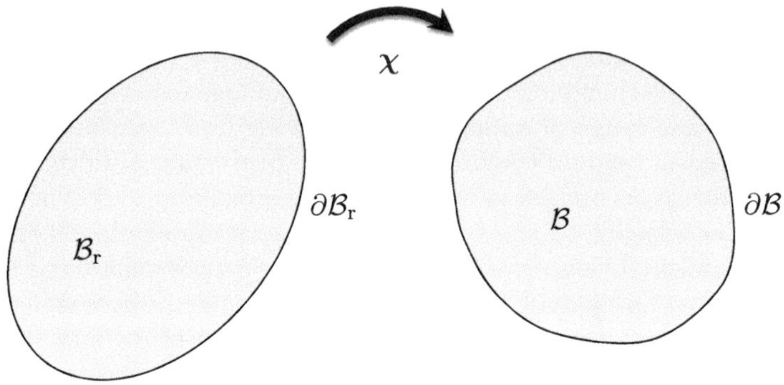

Figure 1. Depiction of a body in the reference configuration $\mathcal{B}_r$ with boundary $\partial\mathcal{B}_r$, which transforms into the deformed configuration $\mathcal{B}$ with boundary $\partial\mathcal{B}$ under the deformation χ.

In this section and subsequently we use the notations Grad, Div and grad, div to denote the gradient and divergence operators with respect to $\mathbf{X}$ and $\mathbf{x}$, respectively, $(\bullet)^{\mathrm{T}}$ to indicate the transpose and $\mathrm{tr}\,(\bullet)$ to denote the trace of a second order tensor $(\bullet)$.

The gradient of χ, denoted $\mathbf{F}$, is defined by

$$\mathbf{F} = \mathrm{Grad}\,\mathbf{x} \tag{2}$$

and referred to as the *deformation gradient tensor*. Its Cartesian components are given by $F_{i\alpha} = \partial x_i/\partial X_\alpha$, where x_i and X_α are the components of $\mathbf{x}$ and $\mathbf{X}$, respectively, for $i, \alpha \in \{1, 2, 3\}$. We use Greek and Roman indices for quantities associated with $\mathcal{B}_r$ and $\mathcal{B}$, respectively. In the absence of deformation $\mathcal{B}$ coincides with $\mathcal{B}_r$ so that $\mathbf{x} = \mathbf{X}$ and $\mathbf{F} = \mathbf{I}$, the identity tensor, and hence $\det\mathbf{F} = 1$. In general, $\mathbf{F}$ is required to be a non-singular tensor, and by convention it is taken to be positive with the standard notation

$$J \equiv \det\mathbf{F} > 0. \tag{3}$$

For an incompressible material, on which we will focus later, the constraint

$$J \equiv \det\mathbf{F} = 1 \tag{4}$$

should be satisfied.

Associated with $\mathbf{F}$ are two important tensors, the *left Cauchy–Green deformation tensor*, denoted $\mathbf{B}$ and the *right Cauchy–Green deformation tensor*, denoted $\mathbf{C}$, which are given in terms of $\mathbf{F}$ as

$$\mathbf{B} = \mathbf{F}\mathbf{F}^{\mathrm{T}}, \quad \mathbf{C} = \mathbf{F}^{\mathrm{T}}\mathbf{F}. \tag{5}$$

The *principal invariants* of $\mathbf{B}$, equivalently of $\mathbf{C}$, are the coefficients I_1, I_2, I_3 in the Cayley–Hamilton theorem, which may be stated as

$$\mathbf{B}^3 - I_1\mathbf{B}^2 + I_2\mathbf{B} - I_3\mathbf{I} = \mathbf{O}, \tag{6}$$

where $\mathbf{O}$ is the null tensor and $\mathbf{I}$ the identity tensor, in this case in $\mathcal{B}$. These invariants are defined by

$$I_1 = \mathrm{tr}\,(\mathbf{B}), \quad I_2 = \frac{1}{2}[I_1^2 - \mathrm{tr}\,(\mathbf{B}^2)], \quad I_3 = \det\mathbf{B} \equiv J^2, \tag{7}$$

Next we record the *polar decompositions* of $\mathbf{F}$, specifically

$$\mathbf{F} = \mathbf{RU} = \mathbf{VR}, \tag{8}$$

each of which is uniquely defined, $\mathbf{R}$ being a proper orthogonal tensor, while $\mathbf{U}$ and $\mathbf{V}$ are positive definite and symmetric tensors, the *right* and *left stretch tensors*, respectively. Since they are symmetric they can be decomposed in spectral form as

$$\mathbf{U} = \sum_{i=1}^{3} \lambda_i \mathbf{u}^{(i)} \otimes \mathbf{u}^{(i)}, \quad \mathbf{V} = \sum_{i=1}^{3} \lambda_i \mathbf{v}^{(i)} \otimes \mathbf{v}^{(i)}, \tag{9}$$

respectively, where $\lambda_i > 0$, $i \in \{1,2,3\}$, are the *principal stretches*, $\mathbf{u}^{(i)}$ are the (unit) eigenvectors of $\mathbf{U}$ and $\mathbf{v}^{(i)}$, the (unit) eigenvectors of $\mathbf{V}$ and $\otimes$ denotes the tensor product. These are referred to as the *Lagrangian principal axes* and *Eulerian principal axes*, respectively. It follows that J can be expressed in the equivalent general forms

$$J = \det\mathbf{F} = \det\mathbf{U} = \det\mathbf{V} = \lambda_1\lambda_2\lambda_3, \tag{10}$$

with

$$J = \det\mathbf{F} = \det\mathbf{U} = \det\mathbf{V} = \lambda_1\lambda_2\lambda_3 = 1 \tag{11}$$

identically for an incompressible material, and hence $I_3 = 1$.

From these basic measures of deformation can be constructed various Lagrangian and Eulerian measures of strain, which, by definition, vanish in the reference configuration. These include $\mathbf{U} - \mathbf{I}$ and $\log\mathbf{U}$, for example. Another is the *Green strain tensor* (or *Green–Lagrange strain tensor*) defined by

$$\mathbf{E} = \frac{1}{2}(\mathbf{C} - \mathbf{I}) = \frac{1}{2}(\mathbf{U}^2 - \mathbf{I}). \tag{12}$$

A particular interpretation of $\mathbf{F}$ is that is relates any infinitesimal line element of material $\mathrm{d}\mathbf{X}$ based at $\mathbf{X}$ to its counterpart consisting of the same material $\mathrm{d}\mathbf{x}$ at $\mathbf{x}$ via the linear transformation $\mathrm{d}\mathbf{x} = \mathbf{F}\mathrm{d}\mathbf{X}$. If we denote by $\mathbf{M}$ a unit vector in the direction $\mathrm{d}\mathbf{X}$, then the ratio $|\mathrm{d}\mathbf{x}|/|\mathrm{d}\mathbf{X}|$ identifies the ratio of lengths of the line element and is referred to as the *stretch in the direction* $\mathbf{M}$, denoted $\lambda(\mathbf{M})$ and given by

$$\lambda(\mathbf{M}) = |\mathrm{d}\mathbf{x}|/|\mathrm{d}\mathbf{X}| = [(\mathbf{F}^{\mathrm{T}}\mathbf{F}\mathbf{M}) \cdot \mathbf{M}]^{1/2}. \tag{13}$$

The principal stretches λ_i, $i = 1, 2, 3$, introduced above are particular cases of this corresponding to $\mathbf{M} = \mathbf{u}^{(i)}$.

2.2 Stress Tensors and Equilibrium

We now consider an arbitrary subdomain of $\mathcal{B}_{\mathrm{r}}$, denoted $\mathcal{D}_{\mathrm{r}}$, with boundary $\partial\mathcal{D}_{\mathrm{r}}$. Let the corresponding domain in $\mathcal{B}$ be $\mathcal{D}$ with boundary $\partial\mathcal{D}$. Consider the element of surface area $\mathrm{d}a$ on $\partial\mathcal{D}$ with unit normal $\mathbf{n}$ as depicted in Figure 2 (by convention area elements are taken to be positive). The contact force acting on this area arising from the surrounding material due to applied loading on $\partial\mathcal{B}$ is written as $\boldsymbol{\sigma}^{\mathrm{T}}\mathbf{n}\mathrm{d}a$, where $\boldsymbol{\sigma}$ is the *Cauchy stress tensor*. Suppose the deformation is caused by the combination of surface tractions on $\partial\mathcal{B}$ and a body force $\mathbf{b}$ per unit mass. For equilibrium of each domain $\mathcal{D}$, the total force on it must vanish, i.e.

$$\int_{\partial\mathcal{D}} \boldsymbol{\sigma}^{\mathrm{T}}\mathbf{n}\,\mathrm{d}a + \int_{\mathcal{D}} \rho\mathbf{b}\,\mathrm{d}v \equiv \int_{\mathcal{D}} (\mathrm{div}\,\boldsymbol{\sigma} + \rho\mathbf{b})\,\mathrm{d}v = \mathbf{0}, \tag{14}$$

where $\mathrm{d}v$ is an element of volume in $\mathcal{D}$ and the divergence theorem has been used to convert the surface integral to a volume integral.

Since $\mathcal{D}$ is arbitrary we deduce from equation (14) that, provided the integrand is continuous, the local form of the equilibrium equation is

$$\mathrm{div}\,\boldsymbol{\sigma} + \rho\mathbf{b} = \mathbf{0}. \tag{15}$$

Equilibrium of moments (under the assumption that there are no intrinsic couple stresses) then ensures that $\boldsymbol{\sigma}$ is symmetric, i.e. $\boldsymbol{\sigma}^{\mathrm{T}} = \boldsymbol{\sigma}$, and the transpose on $\boldsymbol{\sigma}$ can be omitted henceforth.

An alternative expression for the equilibrium equation can be given by reference to $\mathcal{D}_{\mathrm{r}}$ and its boundary. One way to arrive at this is to consider the transformation of the directed area element $\mathbf{N}\mathrm{d}A$ on $\partial\mathcal{D}_{\mathrm{r}}$ to $\mathbf{n}\mathrm{d}a$ on $\partial\mathcal{D}$. The transformation is given by *Nanson's formula* $\mathbf{n}\mathrm{d}a = J\mathbf{F}^{-\mathrm{T}}\mathbf{N}\mathrm{d}A$, where $\mathbf{F}^{-\mathrm{T}} = (\mathbf{F}^{-1})^{\mathrm{T}} = (\mathbf{F}^{\mathrm{T}})^{-1}$. This allows equation (14) to be written as

$$\int_{\partial\mathcal{D}_{\mathrm{r}}} \mathbf{S}^{\mathrm{T}}\mathbf{N}\,\mathrm{d}A + \int_{\mathcal{D}_{\mathrm{r}}} \rho_{\mathrm{r}}\mathbf{b}\,\mathrm{d}V \equiv \int_{\mathcal{D}_{\mathrm{r}}} (\mathrm{Div}\,\mathbf{S} + \rho_{\mathrm{r}}\mathbf{b})\,\mathrm{d}V = \mathbf{0}, \tag{16}$$

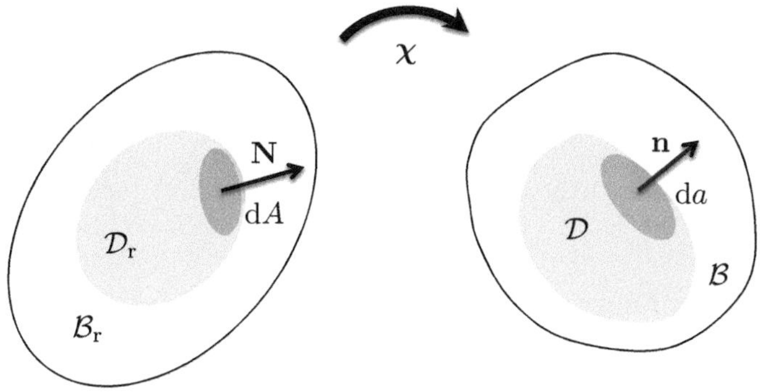

Figure 2. Transformation of an area element dA with unit outward normal $\mathbf{N}$ in the reference domain $\mathcal{D}_r \subset \mathcal{B}_r$ with boundary $\partial\mathcal{D}_r$ into the area element $\mathbf{n}da$ with unit outward normal $\mathbf{n}$ in the deformed domain $\mathcal{D} \subset \mathcal{B}$ with boundary $\partial\mathcal{D}$ under the deformation $\boldsymbol{\chi}$.

where dV is a volume element in $\mathcal{D}_r$, with the connection $dv = JdV$, the divergence theorem (this time in the reference configuration using Div) has again been used and the notation

$$\mathbf{S} = J\mathbf{F}^{-1}\boldsymbol{\sigma} \tag{17}$$

has been introduced for the so-called *nominal stress tensor*, which measures the contact force on $\partial\mathcal{D}$ per unit area of the reference boundary $\partial\mathcal{D}_r$. We have also introduced the density ρ_r in $\mathcal{D}_r$, which, by conservation of mass, is related to ρ by

$$\rho_r = J\rho. \tag{18}$$

Note that for an isochoric deformation ($J = 1$) we have $\rho = \rho_r$, which is always in force for an incompressible material.

The local form of (16), the counterpart of (15), is then given as

$$\mathrm{Div}\,\mathbf{S} + \rho_r\mathbf{b} = \mathbf{0}. \tag{19}$$

Note that, unlike $\boldsymbol{\sigma}$, $\mathbf{S}$ is not in general symmetric but it follows from (17) that

$$\mathbf{S}^T\mathbf{F}^T = \mathbf{F}\mathbf{S}. \tag{20}$$

The transpose $\mathbf{S}^T$ of $\mathbf{S}$ is the *first Piola–Kirchhoff stress tensor*.

We mention here two other stress tensors in common use: the *second Piola–Kirchhoff stress tensor* and the *Biot stress tensor*, here denoted $\mathbf{P}$

and $\mathbf{T}$, respectively, which are both symmetric. They are related to $\boldsymbol{\sigma}$, $\mathbf{S}$ and each other via

$$\mathbf{P} = J\mathbf{F}^{-1}\boldsymbol{\sigma}\mathbf{F}^{-\mathrm{T}} = \mathbf{S}\mathbf{F}^{-\mathrm{T}}, \quad \mathbf{T} = \frac{1}{2}(\mathbf{P}\mathbf{U} + \mathbf{U}\mathbf{P}) = \frac{1}{2}(\mathbf{S}\mathbf{R} + \mathbf{R}^{\mathrm{T}}\mathbf{S}^{\mathrm{T}}), \quad (21)$$

where $\mathbf{R}$ is the proper orthogonal tensor that appears in the polar decomposition (8).

2.3 The Strain-energy Function of an Elastic Material

We are concerned with fibre-reinforced materials that are capable of large elastic deformations, and we restrict attention to purely *elastic* properties of these materials. Their properties can then be characterized in terms of a *strain-energy function*, which we denote by W and define per unit reference volume. For a homogeneous material W depends only on the deformation gradient $\mathbf{F}$ and we write $W = W(\mathbf{F})$. If the material is inhomogeneous then its properties depend separately on $\mathbf{X}$ as well as on dependence on $\mathbf{X}$ through $\mathbf{F}$. For inhomogeneous materials, which we shall consider in Section 3 when dealing with residually stressed materials, we shall omit the explicit dependence on $\mathbf{X}$.

We first consider an elastic material without any internal constraint. Then the nominal stress is work conjugate to the deformation gradient, i.e. $\mathrm{d}W = \mathrm{tr}\,(\mathbf{S}\mathrm{d}\mathbf{F})$ (in components with the usual summation convention for repeated indices, $S_{\alpha i}\mathrm{d}F_{i\alpha}$), and given by

$$\mathbf{S} = \frac{\partial W}{\partial \mathbf{F}}, \quad S_{\alpha i} = \frac{\partial W}{\partial F_{i\alpha}}. \quad (22)$$

Particular note should be made of the convention for the order of the indices used in the latter expression.

When the incompressibility constraint is in force equation $(22)_1$ is replaced by

$$\mathbf{S} = \frac{\partial W}{\partial \mathbf{F}} - p\mathbf{F}^{-1}, \quad \det \mathbf{F} = 1, \quad (23)$$

where p is a Lagrange multiplier associated with the incompressibility constraint.

The corresponding Cauchy stress tensor is obtained from (17) as

$$\boldsymbol{\sigma} = J^{-1}\mathbf{F}\frac{\partial W}{\partial \mathbf{F}}, \quad \boldsymbol{\sigma} = \mathbf{F}\frac{\partial W}{\partial \mathbf{F}} - p\mathbf{I}, \quad (24)$$

for unconstrained and incompressible materials, respectively, with $J = \det \mathbf{F} = 1$ in the latter.

Since we have assumed that the reference configuration $\mathcal{B}_r$, in which $\mathbf{F} = \mathbf{I}$, is stress free then

$$\frac{\partial W}{\partial \mathbf{F}}(\mathbf{I}) = \mathbf{O}, \quad \frac{\partial W}{\partial \mathbf{F}}(\mathbf{I}) - p_0\,\mathbf{I} = \mathbf{O}, \tag{25}$$

for unconstrained and incompressible materials, respectively, where p_0 is the value of p in $\mathcal{B}_r$. We may also take the strain energy to be measured from $\mathcal{B}_r$ and therefore impose the condition

$$W(\mathbf{I}) = 0. \tag{26}$$

The configuration $\mathcal{B}_r$ is referred to as a *natural configuration* if $(25)_{\{1 \text{ or } 2\}}$ and (26) hold together.

When the body is stressed in $\mathcal{B}_r$, so that (25) does not hold, and there are no body forces or applied tractions, i.e. the body is unloaded, then $\mathcal{B}_r$ is referred to as a *residually stressed configuration.* Since there are no applied tractions, the surface force must vanish at every point of the boundary $\partial\mathcal{B}_r$, implying, in particular, that the residual stress is necessarily inhomogeneous. The article by Ogden (2003) reviews relevant aspects of residual stresses, and we shall discuss the formulation of constitutive equations for residually stressed elastic materials in Section 3 together with an application involving the extension and inflation of a residually stress circular cylindrical tube.

Objectivity. Thus far no restrictions have been placed on the strain-energy function other than the fact that it depends locally on $\mathbf{F}$. An important consideration is that the material properties should be independent of observer. In the present context this can be interpreted as the requirement that strain energy should be unaffected by arbitrary superimposed rigid translations and rotations *after deformation*, i.e. in $\mathcal{B}$. Let $\mathbf{Q}$ be the (arbitrary) rotation in question ($\mathbf{Q}$ is a proper orthogonal tensor). When this rotation is superimposed on the existing deformation the deformation gradient changes from $\mathbf{F}$ to $\mathbf{QF}$. Thus, W must be invariant under this transformation, i.e.

$$W(\mathbf{QF}) = W(\mathbf{F}) \tag{27}$$

for *arbitrary* proper orthogonal $\mathbf{Q}$ for any deformation gradient $\mathbf{F}$. This requirement is referred to as (the principle of) *objectivity* or material frame indifference. We use only the former of these terminologies here. It is assumed henceforth that W is objective. Since $\mathbf{Q}$ is arbitrary it may be chosen to be the transpose of the rotation $\mathbf{R}$ occurring in the polar decomposition

(8). Thus, substituting $\mathbf{Q} = \mathbf{R}^{\mathrm{T}}$ in (27) and using the polar decomposition $(8)_1$, we obtain

$$W(\mathbf{F}) = W(\mathbf{U}). \tag{28}$$

This shows that W is independent of the rotational part $\mathbf{R}$ of the deformation gradient and depends on $\mathbf{F}$ only through the stretch tensor $\mathbf{U}$ and may be defined on the class of positive definite symmetric tensors $\mathbf{U}$. Equivalently, in view of $(5)_2$, it may be regarded as a function of $\mathbf{C}$, or, by the definition (12), as a function of the Green strain $\mathbf{E}$ or any other appropriate Lagrangian strain tensor. We remark that W is automatically objective when written as a function of either $\mathbf{U}, \mathbf{C}$ or $\mathbf{E}$, and note that we use the same notation W for the strain-energy function whichever of these variables is used in its argument.

In the same way that $\mathbf{S}$ is work conjugate to $\mathbf{F}$, the Biot stress $\mathbf{T}$ is work conjugate to $\mathbf{U}$ and the dependence (28) enables the Biot stress tensor to be written as

$$\mathbf{T} = \frac{\partial W}{\partial \mathbf{U}}, \quad \mathbf{T} = \frac{\partial W}{\partial \mathbf{U}} - p\,\mathbf{U}^{-1}, \tag{29}$$

for unconstrained and incompressible materials, respectively, with, by (11), $\det \mathbf{U} = 1$ in the latter.

Similarly, the second Piola–Kirchhoff stress tensor is work conjugate to $\mathbf{E}$, and hence

$$\mathbf{P} = \frac{\partial W}{\partial \mathbf{E}}, \quad \mathbf{P} = \frac{\partial W}{\partial \mathbf{E}} - p\,(\mathbf{I} + 2\mathbf{E})^{-1} \tag{30}$$

for unconstrained and incompressible materials, respectively. In this case the incompressibility constraint is $\det \mathbf{C} \equiv \det(\mathbf{I} + 2\mathbf{E}) = 1$.

2.4 Material Symmetry

Apart from the restrictions imposed by objectivity, which are not severe, there are both physical and mathematical considerations that should be taken into account and affect the form of the strain-energy function. In particular, its predictions must be consistent with the data from simple experimental tests involving homogeneous deformations. For non-homogeneous deformations the governing boundary-value problems must be well posed and guarantee, for example, that one or more solutions exist, whatever the material properties. Then the properties of the material have an influence. In particular, many materials possess intrinsic symmetries which may be identified in their natural configuration. An important aspect of material symmetry is that the elastic properties of the material are unaffected by a change of reference configuration corresponding to a symmetry transformation.

Mathematically, this can be stated in the following way. Consider two reference configurations, denoted $\mathcal{B}_r$ and $\mathcal{B}_r^*$, and let a typical material point in these configurations have position vectors $\mathbf{X}$ and $\mathbf{X}^*$, respectively. Suppose $\mathcal{B}_r$ and $\mathcal{B}_r^*$ are related by the deformation gradient $\mathbf{Q}^* = \mathrm{Grad}^*\mathbf{X}$, where Grad^* denotes the gradient with respect to $\mathbf{X}^*$. Then the deformation gradients relative to $\mathcal{B}_r$ and $\mathcal{B}_r^*$, respectively, are $\mathbf{F}$ and $\mathbf{F}^* \equiv \mathbf{F}\mathbf{Q}^*$. If W is unaffected by such a change of reference configuration, then

$$W(\mathbf{F}\mathbf{Q}^*) = W(\mathbf{F}) \tag{31}$$

for all deformation gradients $\mathbf{F}$. Such a $\mathbf{Q}^*$ identifies a symmetry of the material. The full set of such $\mathbf{Q}^*$ satisfying (31) for all deformation gradients $\mathbf{F}$ forms a group of transformations, called the *symmetry group of the material relative to the reference configuration* $\mathcal{B}_r$. In general, the group consists of unimodular transformations, i.e. transformations with $\det \mathbf{Q}^* = 1$, not just proper orthogonal transformations. However, in this chapter we consider only groups of (proper) orthogonal transformations, and we emphasize that the rotation $\mathbf{Q}^*$ is applied in $\mathcal{B}_r$, whereas $\mathbf{Q}$, which is independent of $\mathbf{Q}^*$, is applied in $\mathcal{B}$.

In the following we consider some particular examples of symmetries, specifically *isotropy*, *transverse isotropy* and *orthotropy*.

Isotropy. We first consider *isotropic elastic materials*, for which, as indicated above, the symmetry group is the full *proper orthogonal group*. Thus, (31) should hold for *arbitrary* proper orthogonal $\mathbf{Q}^*$. By combining the independent rotations $\mathbf{Q}$ and $\mathbf{Q}^*$ in (27) and (31), we obtain

$$W(\mathbf{Q}^*\mathbf{U}\mathbf{Q}^{*\mathrm{T}}) = W(\mathbf{U}) \tag{32}$$

for *all* rotations $\mathbf{Q}^*$, or, equivalently, $W(\mathbf{Q}\mathbf{V}\mathbf{Q}^{\mathrm{T}}) = W(\mathbf{V})$. Now the restriction to *proper* orthogonal $\mathbf{Q}^*$, or $\mathbf{Q}$, can be dropped. Equation (32) implies that W is an *isotropic function* of $\mathbf{U}$. We now examine the consequences of this isotropy.

The spectral decomposition (9) enables us to deduce that W depends on $\mathbf{U}$ only through its eigenvalues, the principal stretches $\lambda_1, \lambda_2, \lambda_3$, and is a symmetric function of them: $W = W(\lambda_1, \lambda_2, \lambda_3)$, in which any pair of the stretches can be interchanged. An important consequence of isotropy is that the Cauchy stress tensor $\boldsymbol{\sigma}$ is coaxial with $\mathbf{V}$, i.e. has the same eigenvectors as $\mathbf{V}$, and the Biot stress tensor $\mathbf{T}$ is coaxial with $\mathbf{U}$. It follows that $\boldsymbol{\sigma}$ and $\mathbf{T}$ can be written in spectral form similarly to (9), i.e.

$$\boldsymbol{\sigma} = \sum_{i=1}^{3} \sigma_i \mathbf{v}^{(i)} \otimes \mathbf{v}^{(i)}, \quad \mathbf{T} = \sum_{i=1}^{3} t_i \mathbf{u}^{(i)} \otimes \mathbf{u}^{(i)}, \tag{33}$$

where σ_i are the *principal Cauchy stresses* and t_i, the *principal Biot stresses.* In fact, the triad (t_1, t_2, t_3) is work conjugate to $(\lambda_1, \lambda_2, \lambda_3)$, i.e. $\mathrm{d}W = \sum_{i=1}^{3} t_i \mathrm{d}\lambda_i$, and hence

$$t_i = \frac{\partial W}{\partial \lambda_i}, \quad J\sigma_i = \lambda_i \frac{\partial W}{\partial \lambda_i}, \quad i = 1, 2, 3, \tag{34}$$

for an unconstrained material, the latter following on use of the connections (21) with $\mathbf{v}^{(i)} = \mathbf{R}\mathbf{u}^{(i)}$, $i = 1, 2, 3$. For an incompressible materials these are adjusted to

$$t_i = \frac{\partial W}{\partial \lambda_i} - p\lambda_i^{-1}, \quad \sigma_i = \lambda_i \frac{\partial W}{\partial \lambda_i} - p, \quad \lambda_1\lambda_2\lambda_3 = 1, \quad i = 1, 2, 3. \tag{35}$$

Similarly, expressions for the second Piola–Kirchhoff stress and the nominal stress referred to principal axes may be obtained from (17) with (21) and (33) and the connection $\mathbf{v}^{(i)} = \mathbf{R}\mathbf{u}^{(i)}$, $i = 1, 2, 3$. These are

$$\mathbf{P} = \sum_{i=1}^{3} p_i \mathbf{u}^{(i)} \otimes \mathbf{u}^{(i)}, \quad \mathbf{S} = \sum_{i=1}^{3} t_i \mathbf{u}^{(i)} \otimes \mathbf{v}^{(i)}, \tag{36}$$

where

$$p_i = \lambda_i^{-1} \frac{\partial W}{\partial \lambda_i}, \quad p_i = \lambda_i^{-1} \frac{\partial W}{\partial \lambda_i} - p\lambda_i^{-2}, \quad i = 1, 2, 3, \tag{37}$$

for an unconstrained and incompressible material, respectively. Note that $p_i = \partial W/\partial e_i$, respectively $p_i = \partial W/\partial e_i - p(1 + 2e_i)^{-1}$, $i = 1, 2, 3$, where e_i, $i = 1, 2, 3$, are the *principal Green strains* given by $e_i = (\lambda_i^2 - 1)/2$, i.e. the principal components of $\mathbf{E}$.

The principal invariants I_1, I_2, I_3 defined by (7) are expressed in terms of the principal stretches as

$$I_1 = \lambda_1^2 + \lambda_2^2 + \lambda_3^2, \quad I_2 = \lambda_2^2\lambda_3^3 + \lambda_3^2\lambda_1^2 + \lambda_1^2\lambda_2^2, \quad I_3 = \lambda_1^2\lambda_2^2\lambda_3^2, \tag{38}$$

which are (in general) independent functions of the stretches. Since they are symmetric the dependence of W on the principal stretches can be captured by taking W to be a function of these principal invariants. Then from $(34)_2$ we obtain

$$J\sigma_i = 2\lambda_i^2 \frac{\partial W}{\partial I_1} + 2\lambda_i^2 (I_1 - \lambda_i^2) \frac{\partial W}{\partial I_2} + 2I_3 \frac{\partial W}{\partial I_3}, \tag{39}$$

for an unconstrained material, and, by coaxiality, the Cauchy stress tensor can be written as

$$\boldsymbol{\sigma} = \alpha_0 \mathbf{I} + \alpha_1 \mathbf{B} + \alpha_2 \mathbf{B}^2, \tag{40}$$

where $\mathbf{B}$ is the left Cauchy–Green deformation tension defined in $(5)_1$ and the coefficients $\alpha_0, \alpha_1, \alpha_2$ are given by

$$\alpha_0 = 2I_3^{1/2}W_3, \quad \alpha_1 = 2I_3^{-1/2}(W_1 + I_1W_2), \quad \alpha_2 = -2I_3^{-1/2}W_2, \tag{41}$$

where $W_i = \partial W/\partial I_i$, $i = 1, 2, 3$, with W is now regarded as a function of I_1, I_2, I_3.

Equation (40) may also be obtained directly by using the formula $(24)_1$ in the form

$$\boldsymbol{\sigma} = J^{-1}\mathbf{F}\frac{\partial W}{\partial \mathbf{F}} = J^{-1}\mathbf{F}\sum_{i=1}^{3}\frac{\partial W}{\partial I_i}\frac{\partial I_i}{\partial \mathbf{F}} \tag{42}$$

based on the derivatives

$$\frac{\partial I_1}{\partial \mathbf{F}} = 2\mathbf{F}^{\mathrm{T}}, \quad \frac{\partial I_2}{\partial \mathbf{F}} = 2(I_1\mathbf{F}^{\mathrm{T}} - \mathbf{C}\mathbf{F}^{\mathrm{T}}), \quad \frac{\partial I_3}{\partial \mathbf{F}} = 2I_3\mathbf{F}^{-1}. \tag{43}$$

Since $I_3 \equiv 1$ for an incompressible material, W depends only on I_1 and I_2 and, instead of (40), we have

$$\boldsymbol{\sigma} = \alpha_1\mathbf{B} + \alpha_2\mathbf{B}^2 - p\,\mathbf{I}, \tag{44}$$

where p is the arbitrary hydrostatic pressure, α_1 and α_2 are again given by (41), but with $I_3 = 1$.

An alternative representation for $\boldsymbol{\sigma}$ may be obtained by using the Cayley–Hamilton theorem (6) with $I_3 = 1$ in the form

$$\mathbf{B}^2 = I_1\mathbf{B} - I_2\mathbf{I} + \mathbf{B}^{-1} \tag{45}$$

to give

$$\boldsymbol{\sigma} = \bar{\alpha}_1\mathbf{B} + \bar{\alpha}_{-1}\mathbf{B}^{-1} - \bar{p}\,\mathbf{I}, \tag{46}$$

where $\bar{\alpha}_1 = \alpha_1 + I_1\alpha_2 = 2W_1$, $\bar{\alpha}_{-1} = \alpha_2 = -2W_2$, $\bar{p} = p + I_2\alpha_2$.

Transverse isotropy. Transversely isotropic materials are distinguished by the presence of a specific direction (in the reference configuration), referred to as a *preferred direction*, in such a way that the properties of the material in the transverse direction are isotropic. Let $\mathbf{M}$ be a unit vector in the preferred direction in the reference configuration $\mathcal{B}_{\mathrm{r}}$ at the point $\mathbf{X}$. Note that in general $\mathbf{M}$ may depend on $\mathbf{X}$, in which case the material is inhomogeneous. However, material symmetry is a local property and in considering transverse isotropy we suppress the dependence on $\mathbf{X}$.

The material properties are invariant with respect to an arbitrary rotation about the direction $\mathbf{M}$ and to reversal of $\mathbf{M}$, i.e. to rotations $\mathbf{Q}^*$

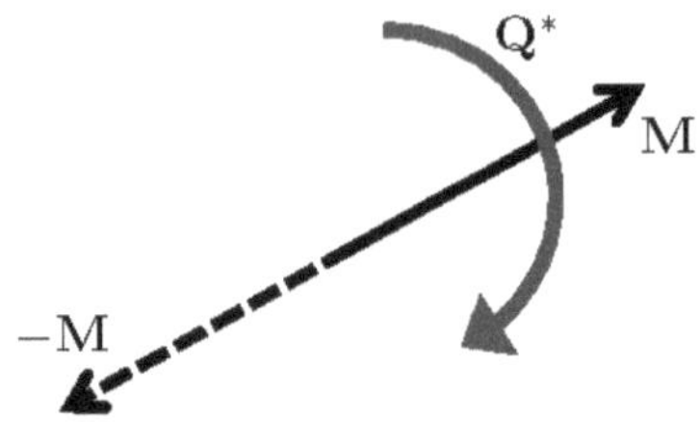

Figure 3. Transverse isotropy with preferred direction $\mathbf{M}$ in the reference configuration. The material properties are independent of rotations $\mathbf{Q}^*$ about $\mathbf{M}$ and reversal of the direction of $\mathbf{M}$: $\mathbf{M} \to -\mathbf{M}$.

such that $\mathbf{Q}^*\mathbf{M} = \pm\mathbf{M}$, as depicted in Figure 3. From the mathematical point of view the strain-energy function must satisfy $W(\mathbf{F}\mathbf{Q}^*) = W(\mathbf{F})$ for all such $\mathbf{Q}^*$. However, by objectivity we can treat W as a function of the right Cauchy–Green tensor $\mathbf{C}$ defined by $(5)_2$ and then the symmetry property can be cast as $W(\mathbf{Q}^*\mathbf{C}\mathbf{Q}^{*\mathrm{T}}) = W(\mathbf{C})$ for all rotations $\mathbf{Q}^*$ satisfying $\mathbf{Q}^*\mathbf{M} = \pm\mathbf{M}$.

An equivalent characterization of transverse isotropy involves regarding W as an *isotropic function* of the two symmetric tensors $\mathbf{C}$ and $\mathbf{M} \otimes \mathbf{M}$, i.e. $W = W(\mathbf{C}, \mathbf{M} \otimes \mathbf{M})$ (see Liu , 1982; see also Spencer, 1972, 1984 for general background material on transverse isotropy). Then the symmetry requirement is

$$W(\mathbf{Q}^*\mathbf{C}\mathbf{Q}^{*\mathrm{T}}, \mathbf{Q}^*\mathbf{M} \otimes \mathbf{Q}^*\mathbf{M}) = W(\mathbf{C}, \mathbf{M} \otimes \mathbf{M}) \tag{47}$$

for *all* orthogonal $\mathbf{Q}^*$. In this case it is not necessary to restrict attention to *proper* orthogonal $\mathbf{Q}^*$. The tensor $\mathbf{M} \otimes \mathbf{M}$ is referred to as a *structure tensor* and it is invariant under the transformation $\mathbf{M} \to -\mathbf{M}$.

For an unconstrained material, the symmetry (47) is captured by dependence of W on five invariants, namely I_1, I_2, I_3, as defined by (7), and two additional invariants, typically denoted I_4 and I_5 and sometimes referred to as quasi-invariants, that depend on $\mathbf{C}$ and $\mathbf{M}$. These are defined by

$$I_4 = \mathbf{M} \cdot (\mathbf{C}\mathbf{M}), \quad I_5 = \mathbf{M} \cdot (\mathbf{C}^2\mathbf{M}). \tag{48}$$

In three dimensions these five invariants provide a complete set of independent invariants for the case of transverse isotropy. We note that I_4 has a clear kinematical interpretation since it defines the square of the stretch in the direction $\mathbf{M}$ according to (13), but I_5 has no such similar interpretation in general. An alternative to the invariant I_5, denoted I_5^*, which does have a physical interpretation but is not so convenient in use, is based on the adjugate $\mathbf{C}^* = I_3\mathbf{C}^{-1}$ of $\mathbf{C}$ and defined by $I_5^* = \mathbf{M}\cdot(\mathbf{C}^*\mathbf{M})$. Based on Nanson's formula, I_5^* represents the square of the ratio of deformed to undeformed area elements initially normal to $\mathbf{M}$ (Merodio and Ogden, 2002).

For particular deformations the number of independent invariants is reduced. For example, in plane strain with $\mathbf{M}$ in the considered plane the connections

$$I_2 = I_1 + I_3 - 1, \quad I_5 = (I_1 - 1)I_4 - I_3 \tag{49}$$

are easily established, and only three independent invariants remain.

By expanding out the formula $(24)_1$ based on the invariants $I_1, \ldots, I_5$ using again the derivatives (43) and additionally the derivatives

$$\frac{\partial I_4}{\partial \mathbf{F}} = 2\mathbf{M}\otimes\mathbf{FM}, \quad \frac{\partial I_5}{\partial \mathbf{F}} = 2(\mathbf{M}\otimes\mathbf{BFM} + \mathbf{CM}\otimes\mathbf{FM}) \tag{50}$$

the Cauchy stress tensor of a transversely isotropic materials can be expressed as

$$\begin{aligned}\boldsymbol{\sigma} &= J^{-1}[2W_1\mathbf{B} + 2W_2(I_1\mathbf{B} - \mathbf{B}^2) + 2I_3W_3\mathbf{I} + 2W_4\mathbf{m}\otimes\mathbf{m} \\ &+ 2W_5(\mathbf{m}\otimes\mathbf{Bm} + \mathbf{Bm}\otimes\mathbf{m})],\end{aligned} \tag{51}$$

where $\mathbf{m} = \mathbf{FM}$ is the 'push-forward' of $\mathbf{M}$ from $\mathcal{B}_r$ to $\mathcal{B}$ and $W_i = \partial W/\partial I_i$, $i = 1, \ldots, 5$. This reduces to the formula (40) with (41) for an isotropic material on setting $W_4 = W_5 = 0$.

Corresponding formulas for the stress tensors $\mathbf{S}, \mathbf{P}$ and $\mathbf{T}$ can be obtained from (51) by using the connections (21) as required.

For an incompressible material $I_3 \equiv 1$, and (51) is replaced by

$$\begin{aligned}\boldsymbol{\sigma} &= 2W_1\mathbf{B} + 2W_2(I_1\mathbf{B} - \mathbf{B}^2) + 2W_4\mathbf{m}\otimes\mathbf{m} \\ &+ 2W_5(\mathbf{m}\otimes\mathbf{Bm} + \mathbf{Bm}\otimes\mathbf{m}) - p\mathbf{I},\end{aligned} \tag{52}$$

equivalently

$$\begin{aligned}\boldsymbol{\sigma} &= 2W_1\mathbf{B} - 2W_2\mathbf{B}^{-1} + 2W_4\mathbf{m}\otimes\mathbf{m} \\ &+ 2W_5(\mathbf{m}\otimes\mathbf{Bm} + \mathbf{Bm}\otimes\mathbf{m}) - \bar{p}\mathbf{I},\end{aligned} \tag{53}$$

where W is now a function of (I_1, I_2, I_4, I_5) and $\bar{p}$ is defined following equation (46).

Orthotropy. An extension of the analysis for a transversely isotropic material arises if there are two preferred directions in the reference configuration $\mathcal{B}_r$ of the material, and we now consider this case. Let the unit vectors be $\mathbf{M}$ and $\mathbf{M}'$ correspond to the two preferred directions in $\mathcal{B}_\mathrm{r}$. Then, the material can be characterized in terms of a strain-energy function that depends on $\mathbf{C}$ and the two structure tensors $\mathbf{M}\otimes\mathbf{M}$ and $\mathbf{M}'\otimes\mathbf{M}'$, and we write $W(\mathbf{C},\mathbf{M}\otimes\mathbf{M},\mathbf{M}'\otimes\mathbf{M}')$, and W is then an isotropic function of the combination of the three tensors and, for any right Cauchy–Green tensor $\mathbf{C}$, must satisfy

$$W(\mathbf{Q}^*\mathbf{C}\mathbf{Q}^{*\mathrm{T}},\mathbf{Q}^*\mathbf{M}\otimes\mathbf{Q}^*\mathbf{M},\mathbf{Q}^*\mathbf{M}'\otimes\mathbf{Q}^*\mathbf{M}')=W(\mathbf{C},\mathbf{M}\otimes\mathbf{M},\mathbf{M}'\otimes\mathbf{M}') \tag{54}$$

for arbitrary orthogonal $\mathbf{Q}^*$.

This symmetry requirement, as for a transversely isotropic material, can be expressed in terms of invariants. In this case W depends not only on the invariants $I_1,\ldots,I_5$, but now also on the invariants, denoted I_6, I_7, analogous to I_4, I_5 but associated with $\mathbf{M}'$, and a coupling invariant, denoted I_8. These are defined by

$$I_6=\mathbf{M}'\cdot(\mathbf{C}\mathbf{M}'),\quad I_7=\mathbf{M}'\cdot(\mathbf{C}^2\mathbf{M}'),\quad I_8=\mathbf{M}\cdot(\mathbf{C}\mathbf{M}'). \tag{55}$$

Note that the definition of I_8, as adopted here, is not strictly invariant since it changes sign if either $\mathbf{M}\to-\mathbf{M}$ or $\mathbf{M}'\to-\mathbf{M}'$, and is therefore sometimes defined with the extra factor $\mathbf{M}\cdot\mathbf{M}'$ so that it is properly invariant. For simplicity, however, we use the definition given in $(55)_3$ since the resulting expressions for the stress are somewhat simpler than would otherwise be the case. The full symmetry is then accounted for in the properties of W. The invariants $I_1,\ldots,I_8$ form a complete independent set of invariants characterizing the symmetry embodied in (54) in three dimensions. As for the case of transverse isotropy, the number of independent invariants is reduced for plane strain, and we have the connections given in (49) together with

$$I_7=(I_1-1)I_4-I_3,\quad I_8^2=I_4I_6-I_3|\mathbf{M}\times\mathbf{M}'|^2. \tag{56}$$

The counterpart of the expression (51) for the Cauchy stress tensor can be obtained similarly by using the additional formula

$$\frac{\partial I_8}{\partial\mathbf{F}}=\mathbf{M}\otimes\mathbf{F}\mathbf{M}'+\mathbf{M}'\otimes\mathbf{F}\mathbf{M} \tag{57}$$

to obtain

$$\begin{aligned}\boldsymbol{\sigma}&=J^{-1}[2W_1\mathbf{B}+2W_2(I_1\mathbf{B}-\mathbf{B}^2)+2I_3W_3\mathbf{I}+2W_4\mathbf{m}\otimes\mathbf{m}\\&+2W_5(\mathbf{m}\otimes\mathbf{B}\mathbf{m}+\mathbf{B}\mathbf{m}\otimes\mathbf{m})+2W_6\mathbf{m}'\otimes\mathbf{m}'\\&+2W_7(\mathbf{m}'\otimes\mathbf{B}\mathbf{m}'+\mathbf{B}\mathbf{m}'\otimes\mathbf{m}')+W_8(\mathbf{m}\otimes\mathbf{m}'+\mathbf{m}'\otimes\mathbf{m})],\end{aligned} \tag{58}$$

where the notation $W_i = \partial W/\partial I_i$ now applies for $i = 1, \ldots, 8$ and $\mathbf{m}' = \mathbf{FM}'$ as well as $\mathbf{m} = \mathbf{FM}$.

For an incompressible material the list of invariants reduces by one to $I_1, I_2, I_4, \ldots, I_8$ and the Cauchy stress is given by

$$\begin{aligned}\boldsymbol{\sigma} &= 2W_1\mathbf{B} + 2W_2(I_1\mathbf{B} - \mathbf{B}^2) + 2W_4\mathbf{m} \otimes \mathbf{m} + 2W_6\mathbf{m}' \otimes \mathbf{m}' \\ &+ 2W_5(\mathbf{m} \otimes \mathbf{Bm} + \mathbf{Bm} \otimes \mathbf{m}) + 2W_7(\mathbf{m}' \otimes \mathbf{Bm}' + \mathbf{Bm}' \otimes \mathbf{m}') \\ &+ W_8(\mathbf{m} \otimes \mathbf{m}' + \mathbf{m}' \otimes \mathbf{m}) - p\mathbf{I}, \end{aligned} \tag{59}$$

the counterpart of (52), the notation $W_i = \partial W/\partial I_i$ now applying for $i = 1, 2, 4, \ldots, 8$.

The above formulas apply for materials with two distinct preferred directions. There are two situations in which they describe *orthotropic* material behaviour. One is if the directions are orthogonal and the other is if the preferred directions are mechanically equivalent (Spencer, 1972), i.e. in the latter, in the case of a fibre-reinforced material, the mechanical properties associated with the two fibre directions are the same.

2.5 Application to Homogeneous Deformations

Homogeneous deformations, or, more strictly, deformations which are approximately homogeneous, are often used in experimental tests as a means of determining material characteristics. We therefore consider such deformations in this subsection. For a homogeneous deformation the deformation gradient is independent of $\mathbf{X}$ and the deformation can therefore be written $\mathbf{x} = \mathbf{FX} + \mathbf{c}$, where $\mathbf{c}$ is an inessential constant which may be omitted (it corresponds to a uniform translation of the whole material). Here we consider a particular example of a homogeneous deformation, namely *pure homogeneous strain*, which is defined by

$$x_1 = \lambda_1 X_1, \quad x_2 = \lambda_2 X_2, \quad x_3 = \lambda_3 X_3, \tag{60}$$

where the principal stretches $\lambda_1, \lambda_2, \lambda_3$ are constants, i.e. independent of the reference (rectangular Cartesian) coordinates (X_1, X_2, X_3). For this deformation $\mathbf{F} = \mathbf{U} = \mathbf{V}$, $\mathbf{R} = \mathbf{I}$ and the principal axes of the deformation are therefore unchanged during the deformation and coincide with the Cartesian coordinate directions. The matrix of components of $\mathbf{F}$ has the diagonal form diag$[\lambda_1, \lambda_2, \lambda_3]$.

In the following we focus mainly on incompressible materials so that the constraint

$$\lambda_1\lambda_2\lambda_3 = 1 \tag{61}$$

is satisfied.

Isotropy. To set the scene we consider first isotropic materials. In order to determine the stresses that are required to maintain the deformation (60) we write (44) on principal axes as

$$\sigma_i = 2W_1\lambda_i^2 + 2W_2(I_1 - \lambda_i^2)\lambda_i^2 - p, \tag{62}$$

with W treated as a function of I_1 and I_2, and eliminate p by considering the differences

$$\sigma_1 - \sigma_3 = 2\lambda_1^{-2}\lambda_2^{-2}(\lambda_1^4\lambda_2^2 - 1)(W_1 + \lambda_2^2 W_2), \tag{63}$$

$$\sigma_2 - \sigma_3 = 2\lambda_1^{-2}\lambda_2^{-2}(\lambda_1^2\lambda_2^4 - 1)(W_1 + \lambda_1^2 W_2), \tag{64}$$

in which (61) has been used to eliminate λ_3, also from the expressions in (38) for the invariants I_1 and I_2, so that

$$I_1 = \lambda_1^2 + \lambda_2^2 + \lambda_1^{-2}\lambda_2^{-2}, \quad I_2 = \lambda_1^{-2} + \lambda_2^{-2} + \lambda_1^2\lambda_2^2. \tag{65}$$

Equations (63) and (64) may be solved explicitly to give expressions for W_1 and W_2 in terms of the principal stretches λ_1 and λ_2 in the $(1, 2)$ plane and the principal Cauchy stress differences $\sigma_1 - \sigma_3$ and $\sigma_2 - \sigma_3$. The result is

$$W_1 = \frac{1}{2}\frac{\lambda_1^2(\sigma_1 - \sigma_3)}{(\lambda_1^2 - \lambda_2^2)(\lambda_1^2 - \lambda_3^3)} - \frac{1}{2}\frac{\lambda_2^2(\sigma_2 - \sigma_3)}{(\lambda_1^2 - \lambda_2^2)(\lambda_2^2 - \lambda_3^3)}, \tag{66}$$

$$W_2 = \frac{1}{2}\frac{(\sigma_2 - \sigma_3)}{(\lambda_1^2 - \lambda_2^2)(\lambda_2^2 - \lambda_3^3)} - \frac{1}{2}\frac{(\sigma_1 - \sigma_3)}{(\lambda_1^2 - \lambda_2^2)(\lambda_1^2 - \lambda_3^3)}, \tag{67}$$

where, for simplicity of expression, we have retained λ_3 rather than having eliminated it using (61).

Since W involves only two independent measures of deformation, I_1 and I_2 in this case, equations (66) and (67) in principle enable the general characters of the functions W_1 and W_2 to be determined from experimental tests involving pure homogeneous strain in which λ_1 and λ_2 are measured (and hence I_1 and I_2 known) provided the stress differences can also be determined from experiment. In practice, experiments are often conducted on thin planar specimens of material for which the transverse stress σ_3 vanishes (the plane stress condition). In the case of biaxial deformations in which λ_1 and λ_2 (or I_1 and I_2) are varied independently the forces applied in the 1 and 2 directions are measured and, knowing the initial cross-sectional areas, the principal Biot stresses t_1 and t_2 can be calculated, and hence the principal Cauchy stresses $\sigma_1 = \lambda_1 t_1$ and $\sigma_2 = \lambda_2 t_2$ for feeding into equations (66) and (67). A schematic of a biaxial tension test on an initially square specimen is shown in Figure 4.

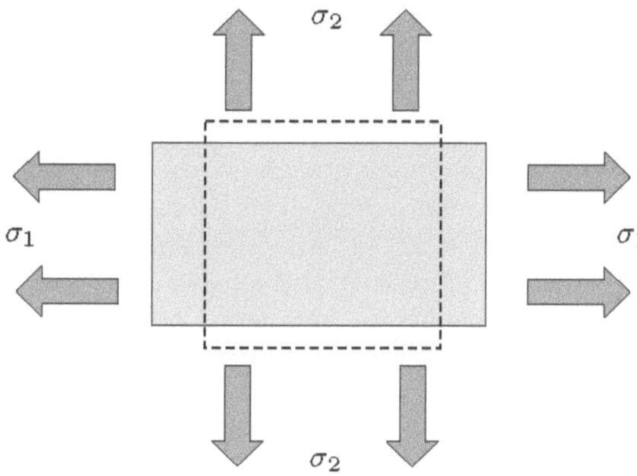

Figure 4. Schematic of a biaxial test on an initially square (shown dashed) thin sheet of isotropic material in the (X_1, X_2) plane with in-plane (principal) Cauchy stresses σ_1 and σ_2 and transverse normal stress $\sigma_3 = 0$.

A general question that arises is "What is the minimum set of experiments required to fully characterize the material properties?" In the case of incompressible isotropic elastic materials (assuming that the materials have first of all been shown to be isotropic), biaxial deformations of the kind discussed above (or their equivalent, such as extension/inflation tests on thin-walled circular cylindrical tubes) are needed and, although only two-dimensional in character, are sufficient to elicit the three-dimensional properties of the material. The tests must be conducted for as wide a range of values of λ_1 and λ_2 (equivalently I_1 and I_2) as possible. Only limited data are obtainable from tests involving only one independent deformation variable, such as simple tension or pure shear.

Thus, as indicated above, biaxial tests can quantify the general characters of W_1 and W_2 as functions of I_1 and I_2 and hence information about the dependence of W on I_1 and I_2 can be deduced. This procedure does not determine the functional form of W, but its functional form can be suggested by the numerical data and then tested further against the data from other tests. For an unconstrained material, of course, biaxial tests alone are not sufficient for characterizing the material properties since W then depends on three independent measures of deformation.

Some soft materials, including natural rubber and some synthetic rubberlike materials, are able to sustain large elastic deformations and can be treated as essentially isotropic relative to their natural configurations, and within certain limitations many soft biological tissues can also be treated as elastic and isotropic. Moreover, both rubberlike materials and most soft tissues can be treated as incompressible, and this enables us to focus on the incompressible theory here.

As we have seen, we may regard W as a function of I_1 and I_2 or equivalently of λ_1 and λ_2. We have also seen how, on the basis of biaxial experimental tests, W may be characterized as a function of I_1 and I_2. The alternative, that of using λ_1 and λ_2, provides a simpler framework than equations (66) and (67) for analyzing the results of biaxial tests, both in structure and interpretation, and we now focus on this option. We recall that for an incompressible material we have the constraint (61) and we therefore use this to cast W as a function of two independent stretches, namely λ_1 and λ_2, alone. For this purpose we define $\hat{W}(\lambda_1, \lambda_2)$, which is symmetric in λ_1 and λ_2, by

$$\hat{W}(\lambda_1, \lambda_2) = W(\lambda_1, \lambda_2, \lambda_1^{-1}\lambda_2^{-1}). \tag{68}$$

This enables p to be eliminated from equations $(35)_2$ by use of the chain rule for partial derivatives to obtain the stress differences

$$\sigma_1 - \sigma_3 = \lambda_1 \frac{\partial \hat{W}}{\partial \lambda_1}, \quad \sigma_2 - \sigma_3 = \lambda_2 \frac{\partial \hat{W}}{\partial \lambda_2}. \tag{69}$$

For the biaxial test discussed above $\sigma_3 = 0$ in equations (69) and then, in terms of the principal Biot stresses, we have simply

$$t_1 = \frac{\partial \hat{W}}{\partial \lambda_1}, \quad t_2 = \frac{\partial \hat{W}}{\partial \lambda_2}. \tag{70}$$

These two equations relate λ_1, λ_2 and t_1, t_2, and therefore provide a basis for characterizing $\hat{W}$ from measured biaxial data. Equations (70) form a simple basis for the interpretation of biaxial data for isotropic elastic materials when one of the stretches (λ_1 or λ_2) is held fixed as the other is varied. For more details of this approach we refer to the review articles by Ogden (1982, 1986).

Thus, we emphasize that once it is determined that a material is isotropic and incompressible, measurements of only two independent deformation variables, I_1 and I_2 or λ_1 and λ_2, for example, in biaxial tests and the associated applied forces, from which the stress components can be calculated, can provide sufficient information to enable the properties of the strain-energy function to be fully characterized. As mentioned above, biaxial tests alone are not sufficient to characterize the properties of a *compressible isotropic* material. Nor are they sufficient for any anisotropic material (compressible or incompressible). For incompressible transverse isotropy, which is the simplest form of anisotropy, this point is illustrated in the following.

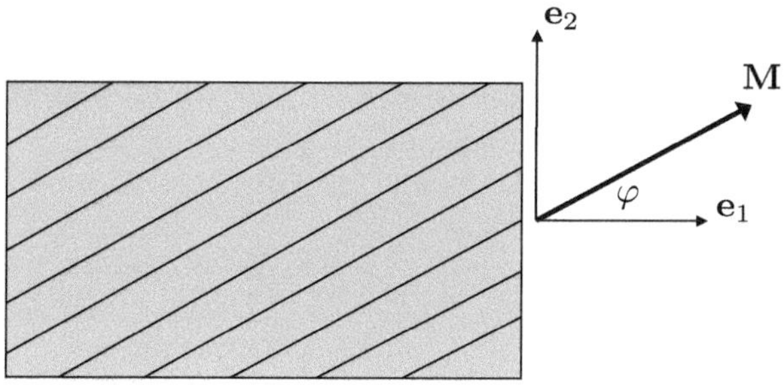

Figure 5. A rectangular sheet of fibre-reinforced material in its reference configuration $\mathcal{B}_r$ corresponding to the (X_1, X_2) plane in terms of rectangular Cartesian axes $\mathbf{e}_1, \mathbf{e}_2$. The fibre direction is $\mathbf{M} = \cos\varphi\,\mathbf{e}_1 + \sin\varphi\,\mathbf{e}_2$.

Fibre-reinforced materials: transverse isotropy. We now consider an isotropic material reinforced with a single family of continuously distributed fibres with direction given by the unit vector $\mathbf{M}$ in $\mathcal{B}_r$ such that the material is incompressible. This is an incompressible transversely isotropic material. Specifically, we consider a thin rectangular sheet of material lying in the $(1,2)$ plane with the fibre direction uniform and aligned also in the $(1,2)$ plane, which is defined by Cartesian basis vectors $\mathbf{e}_1, \mathbf{e}_2$, with $\mathbf{e}_3$ normal to the plane of the sheet. Let $\mathbf{M}$ have components $(\cos\varphi, \sin\varphi, 0)$, where φ is a constant. The geometrical arrangement in $\mathcal{B}_r$ is depicted in Figure 5. Because we are interested in fibre-reinforced materials we shall use the terminologies *fibre direction* and *preferred direction* interchangeably, as already indicated in the Introduction.

For the pure homogeneous strain (60), the matrix of components of $\mathbf{C}$ (also of $\mathbf{B}$) is $\mathrm{diag}[\lambda_1^2, \lambda_2^2, \lambda_3^2]$ and we therefore obtain from the definitions (48)

$$I_4 = \lambda_1^2\cos^2\varphi + \lambda_2^2\sin^2\varphi, \quad I_5 = \lambda_1^4\cos^2\varphi + \lambda_2^4\sin^2\varphi, \tag{71}$$

while I_1 and I_2 are given by (65).

The components of $\mathbf{m} = \mathbf{FM}$ are simply $m_1 = \lambda_1\cos\varphi$, $m_2 = \lambda_2\sin\varphi$. Hence, from (52) we obtain the components of the Cauchy stress $\boldsymbol{\sigma}$ as

$$\sigma_{11} = 2W_1\lambda_1^2 + 2W_2\lambda_1^2(\lambda_2^2+\lambda_3^2) + 2W_4\lambda_1^2\cos^2\varphi + 4W_5\lambda_1^4\cos^2\varphi - p, \tag{72}$$
$$\sigma_{22} = 2W_1\lambda_2^2 + 2W_2\lambda_2^2(\lambda_1^2+\lambda_3^2) + 2W_4\lambda_2^2\sin^2\varphi + 4W_5\lambda_2^4\sin^2\varphi - p, \tag{73}$$
$$\sigma_{12} = 2[W_4 + W_5(\lambda_1^2+\lambda_2^2)]\lambda_1\lambda_2\sin\varphi\cos\varphi, \tag{74}$$
$$\sigma_{33} = 2W_1\lambda_3^2 + 2W_2\lambda_3^2(\lambda_1^2+\lambda_2^2) - p, \tag{75}$$

together with $\sigma_{13} = \sigma_{23} = 0$. Note that if either $\varphi = 0$ or $\varphi = \pi/2$ then $\sigma_{12} = 0$, i.e. no shear stress is required to maintain the deformation. Otherwise, the pure homogeneous strain can only be maintained with the help of the shear stress σ_{12} given in (74).

By eliminating p from equations (72), (73) and (75) we obtain the stress differences

$$\begin{aligned}\sigma_{11} - \sigma_{33} &= 2\lambda_1^{-2}\lambda_2^{-2}(\lambda_1^4\lambda_2^2 - 1)(W_1 + \lambda_2^2 W_2) \\ &+ 2W_4\lambda_1^2\cos^2\varphi + 4W_5\lambda_1^4\cos^2\varphi, \qquad (76)\\ \sigma_{22} - \sigma_{33} &= 2\lambda_1^{-2}\lambda_2^{-2}(\lambda_1^2\lambda_2^4 - 1)(W_1 + \lambda_1^2 W_2) \\ &+ 2W_4\lambda_2^2\sin^2\varphi + 4W_5\lambda_2^4\sin^2\varphi, \qquad (77)\end{aligned}$$

analogous to (63) and (64) for an isotropic material and to which they reduce when $W_4 = W_5 = 0$.

From (65) and (71) we see that for the considered pure homogeneous strain I_1, I_2, I_4, I_5 depend on only two independent deformation variables, namely λ_1 and λ_2. They also depend on the angle φ, but this is not a deformation variable and can be regarded as a material parameter.

We can then adapt the notation $\hat{W}(\lambda_1, \lambda_2)$ for the present situation, but to emphasize the presence of fibre reinforcement here we include φ in the argument of $\hat{W}$ and write

$$\hat{W}(\lambda_1, \lambda_2, \varphi) = W(I_1, I_2, I_4, I_5), \qquad (78)$$

where I_1 and I_2 are given by (65) and I_4 and I_5 by (71). In the isotropic situation $\hat{W}$ is symmetric in λ_1 and λ_2. This is not the case in general for a transversely isotropic material, i.e. $\hat{W}(\lambda_1, \lambda_2, \varphi)$ is *not symmetric* in λ_1 and λ_2, except for the special angles $\varphi = \pi/4$ or $3\pi/4$.

By calculating the derivatives of $\hat{W}$ with respect to λ_1 and λ_2 and using (78) with (65) and (71), we obtain the simple formulas

$$\sigma_{11} - \sigma_{33} = \lambda_1 \frac{\partial \hat{W}}{\partial \lambda_1}, \quad \sigma_{22} - \sigma_{33} = \lambda_2 \frac{\partial \hat{W}}{\partial \lambda_2}, \qquad (79)$$

which are identical in structure to the corresponding equations (69) for an isotropic material, but, we emphasize, there are some differences in their content. First, because $\hat{W}$ is not in general symmetric in λ_1 and λ_2, second because σ_{11} and σ_{22} are not in general principal stresses ($\boldsymbol{\sigma}$ is not in general coaxial with $\mathbf{B}$), and the shear stress σ_{12} is in general required to maintain the deformation.

The strain energy $W(I_1, I_2, I_4, I_5)$ involves four independent (in general) invariants, and the expression (52), or (53), for the Cauchy stress contains four independent constitutive functions, W_1, W_2, W_4, W_5. The equations (74), (76) and (77) associated with the considered biaxial deformation, however, provide only three equations and it is not therefore possible to solve these equations uniquely for W_1, W_2, W_4, W_5 in a similar way to (66) and (67) for the isotropic case, which provide explicit equations for W_1 and W_2.

Mathematically, it is therefore *impossible to fully characterize* the elastic properties of transversely isotropic materials on the basis of biaxial tests alone, and hence it is not possible to deduce a specific mathematical form of W on the basis of experimentally measurable quantities obtained from biaxial tests alone. If, however, one assumes a particular form of W containing only three invariants from the outset then it is in principle possible to determine sufficient information from biaxial tests to be able to characterize the dependence of W on these three invariants.

For more detailed discussion of the issues involved and further analysis of transverse isotropy we refer to Ogden (2009) and Holzapfel and Ogden (2009c).

Fibre-reinforced materials: orthotropy. We now extend the above analysis to the situation in which there are two distinct preferred directions, $\mathbf{M}$ and $\mathbf{M}'$, in the reference configuration. Again, we consider the pure homogeneous strain of a thin sheet with the preferred directions lying in the (X_1, X_2) plane (see Figure 6) and we suppose that they are placed symmetrically with respect to the Cartesian axes and given by

$$\mathbf{M} = \cos\varphi\,\mathbf{e}_1 + \sin\varphi\,\mathbf{e}_2, \quad \mathbf{M}' = \cos\varphi\,\mathbf{e}_1 - \sin\varphi\,\mathbf{e}_2, \tag{80}$$

where, as in the transversely isotropic case, the angle φ is constant and $\mathbf{e}_1, \mathbf{e}_2$ denote the in-plane Cartesian coordinate directions. Let the corresponding vectors in the deformed configuration be denoted by $\mathbf{m}$ and $\mathbf{m}'$, which have components

$$m_1 = \lambda_1 \cos\varphi, \quad m_2 = \lambda_2 \sin\varphi, \quad m_1' = \lambda_1 \cos\varphi, \quad m_2' = -\lambda_2 \sin\varphi. \tag{81}$$

From the definitions (48) and (55) the invariants I_4, I_5, I_6, I_7 and I_8 are found to be

$$I_4 = I_6 = \lambda_1^2 \cos^2\varphi + \lambda_2^2 \sin^2\varphi, \quad I_5 = I_7 = \lambda_1^4 \cos^2\varphi + \lambda_2^4 \sin^2\varphi, \tag{82}$$

$$I_8 = \lambda_1^2 \cos^2\varphi - \lambda_2^2 \sin^2\varphi, \tag{83}$$

and I_1, I_2 are again given by (65).

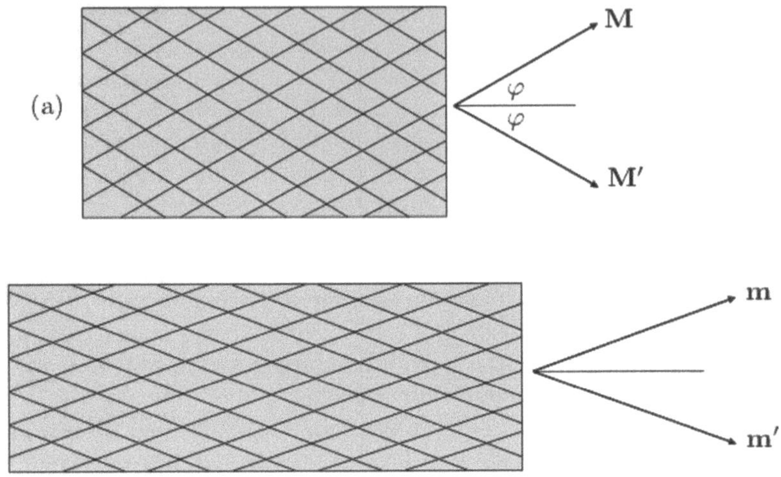

Figure 6. Pure homogeneous strain of a thin sheet of material with two in-plane symmetrically disposed families of fibres with directions $\mathbf{M}$ and $\mathbf{M}'$ at angles φ to the horizontal in the reference configuration (a), and $\mathbf{m}$ and $\mathbf{m}'$ in the deformed configuration (b).

The non-zero components of the Cauchy stress $\boldsymbol{\sigma}$ are now obtained from (59) in the form

$$\begin{aligned}\sigma_{11} &= 2W_1\lambda_1^2 + 2W_2(I_1\lambda_1^2 - \lambda_1^4) + 2(W_4 + W_6 + W_8)\lambda_1^2\cos^2\varphi \\ &\quad + 4(W_5 + W_7)\lambda_1^4\cos^2\varphi - p, &(84)\\ \sigma_{22} &= 2W_1\lambda_2^2 + 2W_2(I_1\lambda_2^2 - \lambda_2^4) + 2(W_4 + W_6 - W_8)\lambda_2^2\sin^2\varphi \\ &\quad + 4(W_5 + W_7)\lambda_2^4\sin^2\varphi - p, &(85)\\ \sigma_{12} &= 2[W_4 - W_6 + (W_5 - W_7)(\lambda_1^2 + \lambda_2^2)]\lambda_1\lambda_2\sin\varphi\cos\varphi, &(86)\\ \sigma_{33} &= 2W_1\lambda_3^2 + 2W_2(I_1\lambda_3^2 - \lambda_3^4) - p. &(87)\end{aligned}$$

These equations extend those for a single fibre family in (72)–(75) to the case of two families of fibres.

We now extend the notation $\hat{W}(\lambda_1, \lambda_2, \varphi)$ used for a transversely isotropic material to allow for the additional invariants and in this case define it by

$$\hat{W}(\lambda_1, \lambda_2, \varphi) = W(I_1, I_2, I_4, I_5, I_6, I_7, I_8), \tag{88}$$

where $I_1, I_2, I_4, \ldots, I_8$ are given by (65), (82) and (83). As in the case of a transversely isotropic material we emphasize that $\hat{W}$ is *not in general symmetric* with respect to interchange of λ_1 and λ_2. Then it is straightforward to show, by using equations (65), (82) and (83), that the stress differences given by the formulas (79) also apply in this case.

Note, in particular, that even though the preferred directions are symmetrically disposed the shear stress σ_{12} given by (86) is not in general zero, and the normal stresses σ_{11} and σ_{22} are not therefore principal stresses. They become principal stresses and the shear stress disappears in the following circumstances: first, if $\varphi = 0$ or $\pi/2$, in which case the two preferred directions coincide; second if both $W_4 = W_6$ and $W_5 = W_7$, which, since $I_4 = I_6$ and $I_5 = I_7$ for the considered deformation, implies that the two families of fibres have the same elastic properties. In the latter case the material is *orthotropic* and the axes of orthotropy coincide with the Cartesian axes.

In general, as is apparent from (84)–(86) and the general form of Cauchy stress (59), there are seven constitutive functions $W_1, W_2, W_4, \ldots, W_8$ that characterize the material properties so that only very limited information can be determined from biaxial experiments. In fact, equation (59), since the Cauchy stress is symmetric, contains only six equations and it is therefore impossible mathematically to solve the six equations uniquely for the seven unknown quantities, irrespective of the amount of stress/deformation data available. In any case experimental data that will distinguish the influences of all the different invariants and constitutive functions are not available, and because data are limited it is usual to restrict the strain-energy function to a subset of the invariants in order to make realistic correlations between theory and experiment. On this basis we shall refer to some particular constitutive models in Subsection 2.7.

2.6 Extension and Inflation of a Circular Cylindrical Tube

Thus far we have considered only homogeneous deformations. In this subsection we consider a basic non-homogeneous deformation that is relevant to arterial wall elasticity, namely the extension and inflation of a thick-walled circular cylindrical tube of elastic material containing two symmetrically disposed families of fibres with equal mechanical properties arranged helically around the tube. We shall then specialize to the approximation of a thin-walled (or membrane) tube.

We consider first a thick-walled circular cylindrical tube with reference geometry defined in cylindrical polar coordinates R, Θ, Z by

$$A \leq R \leq B, \quad 0 \leq \Theta \leq 2\pi, \quad 0 \leq Z \leq L, \tag{89}$$

where A and $B > A$ are the inner and outer radii of the tube and L is its length. Two families of helically arranged fibres, symmetric with respect

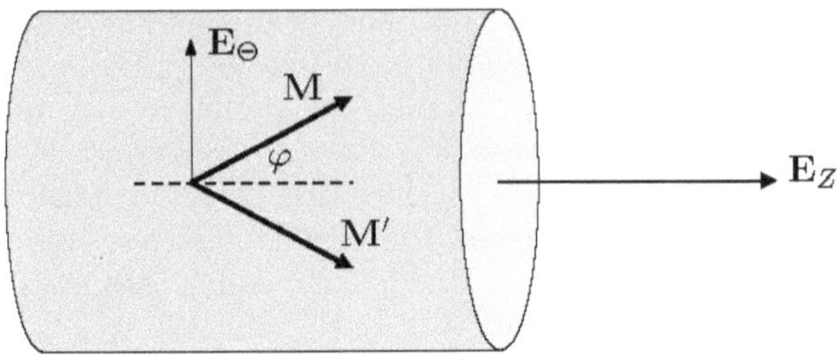

Figure 7. Fibre directions $\mathbf{M}$ and $\mathbf{M}'$ locally in the (Θ, Z) plane at radial distance R from the tube axis.

to the tube axis, are continuously distributed within the tube and have directions defined locally by the unit vectors

$$\mathbf{M} = \cos\varphi\,\mathbf{E}_Z + \sin\varphi\,\mathbf{E}_\Theta, \quad \mathbf{M}' = \cos\varphi\,\mathbf{E}_Z - \sin\varphi\,\mathbf{E}_\Theta, \tag{90}$$

with respect to unit basis vectors $\mathbf{E}_\Theta$ and $\mathbf{E}_Z$ associated with coordinates Θ and Z, as shown in the section of the cylinder at radius R depicted in Figure 7, where φ is the angle that, locally, each fibre makes with the axial direction shown in Figure 7. In general, φ may be allowed to depend on position in the tube, but here, because we wish to maintain symmetry under extension and inflation we allow φ to depend at most on R. We can think of taking the sheet in Figure 6(a) and folding it to form a cylinder so that the straight fibre directions in the sheet become helices in the cylinder.

Subject to appropriate axial and radial loading, the tube is deformed in such a way that its circular cylindrical shape is maintained. The deformation of the tube can then be described by the equations

$$r = f(R), \quad \theta = \Theta, \quad z = \lambda_z Z, \tag{91}$$

where r, θ, z are cylindrical polar coordinates in the deformed configuration associated with basis vectors $\mathbf{e}_r, \mathbf{e}_\theta, \mathbf{e}_z$, f is a function to be determined and λ_z is the (uniform) axial stretch of the tube.

We now assume that the material is incompressible, in which case $f(R)$ is determined explicitly as

$$r = f(R) = \sqrt{a^2 + \lambda_z^{-1}(R^2 - A^2)}, \tag{92}$$

where a is the internal radius of the deformed tube. Thus, $a = f(A)$, and we shall also use the notation $b = f(B)$.

Because of the symmetry, the radial, azimuthal and axial directions are principal directions of both the stretch tensors **U** and **V**, which coincide here, and the associated principal stretches $\lambda_r, \lambda_\theta, \lambda_z$ satisfy the incompressibility condition $\lambda_r\lambda_\theta\lambda_z = 1$. By introducing the notation λ for the azimuthal stretch λ_θ, we then have the connections

$$\lambda_r = \lambda^{-1}\lambda_z^{-1}, \quad \lambda = \lambda_\theta = \frac{r}{R}. \tag{93}$$

It is also convenient to introduce the notations

$$\lambda_a = a/A, \quad \lambda_b = b/B \tag{94}$$

for the values of λ on the inner and outer boundaries, respectively.

From (92) and (93) it can be seen that $(\lambda^2 - \lambda_z^{-1})R^2$ is independent of R, and hence, on use of the definitions (94), we obtain the connections

$$\lambda_a^2\lambda_z - 1 = \frac{R^2}{A^2}(\lambda^2\lambda_z - 1) = \frac{B^2}{A^2}(\lambda_b^2\lambda_z - 1). \tag{95}$$

Thus, $\lambda^2\lambda_z - 1$ has the same sign at every radial location R. In particular, it vanishes if $\lambda^2 = \lambda_z^{-1}$ for *all* $R \in [A, B]$, in which case λ is uniform and the deformation is homogeneous. For an isotropic material this corresponds to the deformation associated with uniaxial tension (with no internal or external radial traction). For the anisotropic material considered below, however, the deformation requires a radial traction since, unlike for an isotropic material, $\lambda\hat{W}_\lambda$, although constant if the fibre angle φ is independent of R, does not vanish, as can be seen, for example, by referring to equations (103) and (125) below. In particular, if there is no external radial traction then a 'pressure' on the inner boundary is required. This is given by equation (104) below and may be positive or negative depending on the values of λ_z and φ.

Suppose the tube is first extended to the fixed stretch λ_z, so that the azimuthal stretch is given by $\lambda^2 = \lambda_z^{-1}$, as above, and then the tube is inflated by an internal pressure. Then, the inequality

$$\lambda^2\lambda_z > 1 \quad \text{for} \quad A \leq R \leq B \tag{96}$$

holds as the inflation proceeds. On the other hand, during deflation $\lambda^2\lambda_z < 1$ for $A \leq R \leq B$ instead provided the circular cylindrical shape is maintained. However, under deflation by external pressure, loss of stability and shape of the tube is very likely in this case (see Zhu et al., 2008 for a discussion of such instability for an isotropic tube).

We have assumed that the two families of fibres have equal mechanical properties and that the cylindrical symmetry is maintained. Thus, there is no need for shear stresses to be applied, as was discussed for the pure homogeneous deformation of a sheet at the end Subsection 2.5. Thus the material response is orthotropic with axes of orthotropy locally coinciding with the cylindrical polar axes. The deformation considered is also locally of the form of a pure homogeneous strain and hence the strain-energy function can be written as $\hat{W}(\lambda, \lambda_z, \varphi)$, as in (88), but with λ_1 and λ_2 replaced by λ and λ_z, respectively.

This means that the stress components $\sigma_{rr}, \sigma_{\theta\theta}, \sigma_{zz}$ are principal stresses and hence, similarly to (79), the stress differences are given by

$$\sigma_{\theta\theta} - \sigma_{rr} = \lambda\frac{\partial\hat{W}}{\partial\lambda}, \quad \sigma_{zz} - \sigma_{rr} = \lambda_z\frac{\partial\hat{W}}{\partial\lambda_z} \tag{97}$$

with an appropriate re-ordering of the components.

As for the homogenous deformation of a thin rectangular sheet, $\hat{W}$ is not in general symmetric in the stretches, in this case λ and λ_z, and now the deformation is in general inhomogeneous since λ depends on R, as do all the terms in (97). The expressions for the invariants are identical to those in (65), (82) and (83) except for the change in notation for the stretches. Thus,

$$I_1 = \lambda^2 + \lambda_z^2 + \lambda^{-2}\lambda_z^{-2}, \quad I_2 = \lambda^{-2} + \lambda_z^{-2} + \lambda^2\lambda_z^2, \tag{98}$$

$$I_4 = I_6 = \lambda_z^2\cos^2\varphi + \lambda^2\sin^2\varphi, \quad I_5 = I_7 = \lambda_z^4\cos^2\varphi + \lambda^4\sin^2\varphi, \tag{99}$$

$$I_8 = \lambda_z^2\cos^2\varphi - \lambda^2\sin^2\varphi, \tag{100}$$

and φ may depend on R, in which case the material is inhomogeneous. In particular, it may be different in each different layer of a tube with multiple concentric layers. The strain-energy function $\hat{W}$ may also be different in each layer if the elastic properties of the layers differ, as is the case in typical arteries (see, or example, Holzapfel et al., 2005). For the remaining discussion in this subsection there is no need to consider specific forms of $\hat{W}$, and it can therefore be considered to be a general function of its arguments.

Because the deformation is inhomogeneous we need to apply the equilibrium equation. Here we adopt equation (15) but without a body force, and, because of the radial symmetry, it reduces to one component equation (the radial equation), namely

$$\frac{d\sigma_{rr}}{dr} + \frac{1}{r}(\sigma_{rr} - \sigma_{\theta\theta}) = 0. \tag{101}$$

With this we associate boundary conditions on the boundaries $r = a$ and $r = b$. We take these to be a pressure $P\,(\geq 0)$ on the inner boundary and zero traction on the outer boundary, and write them as

$$\sigma_{rr} = \begin{cases} -P & \text{on } r = a \\ 0 & \text{on } r = b. \end{cases} \tag{102}$$

Integration of (101) and use of $(97)_1$ and (102) leads to an expression for σ_{rr}:

$$\sigma_{rr} = -\int_r^b \lambda \hat{W}_\lambda \frac{\mathrm{d}r}{r}. \tag{103}$$

Then, by applying (102) again, we obtain an expression for the pressure P as a function of the inner radius a, bearing in mind that b depends on a through the expression $b = f(B)$ obtained from (92). Thus,

$$P = \int_a^b \lambda \hat{W}_\lambda \frac{\mathrm{d}r}{r}, \tag{104}$$

where $\hat{W}_\lambda = \partial \hat{W}/\partial \lambda$. We shall also use the shorthand notation $\hat{W}_{\lambda_z} = \partial \hat{W}/\partial \lambda_z$ in the following.

By using (92) with $\lambda = r/R$ we obtain $1 - \lambda_z^{-1}\lambda^{-2} = (a^2 - \lambda_z^{-1})r^{-2}$ and hence the differential connection

$$\frac{\mathrm{d}\lambda}{\lambda} = -(\lambda^2\lambda_z - 1)\frac{\mathrm{d}r}{r}. \tag{105}$$

Then, on changing the integration variable from r to λ in (104) we obtain the following alternative expression for P:

$$P = \int_{\lambda_b}^{\lambda_a} (\lambda^2\lambda_z - 1)^{-1} \hat{W}_\lambda \mathrm{d}\lambda. \tag{106}$$

This, in particular, highlights the role of the expression $\lambda^2\lambda_z - 1$ appearing in (95), which, as already noted, has the same sign for all $R \in [A, B]$.

The deformation is maintained by the combination of the pressure P on $r = a$ and axial loads on the ends of the the tube. The latter are obtained from the axial stress σ_{zz}, which is given by $(97)_2$ and clearly depends on r. From it we construct the resultant axial load, denoted N, on a tube cross section, which is given by

$$N = 2\pi \int_a^b \sigma_{zz} r \mathrm{d}r, \tag{107}$$

and is independent of z. By using (97) and the equilibrium equation this can be written

$$N = \pi \int_a^b (2\lambda_z \hat{W}_{\lambda_z} - \lambda \hat{W}_\lambda) r \, \mathrm{d}r + \pi \int_a^b \frac{\mathrm{d}}{\mathrm{d}r}(r^2 \sigma_{rr}) \, \mathrm{d}r. \tag{108}$$

The second term can be integrated explicitly to give, with the help of (102), $\pi a^2 P$, which is interpreted as the load contribution on the ends of a tube with closed ends due to the internal pressure. By again changing the integration variable to λ using (105) we obtain

$$N/\pi A^2 = (\lambda_a^2\lambda_z - 1)\int_{\lambda_b}^{\lambda_a}(\lambda^2\lambda_z - 1)^{-2}(2\lambda_z\hat{W}_{\lambda_z} - \lambda\hat{W}_\lambda)\lambda\mathrm{d}\lambda + P\lambda_a^2. \quad (109)$$

The term $N - P\pi a^2$ is the applied part of the axial load and known as the *reduced axial load*, here denoted F.

Typically in tubular soft tissues such as arteries the walls are relatively thin. With this in mind we now consider the thin-walled or membrane approximations for the pressure and axial load. These are based on consideration of the (uniform) wall thickness $B - A$ in the reference configuration, which we denote by H, and introduction of the small parameter $\varepsilon = H/A$. We then linearize the expressions for P and F in ε. First, from (95), we obtain the approximation

$$\lambda_a = \lambda_b + \varepsilon\lambda^{-1}\lambda_z^{-1}(\lambda^2\lambda_z - 1), \quad (110)$$

in which λ can take any value in $[\lambda_b, \lambda_a]$, while application of the mean value theorem to (106) gives

$$P = (\lambda_a - \lambda_b)\frac{\hat{W}_\lambda(\lambda, \lambda_z, \varphi)}{\lambda^2\lambda_z - 1} \quad (111)$$

for some $\lambda \in [\lambda_b, \lambda_a]$. The combination of the latter two equations yields the membrane approximation for P, i.e.

$$P = \varepsilon\lambda^{-1}\lambda_z^{-1}\hat{W}_\lambda(\lambda, \lambda_z, \varphi), \quad (112)$$

where λ is the azimuthal stretch in the membrane.

By the same procedure we obtain the corresponding approximation for the reduced axial load as

$$F = \varepsilon\pi A^2\lambda_z^{-1}[2\lambda_z\hat{W}_{\lambda_z}(\lambda, \lambda_z, \varphi) - \lambda\hat{W}_\lambda(\lambda, \lambda_z, \varphi)]. \quad (113)$$

For specific forms of strain-energy function we shall illustrate these results in the following subsection.

2.7 Constitutive Laws

We now illustrate the theory of the previous subsections by considering first, for an incompressible transversely isotropic material, uniaxial tension in the preferred direction. For this purpose we specialize the strain-energy function $\hat{W}(\lambda_1, \lambda_2, \varphi)$ defined in (78) by setting $\lambda_1 = \lambda$ and hence, by symmetry, $\lambda_2\,(=\lambda_3) = \lambda^{-1/2}$, together with $\varphi = 0$, and define $\tilde{W}(\lambda) = \hat{W}(\lambda, \lambda^{-1/2}, 0)$. The corresponding Biot stress is denoted t and given by $t = \tilde{W}'(\lambda)$.

Illustrative plots of t versus λ are shown in Figure 8 in dimensionless form for two representative strain-energy functions. These are the so-called *standard reinforcing model*

$$W(I_1, I_4) = \frac{1}{2}\mu(I_1 - 3) + \frac{1}{4}\mu\rho(I_4 - 1)^2, \tag{114}$$

and the *Fung-type exponential model*

$$W(I_1, I_4) = \frac{1}{2}\mu(I_1 - 3) + \frac{1}{4}\frac{k_1}{k_2}\{\exp[k_2(I_4 - 1)^2] - 1\}, \tag{115}$$

each of which depends on the two invariants I_1 and I_4, and where μ, ρ, k_1 and k_2 are non-negative material constants. The first term in each case is the isotropic neo-Hookean energy, while the second term quantifies the strength of the anisotropy through the constants ρ and k_1, k_2, respectively. Here we restrict our consideration of constitutive models to these two and their extensions (below) to the case of two fibre families. For discussion of some other constitutive models for fibre-reinforced materials we refer to Ogden (2003).

For the considered deformation, $I_1 = \lambda^2 + 2\lambda^{-1}$ and $I_4 = \lambda^2$ and hence

$$\tilde{W}(\lambda) = \frac{1}{2}\mu(\lambda^2 + 2\lambda^{-1} - 3) + \frac{1}{4}\mu\rho(\lambda^2 - 1)^2 \tag{116}$$

and

$$\tilde{W}(\lambda) = \frac{1}{2}\mu(\lambda^2 + 2\lambda^{-1} - 3) + \frac{1}{4}\frac{k_1}{k_2}\{\exp[k_2(\lambda^2 - 1)^2] - 1\}, \tag{117}$$

respectively for the two models. The corresponding uniaxial stresses are, respectively,

$$t = \tilde{W}'(\lambda) = \mu(\lambda - \lambda^{-2}) + \mu\rho(\lambda^3 - \lambda) \tag{118}$$

and

$$t = \tilde{W}'(\lambda) = \mu(\lambda - \lambda^{-2}) + k_1(\lambda^3 - \lambda)\exp[k_2(\lambda^2 - 1)^2]. \tag{119}$$

The dimensionless stress used in Figure 8 is defined by $t^* = t/\mu$.

The curves in Figure 8(a) correspond to (118) with t^* plotted against λ for several values of the anisotropy parameter ρ, including $\rho = 0$, which

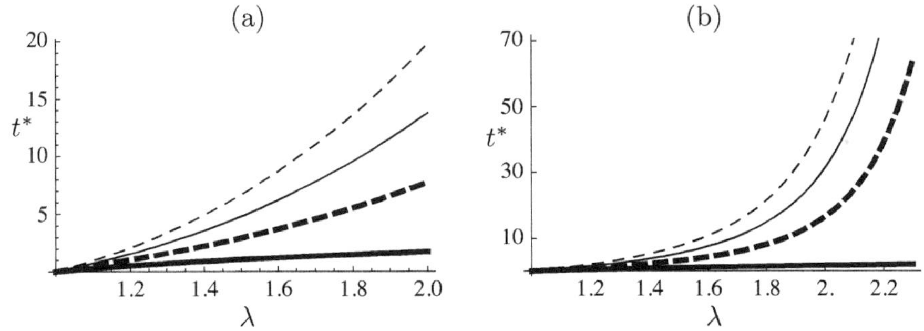

Figure 8. Plots of the dimensionless uniaxial stress $t^* = t/\mu$ versus the stretch λ based on equation (118) in (a) with $\rho = 0, 0.5, 1, 1.5$, respectively the thick continuous, thick dashed, continuous and dashed curves, and equation (119) in (b) with $k_1/\mu = 0, 1, 2, 3$, respectively the thick continuous, thick dashed, continuous and dashed curves, and $k_2 = 0.1$.

gives the result for the (isotropic) neo-Hookean material. As ρ increases the anisotropy becomes stronger and the material response stiffens. Because the model (119) is exponential the response stiffens rapidly, as illustrated in Figure 8(b) for several values of k_1, again including the neo-Hookean case (corresponding to $k_1 = 0$). The constant k_2 is fixed at the low value 0.1 to avoid excessive exponential growth.

We emphasize that the values of the constants chosen are purely illustrative to indicate the dependence on the anisotropy in qualitative terms. They do not necessarily correspond to values for real materials, although the anisotropic curves shown in Figure 8(b) do exhibit the qualitative behaviour of typical soft biological tissues.

Next, to illustrate the effect of fibre orientation we consider two strain-energy functions for a material with two families of fibres with equal mechanical properties. Extending the models (114) and (115) to this case by including the invariant I_6 we have

$$W(I_1, I_4, I_6) = \frac{1}{2}\mu(I_1 - 3) + \frac{1}{4}\mu\rho[(I_4 - 1)^2 + (I_6 - 1)^2] \tag{120}$$

and

$$W(I_1, I_4, I_6) = \frac{1}{2}\mu(I_1 - 3) + \frac{1}{4}\frac{k_1}{k_2}\{\exp[k_2(I_4 - 1)^2] + \exp[k_2(I_6 - 1)^2] - 2\}. \tag{121}$$

Based on the analysis in Subsection 2.6, we then have, since

$$I_4 = I_6 = \lambda^2 \sin^2\varphi + \lambda_z^2 \cos^2\varphi, \tag{122}$$

$$\hat{W}(\lambda, \lambda_z, \varphi) = \frac{1}{2}\mu(\lambda^2 + \lambda_z^2 + \lambda^{-2}\lambda_z^{-2} - 3) + \frac{1}{2}\mu\rho(\lambda^2 \sin^2\varphi + \lambda_z^2 \cos^2\varphi - 1)^2 \tag{123}$$

and

$$\begin{aligned}\hat{W}(\lambda, \lambda_z, \varphi) &= \frac{1}{2}\mu(\lambda^2 + \lambda_z^2 + \lambda^{-2}\lambda_z^{-2} - 3) \\ &+ \frac{1}{2}\frac{k_1}{k_2}\{\exp[k_2(\lambda^2 \sin^2\varphi + \lambda_z^2 \cos^2\varphi - 1)^2] - 1\}.\end{aligned} \tag{124}$$

Then,

$$\hat{W}_\lambda = \mu(\lambda - \lambda^{-3}\lambda_z^{-2}) + 2\mu\rho(\lambda^2 \sin^2\varphi + \lambda_z^2 \cos^2\varphi - 1)\lambda \sin^2\varphi \tag{125}$$

and

$$\begin{aligned}\hat{W}_\lambda &= \mu(\lambda - \lambda^{-3}\lambda_z^{-2}) + 2k_1(\lambda^2 \sin^2\varphi + \lambda_z^2 \cos^2\varphi - 1)\lambda \sin^2\varphi \\ &\times \exp[k_2(\lambda^2 \sin^2\varphi + \lambda_z^2 \cos^2\varphi - 1)^2],\end{aligned} \tag{126}$$

respectively. Similar derivatives are also obtained with respect to λ_z but are not written down explicitly.

In Figure 9, with $\lambda_z = 1$, the dimensionless pressure $P^* = P/\mu\varepsilon$ is plotted against λ for each of the two models with four values of the fibre angle φ in each case and for representative values of the material constants. Results for $\lambda_z > 1$ are qualitatively similar to those for $\lambda_z = 1$ and are not therefore plotted separately. For $\lambda_z < 1$, in the absence of internal pressure, the tube becomes unstable and this case is not therefore considered here. Figure 9 illustrates the strong dependence on fibre orientation. In particular, as the fibres approach the circumferential direction (increasing φ) the pressure required to reach a given circumferential stretch increases, i.e. the fibres have a stronger restraining effect on inflation.

The corresponding reduced axial load F is plotted against λ in Figure 10 in the dimensionless form $F^* = F/(\varepsilon\mu\pi A^2)$, again for the two models, also for $\lambda_z = 1$ and four values of the fibre angle φ. For the smaller values of φ, i.e. for fibres closer to the axial direction than the circumferential direction, there is initially, as λ increases from 1 under pressure, a tendency for the pressure to shorten the tube and an increasing positive (tensile) value of F is required in order to maintain $\lambda_z = 1$. Then, as inflation continues, F reaches a maximum and then becomes negative, so the trend is reversed

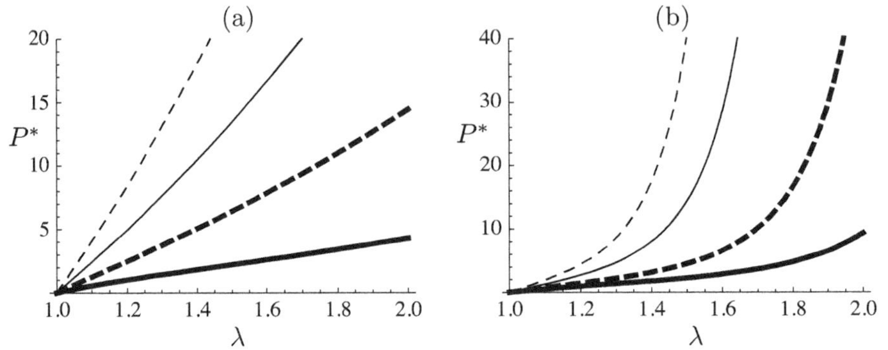

Figure 9. Plots of the dimensionless pressure $P^* = P/\varepsilon\mu$ versus the azimuthal stretch λ based on equation (112) with (125) in (a) with the representative value $\rho = 9$ and $\varphi = \pi/6, \pi/4, \pi/3, \pi/2$, respectively the thick continuous, thick dashed, continuous and dashed curves, and with equation (126) in (b) with $k_1/\mu = 2, k_2 = 1$ and $\varphi = \pi/5, \pi/4, \pi/3, 7\pi/16$, respectively the thick continuous, thick dashed, continuous and dashed curves.

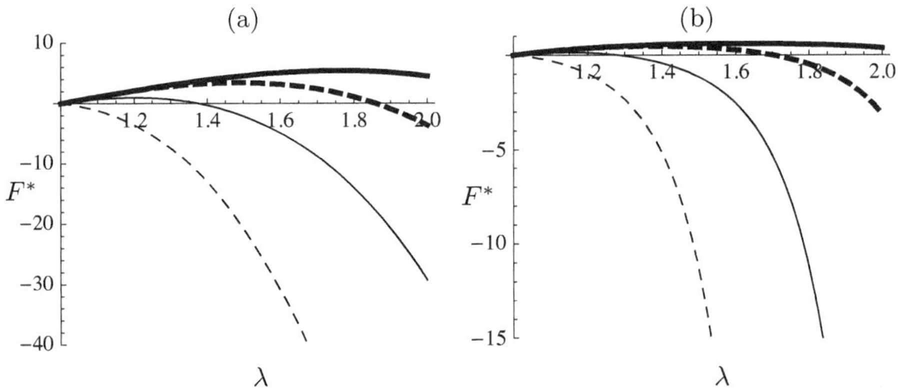

Figure 10. Plots of the dimensionless reduced axial force $F^* = F/(\varepsilon\mu\pi A^2)$ versus the azimuthal stretch λ based on equation (113) for the standard reinforcing model in (a) with the representative value $\rho = 9$ and for the Fung-type model in (b) with $k_1/\mu = 2, k_2 = 1$; in each case the values $\varphi = \pi/6, \pi/5, \pi/4, \pi/3$ are used, respectively the thick continuous, thick dashed, continuous and dashed curves.

and the tube wants to elongate. When the fibres are aligned closer to the circumferential direction, on the other hand, F becomes negative as soon as inflation begins and then decreases rapidly. The behaviour for the two models is qualitatively similar. Such 'switching' in response has been noted previously, both for isotropic materials (Haughton and Ogden, 1979) and for a Fung-type model (Holzapfel and Gasser, 2001).

The results shown in Figures 9 and 10 are for thin-walled (membrane) tubes. However, corresponding plots for thicker-walled tubes are broadly similar and are illustrated in Subsection 3.2. Although the membrane approximation gives a good qualitative picture of the pressure and axial load versus stretch behaviour it cannot account for, for example, the through-thickness stress distribution in artery walls and is not able to support the residual stresses that have an important influence on the mechanical response of arteries. See, for example, the papers by Holzapfel et al. (2000), Ogden and Schulze-Bauer (2000) and Ogden (2003) and references therein for detailed discussion of these features.

With this in mind we shall consider a thick-walled tube in the next section, so that we are able to take account of residual stress, which can be regarded as a structure tensor similar in character to the structure tensor $\mathbf{M}\otimes\mathbf{M}$ considered in respect of the fibre-reinforced materials herein. First, we develop the general theory of residual stress and, in particular, we describe how it is accounted for in the constitutive law of an elastic material.

3 Elasticity with Residual Stress

In the previous section the reference configuration $\mathcal{B}_\mathrm{r}$ was assumed to be stress free. We now drop this assumption and consider $\mathcal{B}_\mathrm{r}$ to be stressed, with residual (Cauchy) stress, which we denote by $\boldsymbol{\tau}$. The deformation is again measured from $\mathcal{B}_\mathrm{r}$ with deformation gradient $\mathbf{F}$, and the associated tensors from Section 2. The source of $\boldsymbol{\tau}$ does not concern us here. It may be associated with some prior material processing, plastic deformation or manufacturing process, for example, and is assumed to be known. It arises in the absence of body forces and surface tractions on the boundary $\partial\mathcal{B}_\mathrm{r}$ of the material body $\mathcal{B}_\mathrm{r}$. It is also assumed that it is not accompanied by intrinsic couple stresses, so that it is symmetric ($\boldsymbol{\tau}^\mathrm{T}=\boldsymbol{\tau}$) and therefore the rotational balance equations are satisfied in $\mathcal{B}_\mathrm{r}$ as well as the equilibrium equation

$$\mathrm{Div}\,\boldsymbol{\tau}=\mathbf{0}, \tag{127}$$

obtained by specializing (19) to the present situation, where Div again denotes the divergence operator with respect to $\mathbf{X}\in\mathcal{B}_\mathrm{r}$. Since there are no surface tractions, it must also satisfy the boundary condition

$$\boldsymbol{\tau}\mathbf{N} = \mathbf{0} \quad \text{on} \quad \partial\mathcal{B}_{\mathrm{r}}. \tag{128}$$

Note that $\boldsymbol{\tau}$ is a residual stress in the sense of Hoger (1985) and is distinguished from other types of initial stress, which may be associated with surface tractions. We emphasize at this point that residual stresses are necessarily non-uniform and the material that carries them is inhomogeneous.

We again consider a strain-energy function W defined per unit volume in $\mathcal{B}_{\mathrm{r}}$, but now W depends not only on the deformation gradient $\mathbf{F}$ but also the residual stress $\boldsymbol{\tau}$, which depends on $\mathbf{X}$. The dependence on $\boldsymbol{\tau}$ is included explicitly in the arguments of W and we write

$$W = W(\mathbf{C}, \boldsymbol{\tau}), \tag{129}$$

which is automatically objective since $\boldsymbol{\tau}$ is unaffected by rotations in the deformed configuration $\mathcal{B}$ and we have now included $\mathbf{C} = \mathbf{F}^{\mathrm{T}}\mathbf{F}$ instead of $\mathbf{F}$.

Note that even if the material has no preferred directions in the sense discussed in Section 2 the elastic properties of the material relative to $\mathcal{B}_{\mathrm{r}}$ are anisotropic, i.e. $\boldsymbol{\tau}$ has an effect on the constitutive law analogous to that of a structure tensor associated with preferred directions. We shall elaborate on this point shortly.

The presence of $\boldsymbol{\tau}$ does not affect the formulas for the nominal and Cauchy stresses (22) and $(24)_1$, except by the dependence of W on $\boldsymbol{\tau}$. They are given by

$$\mathbf{S} = \frac{\partial W}{\partial \mathbf{F}}(\mathbf{F}, \boldsymbol{\tau}), \quad \boldsymbol{\sigma} = J^{-1}\mathbf{F}\mathbf{S} = J^{-1}\mathbf{F}\frac{\partial W}{\partial \mathbf{F}}(\mathbf{F}, \boldsymbol{\tau}) \tag{130}$$

for an unconstrained material, and when these are evaluated in $\mathcal{B}_{\mathrm{r}}$ they each provide the restriction

$$\boldsymbol{\tau} = \frac{\partial W}{\partial \mathbf{F}}(\mathbf{I}, \boldsymbol{\tau}), \tag{131}$$

where $\mathbf{I}$ is the identity tensor in $\mathcal{B}_{\mathrm{r}}$.

For an incompressible material, the corresponding formulas are, from (23) and $(24)_2$,

$$\mathbf{S} = \frac{\partial W}{\partial \mathbf{F}}(\mathbf{F}, \boldsymbol{\tau}) - p\mathbf{F}^{-1}, \quad \boldsymbol{\sigma} = \mathbf{F}\mathbf{S} = \mathbf{F}\frac{\partial W}{\partial \mathbf{F}}(\mathbf{F}, \boldsymbol{\tau}) - p\mathbf{I}, \tag{132}$$

where $\mathbf{I}$ is the identity tensor in $\mathcal{B}$. When $\mathbf{F} = \mathbf{I}$ these both reduce to

$$\boldsymbol{\tau} = \frac{\partial W}{\partial \mathbf{F}}(\mathbf{I}, \boldsymbol{\tau}) - p^{(\mathrm{r})}\mathbf{I}, \tag{133}$$

where $p^{(\mathrm{r})}$ is the value of p in $\mathcal{B}_\mathrm{r}$.

Any symmetric second-order tensor can be expressed in spectral form in terms of its eigenvalues and eigenvectors. In particular, in terms of its eigenvalues, τ_1, τ_2, τ_3 and eigenvectors $\mathbf{M}_1, \mathbf{M}_2, \mathbf{M}_3$, say, $\boldsymbol{\tau}$ can be written

$$\boldsymbol{\tau} = \tau_1 \mathbf{M}_1 \otimes \mathbf{M}_1 + \tau_2 \mathbf{M}_2 \otimes \mathbf{M}_2 + \tau_3 \mathbf{M}_3 \otimes \mathbf{M}_3, \tag{134}$$

and when $\boldsymbol{\tau}$ is included in W, each $\mathbf{M}_i \otimes \mathbf{M}_i$, $i = 1, 2, 3$, has a role similar to the structure tensors discussed in Section 2. They are not all independent, however, and satisfy the identity

$$\mathbf{M}_1 \otimes \mathbf{M}_1 + \mathbf{M}_2 \otimes \mathbf{M}_2 + \mathbf{M}_3 \otimes \mathbf{M}_3 = \mathbf{I}, \tag{135}$$

where $\mathbf{I}$ is the identity tensor in $\mathcal{B}_\mathrm{r}$. Thus, $\boldsymbol{\tau}$ generates invariants, which form a basis for the functional dependence of W. In particular, by specializing $\boldsymbol{\tau}$ to a rank-one tensor, say $\mathbf{M} \otimes \mathbf{M}$, we recover the invariants associated with transverse isotropy. Thus, henceforth, rather than including preferred directions in the sense of Section 2, we now just include the invariants generated by $\mathbf{C}$ and $\boldsymbol{\tau}$.

3.1 Invariant Formulation

With $\boldsymbol{\tau}$ the strain-energy function W must satisfy the symmetry condition

$$W(\mathbf{Q}^*\mathbf{C}\mathbf{Q}^{*\mathrm{T}}, \mathbf{Q}^*\boldsymbol{\tau}\mathbf{Q}^{*\mathrm{T}}) = W(\mathbf{C}, \boldsymbol{\tau}), \tag{136}$$

analogously to (47), for all orthogonal $\mathbf{Q}^*$. This is equivalent to the dependence of W on 10 invariants of $\mathbf{C}$, $\boldsymbol{\tau}$ and their combination. These are typically taken to be, for $\mathbf{C}$,

$$I_1 = \mathrm{tr}\,\mathbf{C}, \quad I_2 = \frac{1}{2}[(\mathrm{tr}\,\mathbf{C})^2 - \mathrm{tr}\,(\mathbf{C}^2)], \quad I_3 = \det \mathbf{C}, \tag{137}$$

as for an isotropic material, for $\boldsymbol{\tau}$ similarly,

$$I_4 \equiv \left\{ \mathrm{tr}\,\boldsymbol{\tau},\ \frac{1}{2}[(\mathrm{tr}\,\boldsymbol{\tau})^2 - \mathrm{tr}\,(\boldsymbol{\tau}^2)],\ \det \boldsymbol{\tau} \right\}, \tag{138}$$

which are collectively denoted I_4 (this duplicates the notation I_4 from Section 2, which is not used in this section). Then, we take the set of independent invariants involving the combination of $\mathbf{C}$ and $\boldsymbol{\tau}$ to be

$$I_5 = \mathrm{tr}\,(\boldsymbol{\tau}\mathbf{C}), \quad I_6 = \mathrm{tr}\,(\boldsymbol{\tau}\mathbf{C}^2), \quad I_7 = \mathrm{tr}\,(\boldsymbol{\tau}^2\mathbf{C}), \quad I_8 = \mathrm{tr}\,(\boldsymbol{\tau}^2\mathbf{C}^2), \tag{139}$$

again duplicating notation from Section 2, but without danger of conflict.

The invariants of $\boldsymbol{\tau}$ are not affected by the deformation, while in the configuration $\mathcal{B}_\mathrm{r}$ the other invariants reduce to

$$I_1 = I_2 = 3, \quad I_3 = 1, \quad I_5 = I_6 = \mathrm{tr}\,\boldsymbol{\tau}, \quad I_7 = I_8 = \mathrm{tr}\,(\boldsymbol{\tau}^2). \tag{140}$$

We emphasize that the above set of 10 invariants, or an equivalent set of alternative invariants, forms a complete set of invariants of $\mathbf{C}$ and $\boldsymbol{\tau}$ in three dimensions. For an incompressible material, since $I_3 = 1$, there are nine independent invariants in general. When the dimension of the considered problem is reduced from three to two, such as for plane strain, the number of independent invariants is reduced, as was illustrated, for example, in Subsection 2.4 for transverse isotropy.

Stress tensors. Expanded expressions for the stress tensors given in (130) and (132) require the calculation of

$$\frac{\partial W}{\partial \mathbf{F}} = \sum_{i \in \mathcal{I}} W_i \frac{\partial I_i}{\partial \mathbf{F}}, \tag{141}$$

where $\mathcal{I}$ is the index set $\{1, 2, 3, 5, 6, 7, 8\}$ for an unconstrained material and $\{1, 2, 5, 6, 7, 8\}$ for incompressible material, and the notation $W_i = \partial W/\partial I_i$, $i \in \mathcal{I}$ has been adopted. Note that the derivative of I_4 with respect to $\mathbf{F}$ vanishes and so is not included in the above expression, but I_4 is included in the functional dependence of W.

For an unconstrained material the nominal and Cauchy stresses are thus obtained from (130) and (141) as

$$\mathbf{S} = \frac{\partial W}{\partial \mathbf{F}} = \sum_{i \in \mathcal{I}} W_i \frac{\partial I_i}{\partial \mathbf{F}}, \quad \boldsymbol{\sigma} = J^{-1}\mathbf{F}\mathbf{S}, \tag{142}$$

and the corresponding expressions for an incompressible material are

$$\mathbf{S} = \frac{\partial W}{\partial \mathbf{F}} - p\mathbf{F}^{-1} = \sum_{i \in \mathcal{I}} W_i \frac{\partial I_i}{\partial \mathbf{F}} - p\mathbf{F}^{-1}, \quad \boldsymbol{\sigma} = \mathbf{F}\mathbf{S}. \tag{143}$$

The expressions for $\partial I_i/\partial \mathbf{F}$, $i = 1, 2, 3$, are given by (43) and for $i = 5, 6, 7, 8$, bearing in mind the symmetry of $\boldsymbol{\tau}$, may be obtained analogously to (50) as

$$\frac{\partial I_5}{\partial \mathbf{F}} = 2\boldsymbol{\tau}\mathbf{F}^\mathrm{T}, \quad \frac{\partial I_6}{\partial \mathbf{F}} = 2(\boldsymbol{\tau}\mathbf{C}\mathbf{F}^\mathrm{T} + \mathbf{C}\boldsymbol{\tau}\mathbf{F}^\mathrm{T}), \tag{144}$$

$$\frac{\partial I_7}{\partial \mathbf{F}} = 2\boldsymbol{\tau}^2\mathbf{F}^\mathrm{T}, \quad \frac{\partial I_8}{\partial \mathbf{F}} = 2(\boldsymbol{\tau}^2\mathbf{C}\mathbf{F}^\mathrm{T} + \mathbf{C}\boldsymbol{\tau}^2\mathbf{F}^\mathrm{T}), \tag{145}$$

by replacing $\mathbf{M}\otimes\mathbf{M}$ in (50) by $\boldsymbol{\tau}$ on noting that $\mathbf{FM}=\mathbf{MF}^{\mathrm{T}}$.

The Cauchy stress in (142) then expands according to

$$\begin{aligned} J\boldsymbol{\sigma} &= 2W_1\mathbf{B}+2W_2(I_1\mathbf{B}-\mathbf{B}^2)+2I_3W_3\mathbf{I}+2W_5\boldsymbol{\Sigma}+2W_6(\boldsymbol{\Sigma}\mathbf{B}+\mathbf{B}\boldsymbol{\Sigma}) \\ &+ 2W_7\boldsymbol{\Sigma}\mathbf{B}^{-1}\boldsymbol{\Sigma}+2W_8(\boldsymbol{\Sigma}\mathbf{B}^{-1}\boldsymbol{\Sigma}\mathbf{B}+\mathbf{B}\boldsymbol{\Sigma}\mathbf{B}^{-1}\boldsymbol{\Sigma}), \end{aligned} \tag{146}$$

in which we have introduced the notation $\boldsymbol{\Sigma}=\mathbf{F}\boldsymbol{\tau}\mathbf{F}^{\mathrm{T}}$ for the Eulerian tensor which is the push-forward of $\boldsymbol{\tau}$. We also recall that $\mathbf{B}=\mathbf{FF}^{\mathrm{T}}$ is the left Cauchy–Green tensor.

The counterpart of (146) for an incompressible material is obtained from (143), yielding

$$\begin{aligned} \boldsymbol{\sigma} &= 2W_1\mathbf{B}+2W_2(I_1\mathbf{B}-\mathbf{B}^2)+2W_5\boldsymbol{\Sigma}+2W_6(\boldsymbol{\Sigma}\mathbf{B}+\mathbf{B}\boldsymbol{\Sigma}) \\ &+ 2W_7\boldsymbol{\Sigma}\mathbf{B}^{-1}\boldsymbol{\Sigma}+2W_8(\boldsymbol{\Sigma}\mathbf{B}^{-1}\boldsymbol{\Sigma}\mathbf{B}+\mathbf{B}\boldsymbol{\Sigma}\mathbf{B}^{-1}\boldsymbol{\Sigma})-p\mathbf{I}. \end{aligned} \tag{147}$$

By evaluating (146) in $\mathcal{B}_{\mathrm{r}}$ we obtain the specialization of (131) in the form

$$\boldsymbol{\tau}=2(W_1+2W_2+W_3)\mathbf{I}+2(W_5+2W_6)\boldsymbol{\tau}+2(W_7+2W_8)\boldsymbol{\tau}^2, \tag{148}$$

where all W_i, $i\in\mathcal{I}$, are evaluated for the invariants given by (140). Thus, following Shams et al. (2011), we obtain the restrictions

$$W_1+2W_2+W_3=0,\quad 2(W_5+2W_6)=1,\quad W_7+2W_8=0 \tag{149}$$

on the strain energy function in $\mathcal{B}_{\mathrm{r}}$, and the corresponding set of restrictions for an incompressible material is

$$2W_1+4W_2-p^{(\mathrm{r})}=0,\quad 2(W_5+2W_6)=1,\quad W_7+2W_8=0, \tag{150}$$

again evaluated in $\mathcal{B}_{\mathrm{r}}$.

Further details of the residual stress formulation are given in Shams et al. (2011) and, for a material also containing a preferred direction, in Ogden and Singh (2011), while an application to a prototype problem involving an inhomogeneous deformation, that of (plane strain) azimuthal shear of a circular cylindrical tube, for a residually-stressed material, is provided by Merodio et al. (2013). In the following we return to the problem of extension and inflation of a thick-walled tube considered in Subsection 2.6, but now including (radial and circumferential) residual stress.

3.2 Extension and Inflation of a Thick-walled Tube with Residual Stress

We consider again the extension and inflation of a thick-walled circular cylindrical tube of incompressible material discussed in Subsection 2.6 with circumferential and axial stretches λ and λ_z, initially without the angle φ associated with the preferred (fibre) directions. Instead, we include radial and circumferential residual stresses τ_{RR} and $\tau_{\Theta\Theta}$, but with no axial residual stress. Thus, the invariants in I_4 in (138) reduce to just $\tau_{RR}+\tau_{\Theta\Theta}$ and $\tau_{RR}\tau_{\Theta\Theta}$, with τ_{RR} and $\tau_{\Theta\Theta}$ satisfying the radial equation of equilibrium

$$\frac{\mathrm{d}\tau_{RR}}{\mathrm{d}R}+\frac{1}{R}(\tau_{RR}-\tau_{\Theta\Theta})=0 \tag{151}$$

and the boundary conditions

$$\tau_{RR}=0 \quad \text{on} \quad R=A \text{ and } B, \tag{152}$$

the specializations of (127) and (128), respectively, to the present geometry.

The invariants I_1 and I_2 are again given by (98), I_4 as above, and

$$I_5=\lambda^{-2}\lambda_z^{-2}\tau_{RR}+\lambda^2\tau_{\Theta\Theta},\quad I_6=\lambda^{-4}\lambda_z^{-4}\tau_{RR}+\lambda^4\tau_{\Theta\Theta}, \tag{153}$$

$$I_7=\lambda^{-2}\lambda_z^{-2}\tau_{RR}^2+\lambda^2\tau_{\Theta\Theta}^2,\quad I_8=\lambda^{-4}\lambda_z^{-4}\tau_{RR}^2+\lambda^4\tau_{\Theta\Theta}^2. \tag{154}$$

Then, by considering the strain-energy function to have the form

$$\hat{W}(\lambda,\lambda_z,\tau_{RR}+\tau_{\Theta\Theta},\tau_{RR}\tau_{\Theta\Theta}) \tag{155}$$

and using (147), we obtain the Cauchy stress differences

$$\sigma_{\theta\theta}-\sigma_{rr}=\lambda\hat{W}_\lambda,\quad \sigma_{zz}-\sigma_{rr}=\lambda_z\hat{W}_{\lambda_z}, \tag{156}$$

similarly to (97), although $\hat{W}$ is different in this case.

At this point we specialize both the form of the residual stress and the strain-energy function in order to provide explicit illustrations of the effect of residual stress on the material response. First, we choose a simple form of τ_{RR} satisfying the boundary conditions (152) and then use (151) to determine $\tau_{\Theta\Theta}$. Specifically, we take

$$\tau_{RR}=\alpha(R-A)(R-B), \tag{157}$$

and hence

$$\tau_{\Theta\Theta}=\alpha[3R^2-2(A+B)R+AB], \tag{158}$$

where $\alpha\,(>0)$ is a constant. Plots of $\tau_{RR}^*=\tau_{RR}/\alpha$ and $\tau_{\Theta\Theta}^*=\tau_{\Theta\Theta}/\alpha$ are shown in Figure 11. Their behaviours are very similar to those arising from the residual stresses calculated for a single layer artery wall from the so-called 'opening angle' method (Ogden, 2003) or the assumption that the circumferential stress at a typical physiological pressure is uniform (Ogden and Schulze-Bauer, 2000), and are therefore realistic.

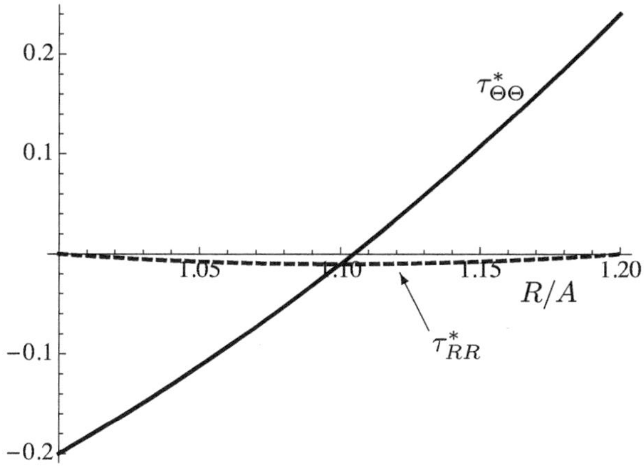

Figure 11. Residual stresses τ^*_{RR} (dashed curve) and $\tau^*_{\Theta\Theta}$ (continuous curve) plotted as functions of R/A for $B/A = 1.2$, a typical value for artery walls.

Next, we specialize the strain-energy function W by including only the invariants I_1 and I_5, with I_4 implicit: $W(I_1, I_5)$. The Cauchy stress (147) then simplifies to

$$\boldsymbol{\sigma} = 2W_1\mathbf{B} + 2W_5\boldsymbol{\Sigma} - p\mathbf{I}, \tag{159}$$

and the restrictions (150) reduce to

$$2W_1 - p^{(\mathrm{r})} = 0, \quad 2W_5 = 1 \tag{160}$$

in $\mathcal{B}_\mathrm{r}$.

We then simplify further by taking

$$W = \frac{1}{2}\mu(I_1 - 3) + \frac{1}{2}(I_5 - \mathrm{tr}\,\boldsymbol{\tau}), \tag{161}$$

so that $W = 0$ and (160) is satisfied with $p^{(\mathrm{r})} = \mu$ in $\mathcal{B}_\mathrm{r}$, where $\mu\,(> 0)$ is a constant.

The strain-energy function (155) then becomes

$$\hat{W} = \frac{1}{2}\mu(\lambda^2+\lambda_z^2+\lambda^{-2}\lambda_z^{-2}-3)+\frac{1}{2}[\lambda^{-2}\lambda_z^{-2}\tau_{RR}+\lambda^2\tau_{\Theta\Theta}-(\tau_{RR}+\tau_{\Theta\Theta})], \tag{162}$$

and the stress differences (156) specialize to

$$\sigma_{\theta\theta} - \sigma_{rr} = \lambda\hat{W}_\lambda = \mu(\lambda^2 - \lambda^{-2}\lambda_z^{-2}) + \lambda^2\tau_{\Theta\Theta} - \lambda^{-2}\lambda_z^{-2}\tau_{RR}, \tag{163}$$

and

$$\sigma_{zz} - \sigma_{rr} = \lambda_z \hat{W}_{\lambda_z} = \mu(\lambda_z^2 - \lambda^{-2}\lambda_z^{-2}) - \lambda^{-2}\lambda_z^{-2}\tau_{RR}. \tag{164}$$

We now evaluate the integrals in the expressions for the pressure P and the reduced axial load $F = N - \pi a^2 P$ from the integrals in (104) and (108) for the case $\lambda_z = 1$. This requires the expressions

$$\lambda\hat{W}_\lambda = (\mu+\alpha AB)(\lambda^2-\lambda^{-2})-\alpha(A+B)R(2\lambda^2-\lambda^{-2})+\alpha R^2(3\lambda^2-\lambda^{-2}) \tag{165}$$

and

$$\begin{aligned} 2\hat{W}_{\lambda_z} - \lambda\hat{W}_\lambda &= 2\mu - (\mu + \alpha AB)(\lambda^2 + \lambda^{-2}) \\ &\quad + \alpha(A+B)R(2\lambda^2 + \lambda^{-2}) - \alpha R^2(3\lambda^2 + \lambda^{-2}), \end{aligned} \tag{166}$$

which are obtained by substituting for the expressions (157) and (158) into (163) and (164) with $\lambda_z = 1$.

The integrals are evaluated, with the help of the definition $\lambda = r/R$, using the radial part of the deformation in the form $r^2 - R^2 = c$ (again for $\lambda_z = 1$), where the notation $c = a^2 - A^2 = b^2 - B^2$ has been introduced. The resulting expressions for P and F, after some manipulation, are obtained as

$$\begin{aligned} P = {} & (\mu + \alpha AB)\log\left(\frac{aB}{Ab}\right) + \frac{1}{2}\mu c\frac{B^2 - A^2}{a^2b^2} + 2\alpha c\log\left(\frac{b}{a}\right) \\ & + \frac{3}{2}\alpha\sqrt{c}(A+B)\left[\tan^{-1}\left(\frac{B}{\sqrt{c}}\right) - \tan^{-1}\left(\frac{A}{\sqrt{c}}\right)\right] \end{aligned} \tag{167}$$

and

$$\begin{aligned} F &= \pi c\left\{(\mu + \alpha AB)\log\left(\frac{bA}{aB}\right) - \alpha c\log\left(\frac{b}{a}\right)\right. \\ &\quad \left. + \alpha\sqrt{c}(A+B)\left[\tan^{-1}\left(\frac{B}{\sqrt{c}}\right) - \tan^{-1}\left(\frac{A}{\sqrt{c}}\right)\right]\right\}. \end{aligned} \tag{168}$$

Inclusion of $\lambda_z \neq 1$ does not alter the qualitative form of the results, so solutions for $\lambda_z \neq 1$ are not given.

For numerical purposes we now non-dimensionalize all the quantities in the above two equations. First, we set $P^* = P/\mu$ and $F^* = F/(\pi\mu A^2)$, which differ from the non-dimensionalizations used for the membrane model in Subsection 2.7. We also introduce the notations

$$\eta = \frac{B}{A}, \quad \alpha^* = \frac{\alpha A^2}{\mu}, \tag{169}$$

and, as a measure of the radial inflation,

$$e = \frac{\sqrt{c}}{A} \equiv \sqrt{\lambda_a^2 - 1}. \tag{170}$$

The connections

$$\frac{b^2}{a^2} = \frac{\eta^2 + e^2}{1 + e^2}, \quad \lambda_b^2 = 1 + \eta^{-2} e^2 \tag{171}$$

then follow and hence P^* and F^* can be written

$$\begin{aligned} P^* &= \frac{1}{2}(1 + \alpha^* \eta) \log \left[\frac{(1 + e^2)\eta^2}{\eta^2 + e^2} \right] + \alpha^* e^2 \log \left(\frac{\eta^2 + e^2}{1 + e^2} \right) \\ &+ \frac{1}{2} \frac{(\eta^2 - 1) e^2}{(\eta^2 + e^2)(1 + e^2)} + \frac{3}{2} \alpha^* (1 + \eta) e \tan^{-1} \left[\frac{(\eta - 1) e}{\eta + e^2} \right], \end{aligned} \tag{172}$$

$$\begin{aligned} F^* &= -\frac{1}{2}(1 + \alpha^* \eta) e^2 \log \left[\frac{(1 + e^2)\eta^2}{\eta^2 + e^2} \right] - \frac{1}{2} \alpha^* e^4 \log \left(\frac{\eta^2 + e^2}{1 + e^2} \right) \\ &+ \alpha^* (1 + \eta) e^3 \tan^{-1} \left[\frac{(\eta - 1) e}{\eta + e^2} \right]. \end{aligned} \tag{173}$$

Illustrative plots of P^* and F^* versus e are shown in Figures 12 and 13, for a residual stress with $\alpha^* = 1$ and $\alpha^* = 15$, respectively. The qualitative nature of P^* is unaffected by the change in the value of α^*, although its value increases significantly (essentially linearly) with α^*. Each term of P^* in (172) is positive. On the other hand, the character of F^* changes because of the competing effects of the positive and negative terms in (173). For $\alpha^* = 1$, F^* is negative for all the considered values of η (a measure of the tube thickness) and increases in magnitude as η increases. Thus, the tube has a tendency to elongate under pressure. But as α^* increases, although its magnitude is hardly affected, F^* turns from negative to positive for the thicker-walled tubes, in which case the tube has a tendency to shorten under pressure.

We next include both the residual stress and the two fibre families based on the standard reinforcing model (120). For the present geometry with $\lambda_z = 1$ the expression for $\lambda \hat{W}_\lambda$ from (125) and the corresponding expression for $\lambda_z \hat{W}_{\lambda_z}$ require the contributions

$$2\mu\rho(\lambda^4 - \lambda^2) \sin^4 \varphi, \quad 2\mu\rho(\lambda^2 - 1) \sin^2 \varphi \cos^2 \varphi, \tag{174}$$

respectively. It is then easily shown that their contributions to the integrals, in dimensionless form, are (with φ taken to be independent of R), for P^*,

$$\rho(1 - \eta^{-2}) e^2 \sin^4 \varphi, \tag{175}$$

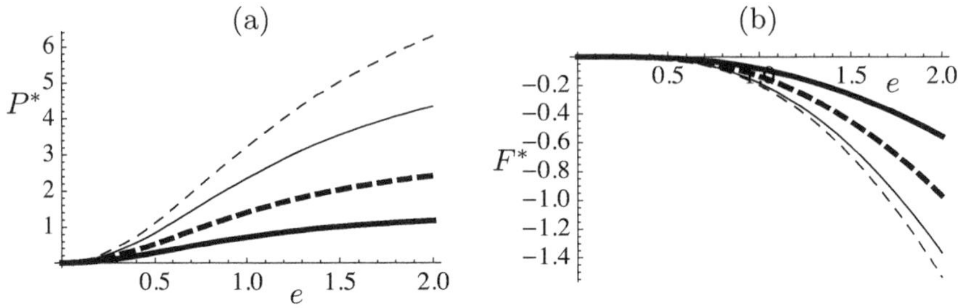

Figure 12. Plots of (a) P^* and (b) F^* versus e based on equations (172) and (173) for $\alpha^* = 1$ and $\eta = 1.2, 1.4, 1.7, 2$, corresponding to the thick continuous, thick dashed, continuous and dashed curves, respectively.

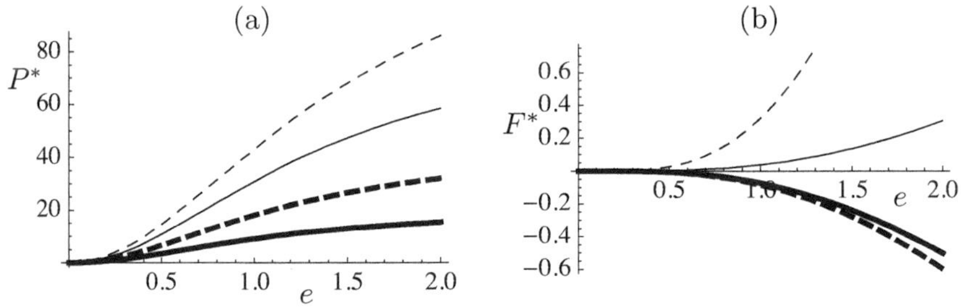

Figure 13. Plots of (a) P^* and (b) F^* versus e based on equations (172) and (173) for $\alpha^* = 15$ and $\eta = 1.2, 1.4, 1.7, 2$, corresponding to the thick continuous, thick dashed, continuous and dashed curves, respectively.

and, for F^*,

$$2\rho \log(\eta) e^2 \sin^2 \varphi (2 - 3 \sin^2 \varphi) - \rho (1 - \eta^{-2}) e^4 \sin^4 \varphi. \qquad (176)$$

With these terms included we provide plots of P^* and F^* in Figure 14 for the representative parameter values $\alpha^* = 4$, $\eta = 1.6$, $\rho = 9$ and for four values of φ, as indicated in the figure caption. For comparison, in Figure 15, we show results for $\alpha^* = 0$ (no residual stress). The latter plots of P^* and F^*, based on the standard reinforcing model, are the thick-walled counterparts of Figures 9(a) and 10(a) for the membrane tube and they show similar features, although the differences in the abscissas should be borne in mind.

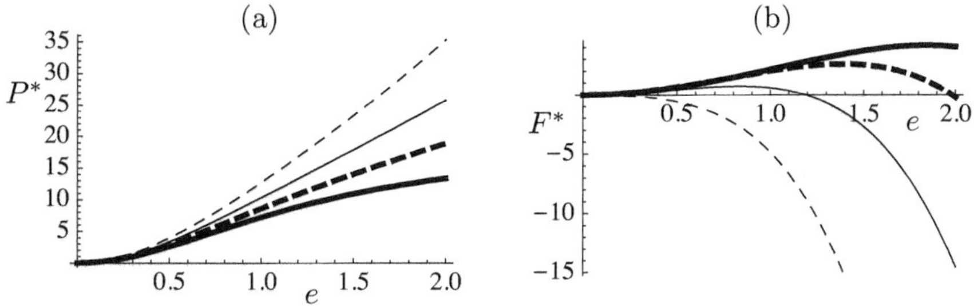

Figure 14. Plots of (a) P^* and (b) F^* versus e based on equations (172) and (173) with the fibre contributions from the additional terms (175) and (176) included. The parameter values are $\alpha^* = 4, \eta = 1.6, \rho = 9$, and the following values of the fibre angle φ: (a) $0, \pi/4, \pi/3, \pi/2$; (b) $\pi/6, \pi/5, \pi/4, \pi/3$. In each of (a) and (b) these correspond to the thick continuous, thick dashed, continuous and dashed curves, respectively.

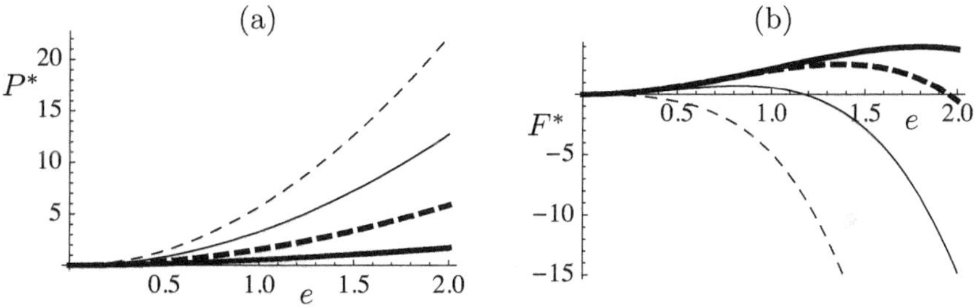

Figure 15. Plots of (a) P^* and (b) F^* versus e based on equations (172) and (173) with the fibre contributions from the additional terms (175) and (176) included. The parameter values are $\alpha^* = 0, \eta = 1.6, \rho = 9$, and the following values of the fibre angle φ: (a) $\pi/6, \pi/4, \pi/3, \pi/2$; (b) $\pi/6, \pi/5, \pi/4, \pi/3$. In each of (a) and (b) these correspond to the thick continuous, thick dashed, continuous and dashed curves, respectively.

Figures 14 and 15 show that, as far as F^* is concerned, the effect of the fibre reinforcement dominates the effect of the residual stress since F^* is virtually identical in the two cases, and this has also been found to be the case for larger values of α^*. On the other hand, the residual stress has a significant effect on P^*. In particular, in the presence of residual stress a much larger value of P^* is required to achieve a given level of inflation than when the residual stress is absent. As the value of α^* is increased the residual stress dominates the behaviour of P^*, which becomes very similar for all values of the fibre orientation (not shown here).

Bibliography

Y. B. Fu and R. W. Ogden. *Nonlinear Elasticity: Theory and Applications.* Cambridge University Press, 2001.

T. C. Gasser, R. W. Ogden, and G. A. Holzapfel. Hyperelastic modelling of arterial layers with distributed collagen fibre orientations. *J. R. Soc. Interface*, 3:15–35, 2006.

D. M. Haughton and R. W. Ogden. Bifurcation of inflated circular cylinders of elastic material under axial loading – I. Membrane theory for thin-walled tubes. *J. Mech. Phys. Solids*, 27:179–212, 1979.

A. Hoger. On the residual stress possible in an elastic body with material symmetry. *Arch. Rational Mech. Anal.*, 88:271–290, 1985.

G. A. Holzapfel. *Nonlinear Solid Mechanics.* Wiley, 2000.

G. A. Holzapfel and T. C. Gasser. A viscoelastic model for fiber-reinforced materials at finite strains: continuum basis, computational aspects and applications. *Comput. Meth. Appl. Mech. Engr.*, 190:4379–4403, 2001.

G. A. Holzapfel, T. C. Gasser, and R. W. Ogden. A new constitutive framework for arterial wall mechanics and a comparative study of material models. *J. Elasticity*, 61:1–48, 2000.

G. A. Holzapfel and R. W. Ogden, editors, *Biomechanics of Soft Tissue in Cardiovascular Systems.* CISM Courses and Lectures no. 441. Springer, Wien, 2003.

G. A. Holzapfel and R. W. Ogden, editors, *Proceedings of the IUTAM Symposium on Mechanics of Biological Tissue*, Graz 2004. Springer, Heidelberg, 2006.

G. A. Holzapfel and R. W. Ogden. Constitutive modelling of passive myocardium. A structurally-based framework for material characterization. *Phil. Trans. R. Soc. Lond. A*, 367:3445–3475, 2009a.

G. A. Holzapfel and R. W. Ogden, editors, *Biomechanical Modelling at the Molecular, Cellular and Tissue Levels.* CISM Courses and Lectures no. 508. Springer, Wien, 2009b.

G. A. Holzapfel and R. W. Ogden. On planar biaxial tests for anisotropic nonlinearly elastic solids. A continuum mechanical framework. *Math. Mech. Solids*, 14:474–489, 2009.

G. A. Holzapfel and R. W. Ogden. Constitutive modelling of arteries. *Proc. R. Soc.Lond. A*, 466:1551–1596, 2010.

G. A. Holzapfel, G. Sommer, T. C. Gasser, and P. Regitnig. Determination of the layer-specific mechanical properties of human coronary arteries with non-atherosclerotic intimal thickening, and related constitutive modelling. *Am. J. Physiol. Heart Circ. Physiol.*, 289:H2048–2058, 2005.

J. D. Humphrey. Mechanics of the arterial wall: review and directions. *Critical Reviews in Biomed. Engr.*, 23:1–162, 1995.

J. D. Humphrey. *Cardiovascular Solid Mechanics. Cells, Tissues, and Organs.* Springer, New York, 2002.

I.-S. Liu. On representations of anisotropic invariants. *Int. J. Eng. Sci.*, 20:1099–1109, 1982.

J. Merodio and R. W. Ogden. Material instabilities in fiber-reinforced nonlinearly elastic solids under plane deformation. *Arch. Mech.*, 54:525–552, 2002.

J. Merodio, R. W. Ogden, and J. Rodriguez. The influence of residual stress on finite deformation elastic response. *Int. J. Non-Linear Mech.*, 56:43–49, 2013.

R. W. Ogden, Elastic deformation of rubberlike solids. In H. G. Hopkins and M. J. Sewell, editors, *Mechanics of Solids, The Rodney Hill 60th Anniversary Volume*, pages 499–537. Pergamon Press, Oxford, 1982.

R. W. Ogden, Recent advances in the phenomenological theory of rubber elasticity. *Rubber Chemistry and Technology*, 59:361–383, 1986.

R. W. Ogden. *Non-linear Elastic Deformations.* Dover, New York, 1997.

R. W. Ogden. Nonlinear elasticity, anisotropy and residual stresses in soft tissue. In G. A. Holzapfel and R. W. Ogden, editors, *Biomechanics of Soft Tissue in Cardiovasular Systems*, CISM Courses and Lectures Series no. 441, pages 65–108. Springer, Wien, 2003.

R. W. Ogden. Anisotropy and nonlinear elasticity in arterial wall mechanics. In G. A. Holzapfel and R. W. Ogden, editors, *Biomechanical Modelling at the Molecular, Cellular and Tissue Levels*, CISM Courses and Lectures Series no. 508, pages 179–258. Springer, Wien, 2009.

R. W. Ogden and C. A. J. Schulze-Bauer. Phenomenological and structural aspects of the mechanical response of arteries. In J. Casey and G. Bao, editors, *Mechanics in Biology.* AMD–Vol. 242/BED–Vol. 46, pages 125–140. ASME, New York, 2000.

R. W. Ogden and B. Singh. Propagation of waves in an incompressible transversely isotropic elastic solid with initial stress: Biot revisited. *J. Mech. Mater. Structures*, 6:453–477, 2011.

M. Shams, M. Destrade, and R. W. Ogden. Initial stresses in elastic solids: constitutive laws and acoustoelasticity. *Wave Motion*, 48:552–567, 2011.

A. J. M. Spencer. *Deformations of Fibre-reinforced Materials.* Oxford University Press, 1972.

A. J. M. Spencer. Constitutive theory for strongly anisotropic solids. In A. J. M. Spencer, editor, *Continuum Theory of the Mechanics of Fibre-reinforced Composites*, CISM Courses and Lectures Series no. 282, pages 1–32. Springer, Wien, 1984.

Y. Zhu, X. Y. Luo, and R. W. Ogden. Asymmetric bifurcations of thick-walled circular cylindrical elastic tubes under axial loading and external pressure. *Int. J. Solids Structures*, 45:3410–3429, 2008.

Porous Materials with Statistically Oriented Reinforcing Fibres

Salvatore Federico

Department of Mechanical and Manufacturing Engineering
The University of Calgary, Calgary, Alberta, Canada

E-mail: salvatore.federico@ucalgary.ca

Abstract. A starting point for modelling biological tissues is to represent them as biphasic continua, with a porous solid phase saturated by an interstitial fluid and reinforced by collagen fibres. In this chapter, modelling techniques for fibre-reinforced composite materials with statistical orientation of the fibres are presented in the monophasic case (solid only), and in the biphasic case (incompressible solid and fluid phases). Both the elastic potential and the permeability are evaluated under large deformations, and hints for the numerical implementation of these models are given throughout.

1 Introduction

The solid matrix of biological tissues is a porous material comprised by various macromolecules, among which collagen fibres, and saturated by an interstitial fluid, essentially water, in which several chemical species are dissolved (Fung, 1993). Often, the fluid flow within the tissue is negligible and the whole system can be represented as comprised exclusively of an incompressible solid. For tissues such articular cartilage, in which the fluid flow plays a central mechanical and biological role, a biphasic, solid-fluid representation is imperative (and a multiphasic representation becomes necessary if the dissolved chemical species are to be taken into account). This chapter is aimed at the description of modelling techniques for biphasic fibre-reinforced materials, with particular attention to the case in which the fibres have a statistical orientation, i.e., an arrangement governed by a given probability distribution of orientation. The focus is on the evaluation of the elastic potential (also called elastic strain energy) and the permeability of the system. The Reader is assumed to be familiar with Linear Algebra,

L. Dorfmann, R. W. Ogden (Eds.), *Nonlinear Mechanics of Soft Fibrous Materials*,
CISM International Centre for Mechanical Sciences
DOI 10.1007/978-3-7091-1838-2_2 © CISM Udine 2015

Differential Calculus and Continuum Mechanics. A review of some fundamental concepts in these fields is given in order to elucidate the notation and justify the level of complexity that has been chosen.

In the most general treatment of Continuum Mechanics, a body is represented as a differentiable manifold and its configurations as a family of embeddings in physical space, parameterised by time. The level chosen here avoids such "extreme" concepts from Differential Geometry, and let bodies dwell in the familiar setting of affine spaces, in the fashion of likewise familiar open sets. Still, the notation and the (fairly long) mathematical prologue most probably need a good deal of effort to be digested. In particular, the choice has been made to keep the distinction between a vector space and its dual or, in one of the many other ways in which this can be said, the distinction between vectors and covectors or, again, the distinction between objects with contravariant and covariant components. While this may seem to be excessive, particularly if Cartesian coordinates are employed, it is very helpful to deeply understand the transformation laws governing all objects in Continuum Mechanics. At any rate, all this suffering was deemed to be a reasonable price to pay to get a little more acquainted with Modern Continuum Mechanics, and the subtleties arising from its sophisticated mathematical structure.

Before getting into the real business, I would like to add a little personal note. I am sure that this chapter contains many mistypings, as well as some (hopefully not too many) plain errors, and many points that could have been explained much better. I shall be grateful to anyone who will point these faults out. May I wish you a good reading.

1.1 Synopsis

Section 2 is a review of Linear Algebra, with the definitions and proofs necessary for the Continuum Mechanics notation employed. Continuum Kinematics is reviewed in Section 3, material symmetries (isotropy and transverse isotropy, in particular) in Section 4, and hyperelasticity in Section 5. Section 6 treats the case of monophasic (solid-only) fibre-reinforced materials, for the case of aligned and statistically oriented fibres. Section 7 recalls the basics of biphasic (solid-fluid) systems in the case of incompressible solid and fluid phases, and reports a methodology for enforcing incompressibility at compaction, i.e., when all the fluid has escaped. Section 8 is devoted to the general case of porous fibre-reinforced material and, aside from putting together the results of Sections 6 and 7 for elasticity, gives full detail of the derivation of the permeability tensor of the system.

1.2 Notation

The Linear Algebra and Continuum Mechanics notation follows generally Marsden and Hughes (1983), although in several instances that from the treatises by Eringen (1980), Ogden (1997) and Epstein (2010) is employed. The Reader shall be given a "warning" whenever an "original" notation (coming mainly from some of our previous works: Federico and Grillo, 2012; Federico, 2012, 2014; Bellini and Federico, 2014) is being used. Einstein's summation convention, according to which a term containing two repeated indices (one high, one low) is summed from one to the dimension of the space, is enforced throughout.

2 Review of Linear Algebra and Tensor Algebra

This section is dedicated to a review of Linear Algebra, as well as to a brief introduction to affine spaces. This permits to study Continuum Mechanics at a sufficient level of detail, without an excessive complexity that would bring us out of the scope of this Course.

The Reader is assumed to be already well acquainted with the concepts and definitions of vector space (we shall focus on finite-dimensional vector spaces on the real numbers $\mathbb{R}$) and basis of a vector space, from which this section shall expand. Please note that there is no claim of extreme rigour nor of completeness: the interested Reader is referred to the textbook by Epstein (2010), which goes far beyond this brief review, and covers all the Differential Geometry needed to get to the highest level of generalisation, constituted by differentiable manifolds.

The notation is preparatory to that employed in Continuum Mechanics. When talking about one space only, the notation shall resemble that of the spatial picture of Mechanics, with lowercase symbols representing scalars, lowercase boldface symbols reserved to vectors and covectors (in this section we also make the additional distinction between Latin symbols $\boldsymbol{u}, \boldsymbol{v}, \boldsymbol{w}, \ldots$ for vectors, and Greek symbols $\boldsymbol{\varphi}, \boldsymbol{\psi}, \boldsymbol{\omega}, \ldots$ for covectors), as well as to second-order tensors $(\boldsymbol{a}, \boldsymbol{b}, \boldsymbol{c}, \ldots)$, and "small" blackboard bold symbols reserved to higher-order tensors $(\mathbb{a}, \mathbb{b}, \mathbb{c}, \ldots)$. When maps between two spaces are involved, object in the domain space will be denoted by uppercase letters (e.g., $\boldsymbol{U}, \boldsymbol{V}, \boldsymbol{W} \ldots$, for vectors, $\boldsymbol{\Phi}, \boldsymbol{\Psi}, \boldsymbol{\Omega}, \ldots$ for covectors, $\boldsymbol{A}, \boldsymbol{B}, \boldsymbol{C} \ldots$, for second-order tensors, $\mathbb{A}, \mathbb{B}, \mathbb{C}, \ldots$ for higher-order tensors), as in the material picture of Mechanics, and objects in the codomain space by lowercase letters, as in the spatial picture.

2.1 Linear Maps

Here, we recall the definition of linear map between two vector spaces and show that the set of all such maps constitutes a vector space.

Let $\mathcal{U}$ and $\mathcal{V}$ be two real vector spaces. A map $\boldsymbol{L} : \mathcal{U} \to \mathcal{V}$ is said to be linear if, and only if,

$$\boldsymbol{L}(\boldsymbol{U} + \boldsymbol{V}) = \boldsymbol{L}(\boldsymbol{U}) + \boldsymbol{L}(\boldsymbol{V}), \quad \forall \boldsymbol{U}, \boldsymbol{V} \in \mathcal{U}, \tag{1a}$$

$$\boldsymbol{L}(\lambda \boldsymbol{U}) = \lambda\, \boldsymbol{L}(\boldsymbol{U}), \quad \forall \boldsymbol{U} \in \mathcal{U},\ \forall \lambda \in \mathbb{R}. \tag{1b}$$

It is straightforward to prove that the two properties in Equation (1) can be condensed into the single property

$$\boldsymbol{L}(\lambda \boldsymbol{U} + \mu \boldsymbol{V}) = \lambda\, \boldsymbol{L}(\boldsymbol{U}) + \mu\, \boldsymbol{L}(\boldsymbol{V}), \quad \forall \boldsymbol{U}, \boldsymbol{V} \in \mathcal{U},\ \forall \lambda, \mu \in \mathbb{R}, \tag{2}$$

i.e., *the image of a linear combination of vectors is the linear combination of the images of each vector.* A linear map can also be called a *homomorphism* (from the Greek ὁμός, *the same*, and μορφή, *form*) between vector spaces, as it is a map that preserves the linear structure when mapping from domain to codomain.

The set of all linear maps (or homomorphisms) from $\mathcal{U}$ to $\mathcal{V}$ is denoted $\mathrm{Hom}(\mathcal{U}, \mathcal{V})$, and has the structure of vector space. We can actually prove that a set much more general than $\mathrm{Hom}(\mathcal{U}, \mathcal{V})$ has the structure of vector space, and we can treat $\mathrm{Hom}(\mathcal{U}, \mathcal{V})$ as a particular case. Indeed, the fact that the domain $\mathcal{V}$ is a vector space is *not* utilised in the proof.

Let $\mathcal{A}$ be a non-empty set, and $\mathcal{V}$ a real vector space, let ${}^{\mathcal{A}}\mathcal{V}$ denote the set of all functions mapping elements of $\mathcal{A}$ into elements of $\mathcal{V}$, and let $f, g, \ldots$ denote the elements of ${}^{\mathcal{A}}\mathcal{V}$. Relying on the fact that $\mathcal{V}$ is a vector space, we are allowed to define the sum of elements of ${}^{\mathcal{A}}\mathcal{V}$ and the multiplication of scalars of $\mathbb{R}$ by elements of ${}^{\mathcal{A}}\mathcal{V}$ as

$$(f + g)(x) = f(x) + g(x), \quad (\lambda\, f)(x) = \lambda\, f(x), \quad \forall x \in \mathcal{A}, \tag{3}$$

because $f(x) + g(x)$ is a sum in $\mathcal{V}$ and $\lambda\, f(x)$ is the multiplication of $\lambda \in \mathbb{R}$ with the vector $f(x) \in \mathcal{V}$. It is easy to verify that these operations satisfy the properties of sum and multiplication by scalars in a vector space. Therefore, ${}^{\mathcal{A}}\mathcal{V}$ is a vector space and so is the particular case $\mathrm{Hom}(\mathcal{U}, \mathcal{V})$ in which the generic set $\mathcal{A}$ is replaced by a vector space $\mathcal{U}$ and the functions are all linear.

If $\mathcal{U}$ has dimension m and $\mathcal{V}$ has dimension n, then $\mathrm{Hom}(\mathcal{U}, \mathcal{V})$ has dimension $n \times m$. Let $L \in \mathrm{Hom}(\mathcal{U}, \mathcal{V})$, $\{\boldsymbol{E}_I\}_{I=1}^m$ be a basis of $\mathcal{U}$ and $\{\boldsymbol{e}_i\}_{i=1}^n$ be a basis of $\mathcal{V}$. For the linearity of $\boldsymbol{L}$, we have that

$$\boldsymbol{v} = \boldsymbol{L}(\boldsymbol{U}) = \boldsymbol{L}(U^I\, \boldsymbol{E}_I) = U^I\, \boldsymbol{L}(\boldsymbol{E}_I). \tag{4}$$

Moreover, since $\boldsymbol{v} = \boldsymbol{L}(\boldsymbol{U}) = U^I\, \boldsymbol{L}(\boldsymbol{E}_I) \in \mathcal{V}$, it can be written in the basis $\{\boldsymbol{e}_i\}_{i=1}^n$, and we have

$$v^i\, \boldsymbol{e}_i = [\boldsymbol{L}(\boldsymbol{U})]^i\, \boldsymbol{e}_i = U^I\, [\boldsymbol{L}(\boldsymbol{E}_I)]^i\, \boldsymbol{e}_i. \tag{5}$$

The scalars

$$L^i{}_I = [\boldsymbol{L}(\boldsymbol{E}_I)]^i \tag{6}$$

constitute the $n \times m$ matrix $[\![L^i{}_I]\!]$, called the *matrix of the linear map* $\boldsymbol{L}$ with respect to the bases $\{\boldsymbol{E}_I\}_{I=1}^m$ in the domain and $\{\boldsymbol{e}_i\}_{i=1}^n$ in the codomain. Note that the columns of $[\![L^i{}_I]\!]$ are given by the components, written in the basis $\{\boldsymbol{e}_i\}_{i=1}^n$, of the images through $\boldsymbol{L}$ of the vectors $\{\boldsymbol{E}_I\}_{I=1}^m$ of the basis of the domain. This proves that the dimension of $\mathrm{Hom}(\mathcal{U}, \mathcal{V})$ is $n \times m$, and allows for writing

$$v^i = L^i{}_I\, U^I, \tag{7}$$

which in Engineering texts is often reported in the matrix formalism

$$\underset{n\times 1}{\{v\}} = \underset{n\times m}{[L]}\ \underset{m\times 1}{\{U\}}. \tag{8}$$

The subset of all bijective (i.e., injective and surjective, and therefore invertible) linear maps in $\mathrm{Hom}(\mathcal{U}, \mathcal{V})$ is a vector subspace, the elements of which are called *isomorphisms*. If the domain and codomain space coincide, a linear map is called an *endomorphism* and one writes $\mathrm{Hom}(\mathcal{V}, \mathcal{V}) = \mathrm{End}(\mathcal{V})$. A bijective endomorphism, i.e., a linear map that is at the same time endomorphism and isomorphism, is called an *automorphism*.

2.2 Dual Space of a Vector Space

The notion of dual space of a vector space has fundamental importance in Physics and Mechanics. The elements of the dual space are called covectors and are linear operators whose action on a vector of the "original" vector space yields a scalar. A prime example in Mechanics is the concept of force, which is not (as usually taught, and as it shall also be assumed here) a vector, but a covector that, acting on a (virtual) velocity vector, yields a scalar called (virtual) power.

Let $\mathcal{V}$ be a real vector space of dimension n, and let us denote by $\mathcal{V}^\star$ the space $\mathrm{Hom}(\mathcal{V}, \mathbb{R})$ of all linear maps $\varphi : \mathcal{V} \to \mathbb{R}$, mapping $\mathcal{V}$ into the real numbers. Since $\mathbb{R}$ is a vector space on $\mathbb{R}$, the space $\mathcal{V}^\star = \mathrm{Hom}(\mathcal{V}, \mathbb{R})$ has the structure of vector space, as seen in Section 2.1. The dimension of $\mathcal{V}^\star = \mathrm{Hom}(\mathcal{V}, \mathbb{R})$ is $1 \times n = n$, i.e., $\mathcal{V}^\star$ has the same dimension of $\mathcal{V}$. Although this descends directly from what has been shown in Section 2.1, as the particular case in which the codomain space is $\mathbb{R}$, it is very insightful to explicitly prove that the dimension of $\mathcal{V}^\star$ is equal to that of $\mathcal{V}$.

Let $\{\boldsymbol{e}_i\}_{i=1}^n$ be a basis for $\mathcal{V}$, in which a vector $\boldsymbol{u} \in \mathcal{V}$ is written $\boldsymbol{u} = u^i\, \boldsymbol{e}_i$, and let $\boldsymbol{\varphi} \in \mathcal{V}^\star$. Because of the linearity of $\boldsymbol{\varphi}$,

$$\boldsymbol{\varphi}(\boldsymbol{u}) = \boldsymbol{\varphi}(u^i\, \boldsymbol{e}_i) = u^i\, \boldsymbol{\varphi}(\boldsymbol{e}_i) = u^i\, \varphi_i, \tag{9}$$

were we denoted $\boldsymbol{\varphi}(\boldsymbol{e}_i) = \varphi_i$. Now, let us define the n linear maps

$$\boldsymbol{e}^i : \mathcal{V} \to \mathbb{R} : \boldsymbol{u} \mapsto u^i, \tag{10}$$

called *projections with respect to the basis* $\{\boldsymbol{e}_i\}_{i=1}^n$, each mapping vector $\boldsymbol{u}$ into its i-th component u^i with respect to the basis $\{\boldsymbol{e}_i\}_{i=1}^n$. Note that this definition implies

$$\boldsymbol{e}^i(\boldsymbol{e}_j) = \delta^i{}_j, \tag{11}$$

where $\delta^i{}_j$ is the Kronecker symbol, such that

$$\delta^i{}_j = \begin{cases} 1 & \text{if } i = j, \\ 0 & \text{if } i \neq j. \end{cases} \tag{12}$$

Using the projections in Equation (9), we obtain

$$\boldsymbol{\varphi}(\boldsymbol{u}) = \varphi_i\, \boldsymbol{e}^i(\boldsymbol{u}), \tag{13}$$

which expresses $\boldsymbol{\varphi}(\boldsymbol{u})$ as the linear combination of $\boldsymbol{e}^i(\boldsymbol{u})$ with coefficients φ_i. If we now drop the argument $\boldsymbol{u}$ in both sides of Equation (13), we get

$$\boldsymbol{\varphi} = \varphi_i\, \boldsymbol{e}^i, \tag{14}$$

i.e., every element $\boldsymbol{\varphi} \in \mathcal{V}^\star$ can be expressed as the linear combination of the elements $\{\boldsymbol{e}^i\}_{i=1}^n$, which therefore constitute a basis of $\mathcal{V}^\star$. Hence, $\mathcal{V}^\star$ has dimension n. More importantly, we have defined a basis of $\mathcal{V}^\star$ by means of the projections $\{\boldsymbol{e}^i\}_{i=1}^n$ with respect to the basis vectors $\{\boldsymbol{e}_i\}_{i=1}^n$ of $\mathcal{V}$.

The space $\mathcal{V}^\star$ is called the *dual space* of $\mathcal{V}$, its elements are called *linear forms* or *one-forms*, or *covectors* (we shall normally employ the latter terminology), the covector basis $\{\boldsymbol{e}^i\}_{i=1}^n$ is called the *dual basis* of $\{\boldsymbol{e}_i\}_{i=1}^n$, and the coefficients φ_i in Equation (14) are called the components of $\boldsymbol{\varphi}$ with respect to the basis $\{\boldsymbol{e}^i\}_{i=1}^n$.

The expression $\boldsymbol{\varphi}(\boldsymbol{u})$ of Equation (9) is also called the *contraction* of the covector $\boldsymbol{\varphi}$ and the vector $\boldsymbol{u}$. An alternative notation for this contraction is that due to Dirac,

$$\boldsymbol{\varphi}(\boldsymbol{u}) = \langle \boldsymbol{\varphi} | \boldsymbol{u} \rangle, \tag{15}$$

where, in the convention of Functional Analysis, normally the *bra* on the left is always the covector and the *ket* on the right is always the vector. We shall use the bra-ket notation in a symmetric manner (see below).

It is possible to prove (Theorem of the Canonical Isomorphism) that the space $\mathcal{V}^{\star\star}$ of the linear maps from $\mathcal{V}^{\star}$ into $\mathbb{R}$, called the *bidual* of $\mathcal{V}$, is *canonically isomorphic* to $\mathcal{V}$, i.e., there is a bijective relation between $\mathcal{V}$ and $\mathcal{V}^{\star\star}$ that *does not* depend on the choice of a basis in $\mathcal{V}$. Therefore, we can consider the bidual $\mathcal{V}^{\star\star}$ to coincide with $\mathcal{V}$, i.e., $\mathcal{V}^{\star\star} \equiv \mathcal{V}$.

The most important consequence of this fundamental result is that vectors in $\mathcal{V}$ can be regarded as linear forms on the covectors of $\mathcal{V}^{\star}$ and that the action of a vector on a covector yields the same result as the action of the same covector on the same vector, i.e.,

$$\boldsymbol{u}(\boldsymbol{\varphi}) = \boldsymbol{\varphi}(\boldsymbol{u}). \tag{16}$$

The Theorem of the Canonical Isomorphism also allows for using Dirac's bra-ket notation in the symmetric fashion

$$\langle \boldsymbol{u} | \boldsymbol{\varphi} \rangle = \langle \boldsymbol{\varphi} | \boldsymbol{u} \rangle. \tag{17}$$

While we shall use the bra-ket notation for a particular type of contraction of higher-order tensors (see Section 2.5), we shall normally simply juxtapose the covector and the vector, omitting the parentheses of Equation (16) or the bra-ket notation of Equation (17):

$$\boldsymbol{u}\,\boldsymbol{\varphi} = \boldsymbol{\varphi}\,\boldsymbol{u}. \tag{18}$$

Note also that, if we contract both sides of Equation (14) with the basis vector $\boldsymbol{e}_j$, we have

$$\boldsymbol{e}_j(\boldsymbol{\varphi}) = \boldsymbol{e}_j(\varphi_i\,\boldsymbol{e}^i) = \varphi_i\,\boldsymbol{e}_j(\boldsymbol{e}^i) = \varphi_i\,\delta^i{}_j = \varphi_j, \tag{19}$$

i.e., the vectors of the basis $\{\boldsymbol{e}_i\}_{i=1}^n$ of $\mathcal{V}$ are the projections of the covectors of $\mathcal{V}^{\star}$:

$$\boldsymbol{e}_i : \mathcal{V}^{\star} \to \mathbb{R} : \boldsymbol{\varphi} \mapsto \varphi_i. \tag{20}$$

Remark 2.1. The way in which the contraction of a covector with a vector is indicated in the literature is various. For instance, Marsden and Hughes (1983) prefer a dot, $\boldsymbol{\varphi} \cdot \boldsymbol{u}$, whereas Epstein (2010) opts for a notation similar to the bra-ket one, $\langle \boldsymbol{\varphi}, \boldsymbol{u} \rangle$, which we will use for a different purpose (see Section 2.7). We also remark that, in the traditional Quantum Mechanics notation, Dirac's bra-ket contraction $\langle \cdot | \cdot \rangle$ actually occurs between two vectors, meaning that the bra-ket involves a scalar (or inner) product (see Section 2.7). However, we do not feel too guilty, as the scalar product automatically induces an isomorphism between a vector space and its dual, and therefore a bra can rightfully be considered as a covector in this isomorphism (this will be clarified in Section 2.7). At any rate, for the case of a vector and a covector, we shall normally prefer the simple juxtaposition, which is what one normally does in the elementary multiplication of a row matrix (covector, bra) with a column matrix (vector, ket), or the contraction of a second-order "mixed" tensor with a vector (Section 2.5). Finally, note that some Authors call the contraction of a vector with a covector a "scalar product", for reasons that probably come from the definition of work and power in Mechanics (the covector being the force and the vector being the displacement or velocity). We shall not use this terminology in this context but reserve it to what the same Authors call "inner product".

2.3 Covariance and Contravariance

The concepts of covariance and contravariance yield the fundamental distinction determining the different transformation laws to which vectors and covectors obey. The difference between vectors and covectors is often said to fade away when orthonormal bases (which, by the way, need the definition of a scalar product) are employed. This is neither entirely true nor necessary. Indeed, regardless of whether orthonormal bases are assumed, vectors and covectors transform in the same way *only* under orthogonal transformations (another concept based on the scalar product), but *not* under general transformations. In Continuum Mechanics, deformation is described by general transformations, and therefore it is crucial to distinguish vectors and covectors, in order not to be forced to memorise the transformation laws of the various families of objects.

Let $\mathcal{V}$ be a vector space, and $\{\boldsymbol{e}_i\}_{i=1}^n$ and $\{\boldsymbol{e}'_j\}_{j=1}^n$ two bases of $\mathcal{V}$, related by the change of basis law

$$\boldsymbol{e}'_j = q^i{}_j \, \boldsymbol{e}_i. \tag{21}$$

A fundamental theorem of Linear Algebra ensures that the $n \times n$ matrix $[\![q^i{}_j]\!]$ is non-singular and therefore admits the inverse matrix $[\![p^j{}_i]\!]$, which governs the inverse change of basis

$$\boldsymbol{e}_i = p^j{}_i \, \boldsymbol{e}'_j. \tag{22}$$

Considering Equation (21), a vector $\boldsymbol{u}$ can be written as

$$\boldsymbol{u} = u^i \, \boldsymbol{e}_i = u'^j \, \boldsymbol{e}'_j = q^i{}_j \, u'^j \, \boldsymbol{e}_i, \tag{23}$$

from which

$$u^i = q^i{}_j \, u'^j. \tag{24}$$

Similarly, using Equation (22), one proves that the inverse relation is

$$u'^j = p^j{}_i \, u^i. \tag{25}$$

Comparing Equations (24) and (21) and Equations (25) and (22), we deduce that the components of vectors transform with a matrix that is the *inverse* of the matrix governing the corresponding change of basis. Indeed, to get the "new" components u'^j from the "old" components u^i, we use the matrix $[\![p^j{}_i]\!]$, which is the inverse of the matrix $[\![q^i{}_j]\!]$ that gives the "new" basis vectors $\boldsymbol{e}'_j$ from the "old" ones $\boldsymbol{e}_i$ (and viceversa for the inverse transformations). For this reason, the components of vectors are said to be *contravariant*, where the Latin prefix *contra-*, meaning "against", is used here in the sense that the transformation law to which the components of vectors obey is the inverse (or "goes against") that followed by the basis vectors.

Let now apply the covector $\boldsymbol{\varphi} \in \mathcal{V}^\star$ to the vector $\boldsymbol{u}$, considering the decomposition (23), i.e.,

$$\boldsymbol{\varphi}(\boldsymbol{u}) = \boldsymbol{\varphi}(u'^j\, \boldsymbol{e}'_j) = \boldsymbol{\varphi}(q^i{}_j\, u'^j\, \boldsymbol{e}_i). \tag{26}$$

From the linearity of $\boldsymbol{\varphi}$, it follows that

$$u'^j\, \boldsymbol{\varphi}(\boldsymbol{e}'_j) = q^i{}_j\, u'^j\, \boldsymbol{\varphi}(\boldsymbol{e}_i) \quad \Rightarrow \quad u'^j\, \varphi'_j = q^i{}_j\, u'^j\, \varphi_i, \tag{27}$$

and from the arbitrariness of $\boldsymbol{u}$, which implies that of the coefficients u'^j, we conclude that

$$\varphi'_j = q^i{}_j\, \varphi_i, \tag{28}$$

the inverse of which is

$$\varphi_i = p^j{}_i\, \varphi'_j. \tag{29}$$

Comparing Equations (28) and (21) and Equations (29) and (22), we conclude that the components of a covector in $\mathcal{V}^\star$ transform with the *same* matrix of the change of basis in $\mathcal{V}$. For this reason, the components of covectors are said to be *covariant*, where the Latin prefix *co*(n)-, meaning "with", is used here in the sense that the transformation law to which the components of covectors obey is the same as (or "goes with") that followed by the basis vectors. It is an easy exercise to show that the basis covectors transform as

$$\boldsymbol{e}^i = q^i{}_j\, \boldsymbol{e}'^j, \tag{30}$$

with inverse transformation

$$\boldsymbol{e}'^j = p^j{}_i\, \boldsymbol{e}^i. \tag{31}$$

Remark 2.2. It may sound confusing at first, but notice that the components of covectors are covariant with respect to the change of basis in $\mathcal{V}$ but are *contravariant* with respect to the change of basis in $\mathcal{V}^\star$! This "paradox" clears up as soon as one thinks about $\mathcal{V}^\star$ and its elements alone: the components of the vectors of a given vector space are *contravariant* with respect to the change of basis in the *same* space: compare Equations (28) and (30) and Equations (29) and (31).

2.4 Tensors

Tensors are multilinear forms, i.e., scalar-valued functions that are linear in each of the arguments, which could be vectors and/or covectors. Tensors are fundamental objects in Physics, describing any possible physical quantity, including scalars, which are regarded as zero-order tensors, and vectors and covectors, which are regarded as first-order tensors. We shall start with the intrinsic definition of tensor, followed by the definition of tensor product, which allows to derive the component representation of a tensor, along with the transformation laws to which the components must obey.

A tensor of order m on the n-dimensional vector space $\mathcal{V}$ is a multilinear form, i.e., a mapping

$$\mathbb{A} : \underbrace{\mathcal{V}^\star \times \ldots \times \mathcal{V}^\star}_{r \text{ times}} \times \underbrace{\mathcal{V} \times \ldots \times \mathcal{V}}_{s \text{ times}} \to \mathbb{R}, \tag{32a}$$

$$\mathbb{A} : (\boldsymbol{\varphi}^1, \ldots \boldsymbol{\varphi}^r, \boldsymbol{u}_1, \ldots \boldsymbol{u}_s) \mapsto \mathbb{A}(\boldsymbol{\varphi}^1, \ldots \boldsymbol{\varphi}^r, \boldsymbol{u}_1, \ldots \boldsymbol{u}_s), \tag{32b}$$

where $r + s = m$, that is linear in each of the arguments separately. Since the Cartesian product of vector spaces is a vector space (recall the classical case of $\mathbb{R}^n = \mathbb{R} \times \ldots \times \mathbb{R}$), the space of all tensors of the type (32) is a vector space (see Section 2.1) with dimension n^m and is denoted by

$$\mathcal{V}^r{}_s = \mathrm{Hom}(\underbrace{\mathcal{V}^\star \times \ldots \times \mathcal{V}^\star}_{r \text{ times}} \times \underbrace{\mathcal{V} \times \ldots \times \mathcal{V}}_{s \text{ times}}, \mathbb{R}). \tag{33}$$

The *tensor product* of the r vectors $\boldsymbol{v}_1, \ldots, \boldsymbol{v}_r \in \mathcal{V}$ and the s covectors $\boldsymbol{\psi}^1, \ldots, \boldsymbol{\psi}^s \in \mathcal{V}^\star$ is the tensor $\boldsymbol{v}_1 \otimes \ldots \otimes \boldsymbol{v}_r \otimes \boldsymbol{\psi}^1 \otimes \ldots \otimes \boldsymbol{\psi}^s \in \mathcal{V}^r{}_s$ such that, for every $\boldsymbol{\varphi}^1, \ldots, \boldsymbol{\varphi}^r \in \mathcal{V}^\star$ and for every $\boldsymbol{u}_1, \ldots, \boldsymbol{u}_s \in \mathcal{V}$,

$$[\boldsymbol{v}_1 \otimes \ldots \otimes \boldsymbol{v}_r \otimes \boldsymbol{\psi}^1 \otimes \ldots \otimes \boldsymbol{\psi}^s](\boldsymbol{\varphi}^1, \ldots, \boldsymbol{\varphi}^r, \boldsymbol{u}_1, \ldots, \boldsymbol{u}_s) = \\ = \boldsymbol{v}_1(\boldsymbol{\varphi}^1) \ldots \boldsymbol{v}_r(\boldsymbol{\varphi}^r)\, \boldsymbol{\psi}^1(\boldsymbol{u}_1) \ldots \boldsymbol{\psi}^s(\boldsymbol{u}_s). \tag{34}$$

The vectors $\boldsymbol{v}_1, \ldots, \boldsymbol{v}_r$ are called the *vector feet* of the tensor, and the covectors $\boldsymbol{\psi}^1, \ldots, \boldsymbol{\psi}^s$ are called the *covector feet* of the tensor. A tensor of the type $\boldsymbol{v}_1 \otimes \ldots \otimes \boldsymbol{v}_r \otimes \boldsymbol{\psi}^1 \otimes \ldots \otimes \boldsymbol{\psi}^s$, given by the tensor product of vectors and covectors, is called *decomposable*.

The definition of tensor product allows to derive the component expression of any tensor $\mathbb{A} \in \mathcal{V}^r{}_s$ with respect to a given basis $\{\boldsymbol{e}_i\}_{i=1}^n$ and dual basis $\{e^i\}_{i=1}^n$. Indeed, for the multilinearity of $\mathbb{A}$, we can write

$$\begin{aligned}\mathbb{A}(\varphi^1, ..., \varphi^r, u_1, ..., u_s) &= \mathbb{A}(\varphi^1_{i_1}\, e^{i_1}, ..., \varphi^r_{i_r}\, e^{i_r}, u_1^{j_1}\, e_{j_1}, ..., u_s^{j_s}\, e_{j_s}) \\ &= \varphi^1_{i_1} \cdots \varphi^r_{i_r}\, u_1^{j_1} \dots u_s^{j_s}\, \mathbb{A}(e^{i_1}, ..., e^{i_r}, e_{j_1}, ..., e_{j_s}) \\ &= \varphi^1_{i_1} \cdots \varphi^r_{i_r}\, u_1^{j_1} \dots u_s^{j_s}\, \mathrm{A}^{i_1 \dots i_r}{}_{j_1 \dots j_s},\end{aligned} \tag{35}$$

where

$$\mathrm{A}^{i_1 \dots i_r}{}_{j_1 \dots j_s} = \mathbb{A}(e^{i_1}, ..., e^{i_r}, e_{j_1}, ..., e_{j_s}) \tag{36}$$

are the components of $\mathbb{A}$ with respect to the basis $\{e_i\}_{i=1}^n$ and dual basis $\{e^i\}_{i=1}^n$, and are said to be r times contravariant and s times covariant. Similarly, $i_1 \dots i_r$ are the contravariant indices of $\mathbb{A}$ and $j_1 \dots j_s$ are the covariant indices of $\mathbb{A}$. Now, using the definition of the action of the basis vectors on covectors and of the basis covectors on vectors, and the definition of tensor product in Equation (34), Equation (35) can be written

$$\begin{aligned}&\mathbb{A}(\varphi^1, ..., \varphi^r, u_1, ..., u_s) \\ &= \varphi^1_{i_1} \cdots \varphi^r_{i_r}\, u_1^{j_1} \dots u_s^{j_s}\, \mathrm{A}^{i_1 \dots i_r}{}_{j_1 \dots j_s} \\ &= e_{i_1}(\varphi^1) ... e_{i_r}(\varphi^r)\, e^{j_1}(u_1) ... e^{j_s}(u_s)\, \mathrm{A}^{i_1 \dots i_r}{}_{j_1 \dots j_s} \\ &= \mathrm{A}^{i_1 \dots i_r}{}_{j_1 \dots j_s}\, [e_{i_1} \otimes ... \otimes e_{i_r} \otimes e^{j_1} \otimes ... \otimes e^{j_s}](\varphi^1, ..., \varphi^r, u_1, ..., u_s),\end{aligned} \tag{37}$$

from which, dropping the arguments $\varphi^1, ..., \varphi^r, u_1, ..., u_s$ on both sides, we obtain the component representation of $\mathbb{A}$ in the *tensor basis* $\{e_{i_1} \otimes ... \otimes e_{i_r} \otimes e^{j_1} \otimes ... \otimes e^{j_s}\}_{i_1,...,i_r,j_1,...j_s=1}^n$ of the tensor space $\mathcal{V}^r{}_s$ as

$$\mathbb{A} = \mathrm{A}^{i_1 \dots i_r}{}_{j_1 \dots j_s}\, e_{i_1} \otimes ... \otimes e_{i_r} \otimes e^{j_1} \otimes ... \otimes e^{j_s}. \tag{38}$$

Tensor $\mathbb{A} \in \mathcal{V}^r{}_s$ is said to have r vector feet and s covector feet: each vector foot contracts with a covector argument, and each covector foot contracts with a vector argument. For this reason, the space $\mathcal{V}^r{}_s$ can be thought of as the *tensor product of vector spaces* (Bishop and Goldberg, 1968)

$$\mathcal{V}^r{}_s = \underbrace{\mathcal{V} \otimes \ldots \otimes \mathcal{V}}_{r \text{ times}} \otimes \underbrace{\mathcal{V}^\star \otimes \ldots \otimes \mathcal{V}^\star}_{s \text{ times}}. \tag{39}$$

Note that $\mathcal{V}$ and $\mathcal{V}^\star$ appear to have been switched from Equation (33) to Equation (39). This, however, makes sense as the spaces in Equation (33) are those to which the *arguments* belong, whereas the spaces in (39) are those to which the *feet* belong. Note also that, in general, a tensor such as $\mathbb{A}$ in Equation (38) is *non-decomposable*, as it cannot be written as the tensor product of vectors and covectors, like that in Equation (34), but only as a linear combination of tensor products of vectors and covectors.

Remark 2.3. With a customary abuse of terminology, tensor $\mathbb{A} \in \mathcal{V}^r{}_s$ is often referred to as an "r times contravariant and s times covariant tensor". Rigorously speaking, it is not the tensor to be contravariant and/or covariant, but its components. We shall follow the custom, and forgive ourselves for this venial sin.

The components of tensors of any order transform following the transformation laws for vectors and covectors, i.e., each vector foot transforms like a vector and each covector foot transforms like a covector. Equivalently, one can say that each contravariant index transforms like the components of a vector, and each covariant index transforms like the components of a covector. Given two bases $\{\boldsymbol{e}_i\}_{i=1}^n$ and $\{\boldsymbol{e}'_j\}_{j=1}^n$ in $\mathcal{V}$ and the corresponding dual bases $\{\boldsymbol{e}^i\}_{i=1}^n$ and $\{\boldsymbol{e}'^j\}_{j=1}^n$ in $\mathcal{V}^\star$, and recalling the transformation laws (25) and (28), the components of tensor $\mathbb{A} \in \mathcal{V}^r{}_s$ transform as

$$\mathrm{A}'^{k_1 \dots k_r}{}_{l_1 \dots l_s} = p^{k_1}{}_{i_1} \cdots p^{k_r}{}_{i_r} \, q^{j_1}{}_{l_1} \cdots q^{j_s}{}_{l_s} \, \mathrm{A}^{i_1 \dots i_r}{}_{j_1 \dots j_s} \,. \tag{40}$$

Note that we could define a tensor of order $m = r + s$ with the Cartesian products in Equation (32a) or the tensor products in (39) performed in any order, i.e., the r covector arguments need not be the first r and the s vector argument need not be the last s; equivalently, the r vector feet need not be the first r and the s covector feet need not be the last s.

Considering that, given an order m, the integers r and s may vary from zero to m, subjected to the constraint $m = r + s$, we have that there exist 2^{r+s} spaces of tensors of order $m = r + s$. Tensor bases for each of these spaces can be built starting from a basis $\{\boldsymbol{e}_i\}_{i=1}^n$ in $\mathcal{V}$ and its dual basis $\{\boldsymbol{e}^i\}_{i=1}^n$ in $\mathcal{V}^\star$. For instance, there are $2^4 = 16$ possible fourth-order tensor spaces. The bases and tensor representations in some of these 16 possible spaces are given below as an example:

$$\mathbb{A} \in \mathcal{V}^2{}_2, \qquad \mathbb{A} = \mathrm{A}^{ij}{}_{kl} \; \boldsymbol{e}_i \otimes \boldsymbol{e}_j \otimes \boldsymbol{e}^k \otimes \boldsymbol{e}^l, \tag{41a}$$

$$\mathbb{A} \in \mathcal{V}^1{}_1{}^1{}_1, \quad \mathbb{A} = \mathrm{A}^i{}_j{}^k{}_l \; \boldsymbol{e}_i \otimes \boldsymbol{e}^j \otimes \boldsymbol{e}_k \otimes \boldsymbol{e}^l, \tag{41b}$$

$$\mathbb{A} \in \mathcal{V}^4_0, \qquad \mathbb{A} = \mathrm{A}^{ijkl} \; \boldsymbol{e}_i \otimes \boldsymbol{e}_j \otimes \boldsymbol{e}_k \otimes \boldsymbol{e}_l. \tag{41c}$$

Finally, note that there is only $2^0 = 1$ space of tensors of order zero, i.e., the set of the real numbers, $\mathcal{V}^0_0 \equiv \mathbb{R}$, and there are $2^1 = 2$ spaces of tensors of order one, i.e., the vector space itself $\mathcal{V}^1_0 = \mathcal{V}$ and its dual space $\mathcal{V}^0_1 = \mathcal{V}^\star$.

2.5 Tensors as Linear Maps, Tensor Contractions

A tensor can be seen as a linear map between two tensor spaces. Two simple cases (and those most common in Continuum Mechanics) are those of second- and fourth-order tensors. The simplest example is that of tensors in $\mathcal{V}^1{}_1$, often called "mixed tensors", as their components have one index high and one low, which are equivalent to linear maps in $\mathrm{End}(\mathcal{V}) = \mathrm{Hom}(\mathcal{V}, \mathcal{V})$. Their action on a vector, called a *simple contraction*, is denoted by simple juxtaposition: $\boldsymbol{a}\,\boldsymbol{u} = \boldsymbol{v} \in \mathcal{V}$; in components, $a^i{}_j\, u^j = v^i$. Analogously, mixed tensors in $\mathcal{V}_1{}^1$ (the space to which the transposes of tensors in $\mathcal{V}^1{}_1$ belong) are equivalent to linear maps in $\mathrm{End}(\mathcal{V}^\star)$, and their action on covectors is also denoted by juxtaposition: $\boldsymbol{b}\,\boldsymbol{\varphi} = \boldsymbol{\psi} \in \mathcal{V}^\star$; in components, $b_i{}^j\, \varphi_j = \psi_i$. It is a simple exercise to show that tensors in $\mathcal{V}^2_0$ and $\mathcal{V}^0_2$ can be identified with linear maps in $\mathrm{Hom}(\mathcal{V}^\star, \mathcal{V})$ and in $\mathrm{Hom}(\mathcal{V}, \mathcal{V}^\star)$, respectively.

A tensor $\mathbb{A} \in \mathcal{V}^r{}_s$ can be seen as a linear form on its dual space $\mathcal{V}_r{}^s$, mapping elements of $\mathcal{V}_r{}^s$ into $\mathbb{R}$, and the contraction of $\mathbb{A} \in \mathcal{V}^r{}_s$ and $\mathbb{B} \in \mathcal{V}_r{}^s$ is denoted by the bra-ket $\langle \mathbb{A} | \mathbb{B} \rangle = \mathrm{A}^{i_1 \ldots i_r}{}_{j_1 \ldots j_s}\, \mathrm{B}_{i_1 \ldots i_r}{}^{j_1 \ldots j_s}$. Naturally, by virtue of the Canonical Isomorphism between a space and its bidual, and in the spirit of the bra-ket notation as introduced in Equation (17), one can equivalently say that $\mathbb{B} \in \mathcal{V}_r{}^s$ maps elements of $\mathcal{V}^r{}_s$ into $\mathbb{R}$. For the case of second-order tensors, the *double contraction* is usually denoted by a colon; e.g., for the case of a tensor $\boldsymbol{p} \in \mathcal{V}^2_0$ mapping a tensor $\boldsymbol{q} \in \mathcal{V}^0_2$ into $\mathbb{R}$, we write $\langle \boldsymbol{p} | \boldsymbol{q} \rangle = \boldsymbol{p} : \boldsymbol{q} = p^{ij}\, q_{ij}$.

Fourth-order tensors are often employed as maps between spaces of second-order tensors. For instance, a tensor $\mathbb{T} \in \mathcal{V}^2{}_2$ can be seen as a linear map in $\mathrm{Hom}(\mathcal{V}^2_0, \mathcal{V}^2_0)$, and its action on $\boldsymbol{a} \in \mathcal{V}^2_0$ is denoted by a double contraction: $\mathbb{T} : \boldsymbol{a} = \boldsymbol{b} \in \mathcal{V}^2_0$; in components, $\mathrm{T}^{ij}{}_{kl}\, a^{kl} = b^{ij}$. All other cases of fourth-order tensor are analogous. We conclude by noting that a fourth-order tensor can also be seen as a bilinear form on a space of second-order tensors. For instance, a tensor $\mathbb{L} \in \mathcal{V}^4_0$ can be seen as a bilinear form in $\mathrm{Hom}(\mathcal{V}^0_2 \times \mathcal{V}^0_2, \mathbb{R})$, with two double contractions, on the left and on the right, i.e., $\boldsymbol{\epsilon} : \mathbb{L} : \boldsymbol{\eta} = \epsilon_{ij}\, \mathrm{L}^{ijkl}\, \eta_{kl}$. This is the case of elasticity stiffness tensors. Since elasticity compliance tensors are inverses of elasticity stiffness tensors, they must be such $\mathbb{L} : \mathbb{A} = \mathbb{I}$, where $\mathbb{I} \in \mathcal{V}^2{}_2$ is the fourth-order identity (see, e.g., Federico, 2012), and are therefore tensors in $\mathcal{V}^0_4$.

Remark 2.4. Note that, strictly speaking, a tensor seen as a multilinear form (i.e., a multilinear map into $\mathbb{R}$) and the "same" tensor seen as a linear map between two tensor spaces are two *different* objects. However, it is not worth to keep the distinction, because this identification virtually never causes ambiguity.

2.6 Norm and Metric (Distance)

Although in everyday life we perfectly know that every arrow drawn on a blackboard, possibly representing a vector, has a definite length, in the "bare" definition of vector space, there is no mention whatsoever to the concept of "length of a vector". The only thing one can certainly say is that, if $\boldsymbol{u}, \boldsymbol{v} \in \mathcal{V}$ are such that $\boldsymbol{u} = \lambda\, \boldsymbol{v}$, then the "length" of $\boldsymbol{u}$ is λ times that of $\boldsymbol{v}$, which is a *relative* concept. The absolute concept of "length", or *magnitude*, in a vector space must thus be introduced as an additional feature, which technically is called a *norm*. We shall give the general definition of norm in this section, but then shall only deal with the so-called *Euclidean norm*, which descends from a *scalar product* or *inner product*, and shall be defined in Section 2.7. We shall also give the definition of *distance* or *metric*.

A norm in a real vector space $\mathcal{V}$ is a function

$$\|\cdot\| : \mathcal{V} \to \mathbb{R} : \boldsymbol{u} \mapsto \|\boldsymbol{u}\|, \tag{42}$$

such that, for every $\boldsymbol{u}, \boldsymbol{v} \in \mathcal{V}$ and for every $\lambda \in \mathbb{R}$, it satisfies the properties

$$\begin{aligned}
&\text{positive definiteness:} && \|\boldsymbol{u}\| \geq 0, \quad \text{and} \quad \|\boldsymbol{u}\| = 0 \iff \boldsymbol{u} = \mathbf{0}_{\mathcal{V}}, && \text{(43a)}\\
&\text{absolute homogeneity:} && \|\lambda \boldsymbol{u}\| = |\lambda|\, \|\boldsymbol{u}\|, && \text{(43b)}\\
&\text{sub-additivity:} && \|\boldsymbol{u} + \boldsymbol{v}\| \leq \|\boldsymbol{u}\| + \|\boldsymbol{v}\|, && \text{(43c)}
\end{aligned}$$

where $\mathbf{0}_{\mathcal{V}}$ is the zero vector of $\mathcal{V}$ and Property (43c) is also called *triangular inequality*. A vector space $\mathcal{V}$ equipped with a norm $\|\cdot\|$ is called a *normed vector space*, or simply a *normed space*, and is denoted by $(\mathcal{V}, \|\cdot\|)$.

A distance (or metric) needs not be defined on a vector space and so we shall give the general definition, which can then be specialised to vector spaces. Let $\mathcal{A}$ be a non-empty set. A metric on $\mathcal{A}$ is a function

$$\varrho : \mathcal{A} \times \mathcal{A} \to \mathbb{R} : (x, y) \mapsto \varrho(x, y), \tag{44}$$

such that, for every $x, y, z \in \mathcal{A}$, it satisfies the properties

$$\begin{aligned}
&\text{positive definiteness:} && \varrho(x, y) \geq 0, \quad \text{and} \quad \varrho(x, y) = 0 \iff x = y, && \text{(45a)}\\
&\text{symmetry:} && \varrho(x, y) = \varrho(y, x), && \text{(45b)}\\
&\text{sub-additivity:} && \varrho(x, z) \leq \varrho(x, y) + \varrho(y, z). && \text{(45c)}
\end{aligned}$$

Also in this case, Property (45c) can be called *triangular inequality*. The set $\mathcal{A}$ equipped with the distance ϱ is called a *metric space* and is denoted by $(\mathcal{A}, \varrho)$. It is fairly easy to prove that, given a normed vector space $(\mathcal{V}, \|\cdot\|)$, the norm automatically induces the metric defined by $\varrho(\boldsymbol{u}, \boldsymbol{v}) = \|\boldsymbol{u} - \boldsymbol{v}\|$. This is of crucial importance because it enables us to define open sets (i.e., to define a *topology*), to calculate limits, and determine whether a function is continuous in normed vector spaces.

2.7 Metric Tensor (or Scalar Product, or Inner Product)

A metric tensor in a vector space $\mathcal{V}$ is a tensor $\boldsymbol{g} \in \mathcal{V}^0_2$, i.e., a bilinear form on $\mathcal{V} \times \mathcal{V}$, such that, for every $\boldsymbol{u}, \boldsymbol{v} \in \mathcal{V}$, it satisfies the properties

$$\text{symmetry:} \qquad \boldsymbol{g}(\boldsymbol{u}, \boldsymbol{v}) = \boldsymbol{g}(\boldsymbol{v}, \boldsymbol{u}), \tag{46a}$$

$$\text{positive definiteness:} \quad \boldsymbol{g}(\boldsymbol{u}, \boldsymbol{u}) \geq 0, \ \text{ and } \ \boldsymbol{g}(\boldsymbol{u}, \boldsymbol{u}) = 0 \iff \boldsymbol{u} = \boldsymbol{0}_{\mathcal{V}}. \tag{46b}$$

The application of the metric tensor $\boldsymbol{g}$, to two vectors $\boldsymbol{u}$ and $\boldsymbol{v}$ is called the *scalar product* or *inner product* of $\boldsymbol{u}$ and $\boldsymbol{v}$,

$$\boldsymbol{g}(\boldsymbol{u}, \boldsymbol{v}) = \langle \boldsymbol{u}, \boldsymbol{v} \rangle = \boldsymbol{u}.\boldsymbol{v} = u^a \, g_{ab} \, v^b, \tag{47}$$

where we prefer the use of the "low" dot (Chadwick, 1976), as opposed to the "centred dot" of the typical notation $\boldsymbol{u} \cdot \boldsymbol{v}$. Also, the metric automatically induces the norm $\|\boldsymbol{u}\| = \sqrt{\boldsymbol{u}.\boldsymbol{u}}$, called *Euclidean norm*, which in turn induces the *Euclidean metric* $\varrho(\boldsymbol{u}, \boldsymbol{v}) = \|\boldsymbol{u} - \boldsymbol{v}\| = \sqrt{(\boldsymbol{u} - \boldsymbol{v}).(\boldsymbol{u} - \boldsymbol{v})}$.

Being positive definite, the metric tensor $\boldsymbol{g}$ is invertible and its inverse is the tensor $\boldsymbol{g}^{-1} \in \mathcal{V}^2_0$, defining a scalar product in $\mathcal{V}^\star$, such that

$$\boldsymbol{g}^{-1}(\boldsymbol{\varphi}, \boldsymbol{\psi}) = \langle \boldsymbol{\varphi}, \boldsymbol{\psi} \rangle = \boldsymbol{\varphi}.\boldsymbol{\psi} = \varphi_a \, (\boldsymbol{g}^{-1})^{ab} \, \psi_b, \tag{48}$$

along with the associated norm and metric.

Furthermore, $\boldsymbol{g}$ lowers contravariant indices of tensors of any order (e.g., given a vector $\boldsymbol{v} \in \mathcal{V}$, its associated covector is given by $\boldsymbol{v}^\flat = \boldsymbol{g}\,\boldsymbol{v}$, with components $[\boldsymbol{v}^\flat]_a = v_a = g_{ab} v^b$), and the inverse metric tensor $\boldsymbol{g}^{-1} \in \mathcal{V}^2_0$ raises covariant indices (e.g., given a covector $\boldsymbol{\varphi} \in \mathcal{V}^\star$, its associated vector is $\boldsymbol{\varphi}^\sharp = \boldsymbol{g}^{-1}\boldsymbol{\varphi}$, with components $[\boldsymbol{\varphi}^\sharp]^a = \varphi^a = [\boldsymbol{g}^{-1}]^{ab}\varphi_b$). From this, and the fact that $\boldsymbol{g}^{-1}\boldsymbol{g} = \boldsymbol{i}$ and $\boldsymbol{g}\,\boldsymbol{g}^{-1} = \boldsymbol{i}^T$, where $\boldsymbol{i} \in \mathcal{V}^1{}_1$ is the identity tensor in $\mathcal{V}$ (with components $\delta^i{}_j$) and its transpose $\boldsymbol{i}^T \in \mathcal{V}_1{}^1$ is the identity tensor in $\mathcal{V}^\star$ (with components $\delta_i{}^j$), it is straightforward to show that the components of the inverse metric tensor are obtained by raising both covariant indices of the metric tensor, i.e., $[\boldsymbol{g}^{-1}]^{ab} = g^{ab}$. Indeed, the indices of $\boldsymbol{g}$ are risen by left- and right-contraction with $\boldsymbol{g}^{-1}$, i.e., $\boldsymbol{g}^\sharp = \boldsymbol{g}^{-1}\boldsymbol{g}\,\boldsymbol{g}^{-1} = \boldsymbol{i}\boldsymbol{g}^{-1} = \boldsymbol{g}^{-1}$. Therefore, the scalar product of covectors in Equation (48) can be written, in components, $\varphi_a \, g^{ab} \, \psi_b$.

Following Marsden and Hughes (e.g., 1983), we use the suggestive musical analogy, with the *flat* "$\flat$" lowering indices (analogously to the musical flat, which lowers notes a semitone down) and the *sharp* "$\sharp$" raising indices (analogously to the musical sharp, which raises notes a semitone up). Via the operations of lowering indices of vectors and rising indices of covectors,

$$^\flat : \mathcal{V} \to \mathcal{V}^\star \qquad \boldsymbol{u} \mapsto \boldsymbol{u}^\flat = \boldsymbol{g}\,\boldsymbol{u}, \tag{49a}$$

$$^\sharp : \mathcal{V}^\star \to \mathcal{V} \qquad \boldsymbol{\varphi} \mapsto \boldsymbol{\varphi}^\sharp = \boldsymbol{g}^{-1}\boldsymbol{\varphi}, \tag{49b}$$

the scalar product induces an isomorphism between $\mathcal{V}$ and its dual space $\mathcal{V}^\star$, for the case of finite dimension. Indeed, once a metric $\boldsymbol{g}$ is assigned, to every vector $\boldsymbol{u}$ corresponds a unique covector $\boldsymbol{g}\boldsymbol{u} = \boldsymbol{u}^\flat$, and to every covector $\boldsymbol{\varphi}$ corresponds a unique vector $\boldsymbol{g}^{-1}\boldsymbol{\varphi} = \boldsymbol{\varphi}^\sharp$, such that $\langle \boldsymbol{\varphi} | \boldsymbol{u} \rangle = \langle \boldsymbol{\varphi}^\sharp, \boldsymbol{u} \rangle = \langle \boldsymbol{\varphi}, \boldsymbol{u}^\flat \rangle$, which justifies the notation $\langle \cdot\,, \cdot \rangle$ for the scalar product, recalling the bra-ket $\langle \cdot | \cdot \rangle$.

The scalar product of two tensors $\mathbb{A}, \mathbb{B} \in \mathcal{V}^r{}_s$ is obtained by contracting homologous indices by means of the appropriate metric tensor, i.e.,

$$\langle \mathbb{A}, \mathbb{B} \rangle = \mathrm{A}^{i_1 \ldots i_r}{}_{j_1 \ldots j_s}\, g_{i_1 k_1} \cdots g_{i_r k_r}\, g^{j_1 l_1} \ldots g^{j_s l_s}\, \mathrm{B}^{k_1 \ldots k_r}{}_{l_1 \ldots l_s}. \tag{50}$$

For tensors of order one, i.e., vectors and covectors, we shall normally retain the simple "dot" notation resulting in $\langle \boldsymbol{u}, \boldsymbol{v} \rangle = \boldsymbol{u}.\boldsymbol{v}$ and $\langle \boldsymbol{\varphi}, \boldsymbol{\psi} \rangle = \boldsymbol{\varphi}.\boldsymbol{\psi}$.

Tensors $\boldsymbol{g}$ and $\boldsymbol{g}^{-1}$ are also used for the single contraction of tensors of higher order. For two tensors of any order, such that (the repeated tensor products of spaces in the underbraces could be any permutation with repetition of $\mathcal{V}$ and $\mathcal{V}^\star$)

$$\mathbb{A} \in \underbrace{\mathcal{V} \otimes \ldots \otimes \mathcal{V}^\star}_{n-1 \text{ times}} \otimes \mathcal{V}, \qquad \mathbb{B} \in \mathcal{V} \otimes \underbrace{\mathcal{V} \otimes \ldots \otimes \mathcal{V}^\star}_{m-1 \text{ times}},$$

the single contraction is indicated by a dot "." and obtained by use of the metric tensor $\boldsymbol{g}$, which serves to contract the last index of $\mathbb{A}$ with the first index of $\mathbb{B}$, which are both contravariant: $\mathbb{A}.\mathbb{B} = \mathbb{A}\, \boldsymbol{g}\, \mathbb{B}$. Analogously, for

$$\mathbb{C} \in \underbrace{\mathcal{V} \otimes \ldots \otimes \mathcal{V}^\star}_{n-1 \text{ times}} \otimes \mathcal{V}^\star, \qquad \mathbb{D} \in \mathcal{V}^\star \otimes \underbrace{\mathcal{V} \otimes \ldots \otimes \mathcal{V}^\star}_{m-1 \text{ times}},$$

the single contraction is indicated by $\mathbb{C}.\mathbb{D} = \mathbb{C}\, \boldsymbol{g}^{-1} \mathbb{D}$, where the inverse metric tensor $\boldsymbol{g}^{-1}$ contracts the last index of $\mathbb{C}$ with the first index of $\mathbb{D}$, which are both covariant.

2.8 Affine Spaces

Loosely speaking, an affine space $\mathcal{S}$ is a set at each point x of which one can attach vectors from a given vector space $\mathcal{V}$. The first example that comes to mind is the velocity of a particle, which, at each instant of time, is attached at the current placement x of the particle. Affine spaces provide the simplest way to describe the natural three-dimensional space of Physics, as well as three-dimensional bodies, which are the objects of interest in Continuum Mechanics. We remark that the structure of normed affine space is essential in the basic definitions and theorems of Differential Calculus.

An *affine space* is a structure consisting of a non-empty set $\mathcal{S}$, called the *point space*, a vector space $\mathcal{V}$, called the *supporting* or *modelling* vector space, and a mapping

$$\mathcal{F} : \mathcal{S} \times \mathcal{S} \to \mathcal{V} : (x, y) \mapsto \mathcal{F}(x, y) = \vec{xy} = y - x, \tag{51}$$

satisfying the properties

$$\begin{aligned} &\text{anticommutativity:} && y - x = -[x - y], && \forall x, y \in \mathcal{S} && (52a)\\ &\text{parallelepiped rule:} && y - x = [y - z] + [z - x], && \forall x, y, z \in \mathcal{S} && (52b)\\ &\text{arbitrary origin:} && \forall x \in \mathcal{S},\ \forall \boldsymbol{u} \in \mathcal{V},\ \exists!\, y \in \mathcal{S} : && y - x = \boldsymbol{u}. && (52c) \end{aligned}$$

The vector $y - x = \boldsymbol{u} \in \mathcal{V}$ is called the *oriented segment from x to y*. The map $\mathcal{F}$ is not injective, i.e., there can be more than one pair (x, y) such that $y - x = \boldsymbol{u}$. The map $\mathcal{F}$ determines in fact an *equivalence relation*, in which $y - x$ and $q - p$ are equivalent if they are represented by the same vector $\boldsymbol{u} \in \mathcal{V}$. However, given $x \in \mathcal{S}$ and $\boldsymbol{u} \in \mathcal{V}$, the point $y = x + \boldsymbol{u}$ is unique.

Given a point $x \in \mathcal{S}$, the set of all vectors $\boldsymbol{u}_x$ emanating from x, denoted

$$T_x\mathcal{S} = \{\boldsymbol{u}_x = y - x : y \in \mathcal{S}\}, \tag{53}$$

is a vector space isomorphic to $\mathcal{V}$, called the *tangent space* to $\mathcal{S}$ at point x, in which the operations of sum and product by scalars are defined exactly as in $\mathcal{V}$. A vector $\boldsymbol{u}_x \in T_x\mathcal{S}$ is called a *tangent vector* at x. With a generally accepted abuse of notation, we shall drop the base point x in the expression $\boldsymbol{u}_x$ and just write $\boldsymbol{u}$, as there is seldom danger of confusion. The disjoint union (denoted by $\sqcup$) of all tangent spaces,

$$T\mathcal{S} = \bigsqcup_{x \in \mathcal{S}} T_x\mathcal{S}, \tag{54}$$

is called the *tangent bundle* of $\mathcal{S}$ and, for the case of an affine space, is given by the Cartesian product

$$T\mathcal{S} = \mathcal{S} \times \mathcal{V} = \{\boldsymbol{u}_x \equiv (x, \boldsymbol{u}) : x \in \mathcal{S},\ \boldsymbol{u} \in \mathcal{V}\}. \tag{55}$$

It is extremely important to emphasise that affine spaces are the simplest case in which the tangent bundle is expressible as a Cartesian product. In the most general case of differentiable manifolds, this is *not* true.

Finally, we note that, if the vector space $\mathcal{V}$ on which $\mathcal{S}$ is modelled is normed or Euclidean (i.e., endowed with a metric tensor $\boldsymbol{g}$), then the norm or metric is inherited by $\mathcal{S}$, which is called a *normed affine space* or *Euclidean affine space*, respectively.

3 Review of Continuum Kinematics

This section is mainly aimed at giving some basic definitions in order to elucidate the notation that shall be employed in the remainder of the chapter.

3.1 Basic Definitions

The natural physical space is represented by a Euclidean affine space $\mathcal{S}$, modelled after $\mathbb{R}^3$ and with metric $\boldsymbol{g}$. The metric tensor $\boldsymbol{g}$ belongs to $[T\mathcal{S}]^0_2$ and has covariant components g_{ab}. The inverse metric tensor $\boldsymbol{g}^{-1}$ belongs to $[T\mathcal{S}]^2_0$ and has contravariant components g^{ab}. The contraction $\boldsymbol{g}^{-1}\boldsymbol{g}$ yields the spatial identity tensor $\boldsymbol{i} \in [T\mathcal{S}]^1{}_1$ with components $i^a{}_b = \delta^a{}_b$. In the most elementary treatment, a three-dimensional continuum body (here strings and shells are ruled out) is described by means of one of its placements in space, called a reference configuration and represented by an open subset $\mathcal{B}_R$ of $\mathcal{S}$ in which a metric $\boldsymbol{G}$ is defined. The metric tensor $\boldsymbol{G}$ belongs to $[T\mathcal{B}_R]^0_2$ and has covariant components G_{AB}. The inverse metric tensor $\boldsymbol{G}^{-1}$ belongs to $[T\mathcal{B}_R]^2_0$ and has contravariant components G^{AB}. The contraction $\boldsymbol{G}^{-1}\boldsymbol{G}$ yields the material identity tensor $\boldsymbol{I} \in [T\mathcal{B}_R]^1{}_1$ with components $I^A{}_B = \delta^A{}_B$.

3.2 Motion and Deformation

The motion of the body is represented by a diffeomorphism (i.e., an invertible differentiable map with differentiable inverse)

$$\chi(\,\cdot\,,t) : \mathcal{B}_R \to \mathcal{S} : X \mapsto x = \chi(X,t), \tag{56}$$

called configuration map, which, at each time t maps *material* points $X = (X^1, X^2, X^3)$ into *spatial* points $x = (x^1, x^2, x^3)$. At fixed t, the *tangent map* of χ at point $X \in \mathcal{B}_R$ is the linear map

$$(T\chi)(X,t) = \boldsymbol{F}(X,t) : T_X\mathcal{B}_R \to T_x\mathcal{S}, \tag{57}$$

such that, for every tangent vector $\boldsymbol{W} \in T_X\mathcal{B}_R$,

$$[(T\chi)(X,t)](\boldsymbol{W}) = [\boldsymbol{F}(X,t)]\boldsymbol{W} = \lim_{h\to 0} \frac{\chi(X + h\boldsymbol{W}, t) - \chi(X,t)}{h}, \tag{58}$$

is the directional derivative $(\partial_{\boldsymbol{W}}\chi)(X,t)$ of χ with respect to $\boldsymbol{W}$, which is a tangent vector in the tangent space $T_x\mathcal{S}$ at $x = \chi(X,t)$. $\boldsymbol{F}$ is called the *deformation gradient*, and has components $F^a{}_A = \chi^a{}_{,A}$, where the comma denote partial differentiation. Note that the limit as written in Equation (58) only makes sense in affine spaces, and writing an equivalent expression for the case of differentiable manifolds is a lot more laborious.

By dropping the argument X of $\boldsymbol{F}$, the base points X and $x = \chi(X,t)$ need not be specified, and one can write

$$\boldsymbol{F}(\,\cdot\,,t) : T\mathcal{B}_R \to T\mathcal{S}, \tag{59}$$

where $T\mathcal{B}_R$ and $T\mathcal{S}$ are the tangent bundles of $\mathcal{B}_R$ and $\mathcal{S}$, respectively. As a two-point tensor field, the deformation gradient reads

$$\boldsymbol{F}(\,\cdot\,,t) : \mathcal{B}_R \to T\mathcal{S} \otimes T^\star\mathcal{B}_R, \tag{60}$$

Since χ is a diffeomorphism, then $\boldsymbol{F}$ admits the inverse

$$\boldsymbol{F}^{-1}(\,\cdot\,,t) : \chi(\mathcal{B}_R,t) \to T\mathcal{B}_R \otimes T^\star\mathcal{S}, \tag{61}$$

such that $\boldsymbol{F}^{-1}\boldsymbol{F} = \boldsymbol{I}$ and $\boldsymbol{F}\boldsymbol{F}^{-1} = \boldsymbol{i}$. The algebraic transpose of $\boldsymbol{F}$ is the tensor field

$$\boldsymbol{F}^T(\,\cdot\,,t) : \chi(\mathcal{B}_R,t) \to T^\star\mathcal{B}_R \otimes T\mathcal{S}, \tag{62}$$

such that, for every $\varphi \in T^\star S$ and $\boldsymbol{W} \in T\mathcal{B}_R$, $\varphi\boldsymbol{F}\boldsymbol{W} = \boldsymbol{W}\boldsymbol{F}^T\varphi$ and has components $(\boldsymbol{F}^T)_A{}^a = F^a{}_A$. The metric transpose (Federico, 2012, 2014) of $\boldsymbol{F}$ is the tensor field

$$\boldsymbol{F}^t(\,\cdot\,,t) : \chi(\mathcal{B}_R,t) \to T\mathcal{B}_R \otimes T^\star\mathcal{S}, \tag{63}$$

such that, for every $\boldsymbol{w} \in T\mathcal{S}$ and $\boldsymbol{Y} \in T\mathcal{B}_R$, $\langle \boldsymbol{w}, \boldsymbol{F}\boldsymbol{Y}\rangle_{\boldsymbol{g}} = \langle \boldsymbol{F}^t\boldsymbol{w}, \boldsymbol{Y}\rangle_{\boldsymbol{G}}$, from which $\boldsymbol{F}^t = \boldsymbol{G}^{-1}\boldsymbol{F}\,\boldsymbol{g}$, with components $(\boldsymbol{F}^t)^A{}_a = G^{AB}(\boldsymbol{F}^T)_B{}^b g_{ba}$.

Cauchy's Polar Decomposition Theorem states that, if $\boldsymbol{F}$ is invertible,

$$\boldsymbol{F} = \boldsymbol{R}.\boldsymbol{U} = \boldsymbol{V}.\boldsymbol{R} \quad \Rightarrow \quad \boldsymbol{F} = \boldsymbol{R}\,\boldsymbol{G}^{-1}\boldsymbol{U} = \boldsymbol{V}\boldsymbol{g}\,\boldsymbol{R}, \tag{64}$$

where $\boldsymbol{R}$ is a two-point proper orthogonal tensor field called *rotation* and valued in $T\mathcal{S} \otimes T^\star\mathcal{B}_R$, like $\boldsymbol{F}$, $\boldsymbol{U}$ is a symmetric material tensor field called *right stretch tensor* and valued in $[T\mathcal{B}_R]^0_2$, and $\boldsymbol{V}$ is a spatial symmetric tensor called *left stretch tensor* and valued in $[T\mathcal{S}]^2_0$. Note the use of the metric tensors, allowing to write $\boldsymbol{U}$ as a covariant tensor and $\boldsymbol{V}$ as a contravariant tensor, which is customary in the description of deformation and constitutive modelling. Indeed, in components, we have (Federico, 2014)

$$F^a{}_C = R^a{}_A\, G^{AB}\, U_{BC} = V^{ab}\, g_{bc}\, R^c{}_C. \tag{65}$$

Orthogonality of $\boldsymbol{R}$ means that it is an *isometry* (i.e., preserves the metric), and its inverse equals its metric transpose. Indeed, for every $\boldsymbol{W}, \boldsymbol{Y} \in T\mathcal{B}_R$,

$$\begin{aligned}
\langle \boldsymbol{W}, \boldsymbol{Y}\rangle_{\boldsymbol{G}} = \langle \boldsymbol{R}\boldsymbol{W}, \boldsymbol{R}\boldsymbol{Y}\rangle_{\boldsymbol{g}} &\Rightarrow \boldsymbol{W}\boldsymbol{G}\boldsymbol{Y} = (\boldsymbol{R}\boldsymbol{W})\boldsymbol{g}(\boldsymbol{R}\boldsymbol{Y}) \\
&\Rightarrow \boldsymbol{W}\boldsymbol{G}\boldsymbol{Y} = \boldsymbol{W}(\boldsymbol{R}^T\boldsymbol{g}\,\boldsymbol{R})\boldsymbol{Y} \\
&\Rightarrow \boldsymbol{G} = \boldsymbol{R}^T\boldsymbol{g}\,\boldsymbol{R} \\
&\Rightarrow \boldsymbol{R}^{-1} = \boldsymbol{G}^{-1}\boldsymbol{R}^T\boldsymbol{g} = \boldsymbol{R}^t,
\end{aligned} \tag{66}$$

When there is no deformation, the deformation gradient takes the value $\boldsymbol{F} = \mathbf{1}$, where

$$\mathbf{1} : T\mathcal{B}_R \to T\mathcal{S} \tag{67}$$

is the *shifter*, i.e., the two-point orthogonal tensor in $T\mathcal{S} \otimes T^\star\mathcal{B}_R$ that operates parallel translation of tangent vectors from $T\mathcal{B}_R$ into $T\mathcal{S}$ and can be thought as the "two-point identity tensor". Indeed, in the most elementary case of collinear Cartesian coordinates in $\mathcal{B}_R$ and $\mathcal{S}$, the components of the shifter are simply $\mathbf{1}^a{}_A = \delta^a{}_A$ (Eringen, 1980; Marsden and Hughes, 1983).

3.3 Push-Forward and Pull-Back

Given a material tensor field $\mathbb{P}$ valued in $[T\mathcal{B}_R]^r{}_s$ (as usual, the vector and covector feet could be in any order), its *push-forward* $\chi_*[\mathbb{P}] = \mathrm{p}$ is the tensor field valued in $[T\mathcal{S}]^r{}_s$ obtained by contracting each contravariant index with $\boldsymbol{F}$ and each covariant index with $\boldsymbol{F}^{-T}$, which in components reads $\mathrm{p}^{a\cdots}{}_{\dots b} = F^a{}_A \dots (\boldsymbol{F}^{-1})^B{}_b \, \mathrm{P}^{A\cdots}{}_{\dots B}$. Analogously, given a spatial tensor field q valued in $[T\mathcal{S}]^r{}_s$, its *pull-back* $\chi^*[\mathrm{q}] = \mathbb{Q}$ is the tensor field valued in $[T\mathcal{B}_R]^r{}_s$ obtained by contracting each contravariant index by $\boldsymbol{F}^{-1}$ and each covariant index by $\boldsymbol{F}^T$, i.e., $\mathrm{Q}^{A\cdots}{}_{\dots B} = (\boldsymbol{F}^{-1})^A{}_a \dots F^b{}_B \, \mathrm{q}^{a\cdots}{}_{\dots b}$.

3.4 Measures of Deformation and Strain

The pull-backs and push-forwards of the metric tensors have fundamental importance in Continuum Mechanics. The right and left Cauchy-Green deformation tensors $\boldsymbol{C}$ (valued in $[T\mathcal{B}_R]^0_2$) and $\boldsymbol{b}$ (valued in $[T\mathcal{S}]^2_0$ are defined as the pull-back of the spatial metric $\boldsymbol{g}$ and the push-forward of the inverse material metric $\boldsymbol{G}^{-1}$, respectively,

$$\boldsymbol{C} = \boldsymbol{F}^T\!.\boldsymbol{F} = \boldsymbol{F}^T \boldsymbol{g}\, \boldsymbol{F} \quad = \chi^*[\boldsymbol{g}], \tag{68}$$

$$\boldsymbol{b} = \boldsymbol{F}.\boldsymbol{F}^T = \boldsymbol{F}\, \boldsymbol{G}^{-1} \boldsymbol{F}^T = \chi_*[\boldsymbol{G}^{-1}], \tag{69}$$

and have components

$$C_{AB} = F^a{}_A \, g_{ab} \, F^b{}_B, \tag{70}$$

$$b^{ab} = F^a{}_A \, G^{AB} \, F^b{}_B. \tag{71}$$

Their inverses are the Piola deformation tensor $\boldsymbol{B} = \boldsymbol{C}^{-1}$ (valued in $[T\mathcal{B}_R]^2_0$) and the Finger deformation tensor $\boldsymbol{c} = \boldsymbol{b}^{-1}$ (valued in $[T\mathcal{S}]^0_2$) and are given by the pull-back of the inverse spatial metric $\boldsymbol{g}^{-1}$ and the push-forward of the material metric $\boldsymbol{G}$, respectively,

$$\boldsymbol{B} = \boldsymbol{C}^{-1} = \boldsymbol{F}^{-1}\!.\boldsymbol{F}^{-T} = \boldsymbol{F}^{-1} \boldsymbol{g}^{-1} \boldsymbol{F}^{-T} = \chi^*[\boldsymbol{g}^{-1}], \tag{72}$$

$$\boldsymbol{c} = \boldsymbol{b}^{-1} = \boldsymbol{F}^{-T}\!.\boldsymbol{F}^{-1} = \boldsymbol{F}^{-T} \boldsymbol{G}\, \boldsymbol{F}^{-1} = \chi_*[\boldsymbol{G}], \tag{73}$$

and have components

$$B^{AB} = (\boldsymbol{F}^{-1})^A{}_a \, g^{ab} \, (\boldsymbol{F}^{-1})^B{}_b, \tag{74}$$

$$c_{ab} = (\boldsymbol{F}^{-1})^A{}_a \, G_{AB} \, (\boldsymbol{F}^{-1})^B{}_b. \tag{75}$$

Note that the Polar Decomposition Theorem allows for expressing $\boldsymbol{C}$, $\boldsymbol{b}$, $\boldsymbol{B}$ and $\boldsymbol{c}$ as

$$\boldsymbol{C} = \boldsymbol{U}^2 = \boldsymbol{U}.\boldsymbol{U}, \tag{76}$$

$$\boldsymbol{b} = \boldsymbol{V}^2 = \boldsymbol{V}.\boldsymbol{V}, \tag{77}$$

$$\boldsymbol{B} = \boldsymbol{U}^{-2} = \boldsymbol{U}^{-1}.\boldsymbol{U}^{-1}, \tag{78}$$

$$\boldsymbol{c} = \boldsymbol{V}^{-2} = \boldsymbol{V}^{-1}.\boldsymbol{V}^{-1}. \tag{79}$$

Since $\boldsymbol{C}$ is completely material and $\boldsymbol{b}$ is completely spatial, their determinants are well-defined, with $\det(\boldsymbol{C}) = \det(\boldsymbol{b})$. The determinant of the two-point tensor $\boldsymbol{F}$ is defined as (see, e.g., Federico, 2014)

$$J = \det(\boldsymbol{F}) = \sqrt{\det(\boldsymbol{C})} = \sqrt{\det(\boldsymbol{b})}, \tag{80}$$

and is called volumetric deformation ratio.

The standard material and spatial measures of strain are the Green-Lagrange strain $\boldsymbol{E}$ and the Almansi strain $\boldsymbol{e}$,

$$\boldsymbol{E} = \tfrac{1}{2}\,(\boldsymbol{C} - \boldsymbol{G}), \tag{81}$$

$$\boldsymbol{e} = \tfrac{1}{2}\,(\boldsymbol{g} - \boldsymbol{c}) = \tfrac{1}{2}\,(\boldsymbol{g} - \boldsymbol{b}^{-1}), \tag{82}$$

which are related by a push-forward/pull-back relation

$$\boldsymbol{E} = \chi^*[\boldsymbol{e}] = \boldsymbol{F}^T \boldsymbol{e}\,\boldsymbol{F}, \tag{83}$$

$$\boldsymbol{e} = \chi_*[\boldsymbol{E}] = \boldsymbol{F}^{-T} \boldsymbol{E}\,\boldsymbol{F}^{-1}. \tag{84}$$

3.5 Velocity and Time Rates

The velocity is defined as the spatial vector field $\boldsymbol{v}(\,\cdot\,,t) : \chi(\mathcal{B}_R, t) \to T\mathcal{S}$ such that, for every $x = \chi(X,t) \in \chi(\mathcal{B}_R, t)$,

$$\boldsymbol{v}(x,t) = \dot{\chi}(X,t) = \lim_{h \to 0} \frac{\chi(X, t+h) - \chi(X,t)}{h}. \tag{85}$$

Note that, similarly to the case of the definition (58) of deformation gradient, the limit written as in Equation (85) only makes sense in affine spaces. The velocity gradient is the tensor field valued in $[T\mathcal{S}]^1{}_1$, defined as

$$\boldsymbol{l} = \operatorname{grad} \boldsymbol{v}, \tag{86}$$

and is related to the time derivative of the deformation gradient by

$$\dot{\boldsymbol{F}} = \boldsymbol{l}\,\boldsymbol{F}. \tag{87}$$

The fully covariant velocity gradient, the tensor field $\boldsymbol{l}^\flat = \boldsymbol{g}\,\boldsymbol{l}$ valued in $[T\mathcal{S}]^0_2$, can be decomposed into the symmetric part $\boldsymbol{d}$, called deformation rate, and the antisymmetric part $\boldsymbol{w}$, called spin tensor:

$$\boldsymbol{l}^\flat = \tfrac{1}{2}(\boldsymbol{g}\,\boldsymbol{l} + \boldsymbol{l}^T\boldsymbol{g}) + \tfrac{1}{2}(\boldsymbol{g}\,\boldsymbol{l} - \boldsymbol{l}^T\boldsymbol{g}) = \boldsymbol{d} + \boldsymbol{w}. \tag{88}$$

It is easy to show that the pull-back of the deformation rate $\boldsymbol{d}$ is the rate $\dot{\boldsymbol{E}}$ of the Green-Lagrange strain tensor $\boldsymbol{E}$, or one-half of the rate $\dot{\boldsymbol{C}}$ of the right Cauchy-Green deformation tensor $\boldsymbol{C}$ (note that $\dot{\boldsymbol{G}}$ vanishes identically):

$$\chi^*[\boldsymbol{d}] = \boldsymbol{F}^T\boldsymbol{d}\,\boldsymbol{F} = \dot{\boldsymbol{E}} = \tfrac{1}{2}\,\dot{\boldsymbol{C}}. \tag{89}$$

4 Review of Material Symmetries

The material symmetries of a body must be studied in the reference configuration (Eringen, 1980; Marsden and Hughes, 1983), and this is indeed what is customarily done in Elasticity. Here, for the case of the determination of the permeability, because of the upscaling method that we shall employ, we shall show that the study of the symmetries in the current configuration (spatial picture) yields physically admissible results, which are a particular case of the most general, invariant-based material theory. In this section, we review two fundamental material symmetries, isotropy and transverse isotropy, viewed as purely geometrical concepts that can, thus, be studied in the material or the spatial picture. We shall also remark how, in general, a material symmetry does *not* imply the corresponding spatial symmetry. Note that, in order to speak about invariance under a certain class of isometries (i.e., rotations), it is necessary to assume that a metric tensor is available, $\boldsymbol{G}$ in $\mathcal{B}_R$ and $\boldsymbol{g}$ in $\mathcal{S}$ (see, e.g., Marsden and Hughes, 1983; Federico, 2014).

With respect to the metric $\boldsymbol{G}$ in $\mathcal{B}_R$, the collection of all possible directions in the tangent space $T_X\mathcal{B}_R$ at the material point $X \in \mathcal{B}_R$ is represented by the material unit sphere

$$\mathbb{S}^2_X\mathcal{B}_R = \{\boldsymbol{M} \in T_X\mathcal{B}_R : \|\boldsymbol{M}\| = 1\}, \tag{90}$$

where $\|\boldsymbol{M}\| = \sqrt{\boldsymbol{M}.\boldsymbol{M}}$ is the Euclidean norm induced by $\boldsymbol{G}$. The material bundle unit sphere $\mathbb{S}^2\mathcal{B}_R = \{\boldsymbol{M} \in T\mathcal{B}_R : \|\boldsymbol{M}\| = 1\}$ refers to unit vectors in the tangent bundle $T\mathcal{B}_R$. Similarly, given a metric $\boldsymbol{g}$ in $\mathcal{S}$, the set of all possible directions in the tangent space $T_x\mathcal{S}$ at a point $x \in \mathcal{S}$ is the spatial unit sphere

$$\mathbb{S}^2_x\mathcal{S} = \{\boldsymbol{m} \in T_x\mathcal{S} : \|\boldsymbol{m}\| = 1\}, \tag{91}$$

where $\|\boldsymbol{m}\| = \sqrt{\boldsymbol{m}.\boldsymbol{m}}$ is the Euclidean norm induced by $\boldsymbol{g}$. The spatial bundle unit sphere $\mathbb{S}^2\mathcal{S} = \{\boldsymbol{m} \in T\mathcal{S} : \|\boldsymbol{m}\| = 1\}$ refers to unit vectors in the tangent bundle $T\mathcal{S}$.

4.1 Isotropic Second-Order Tensors

Isotropy in $\mathcal{B}_R$ is the symmetry defined as the the invariance under arbitrary rotations (i.e., material orthogonal tensors $\boldsymbol{Q} \in [T\mathcal{B}_R]^1{}_1$). Material second-order isotropic (or spherical) tensors in $[T\mathcal{B}_R]^2_0$ ("contravariant"), $[T\mathcal{B}_R]^0_2$ ("covariant"), in $[T\mathcal{B}_R]^1{}_1$ and $[T\mathcal{B}_R]_1{}^1$ ("mixed") are all tensors proportional to the inverse material metric $\boldsymbol{G}^{-1}$, the material metric $\boldsymbol{G}$, the material identity $\boldsymbol{I}$, and the transpose material identity $\boldsymbol{I}^T$, respectively. Their push-forwards are spatial tensors proportional to the left Cauchy-Green deformation $\boldsymbol{b}$ valued in $[T\mathcal{S}]^2_0$, the Finger deformation $\boldsymbol{c} = \boldsymbol{b}^{-1}$ valued in $[T\mathcal{S}]^0_2$, the spatial identity $\boldsymbol{i}$ in $[T\mathcal{S}]^1{}_1$, and the transpose identity $\boldsymbol{i}^T$ in $[T\mathcal{S}]_1{}^1$, respectively. Analogously, spatial second-order isotropic (or spherical) tensors in $[T\mathcal{S}]^2_0$ ("contravariant"), $[T\mathcal{S}]^0_2$ ("covariant"), in $[T\mathcal{S}]^1{}_1$ and $[T\mathcal{S}]_1{}^1$ ("mixed") are all tensors proportional to the inverse spatial metric $\boldsymbol{g}^{-1}$, the spatial metric $\boldsymbol{g}$, the spatial identity $\boldsymbol{i}$, and the transpose spatial identity $\boldsymbol{i}^T$, respectively. Their pull-backs are material tensors proportional to the Piola deformation $\boldsymbol{B} = \boldsymbol{C}^{-1}$ valued in $[T\mathcal{B}_R]^2_0$, the right Caughy-Green deformation $\boldsymbol{C}$ valued in $[T\mathcal{B}_R]^0_2$, the material identity $\boldsymbol{I}$ in $[T\mathcal{B}_R]^1{}_1$, and the transpose of the material identity $\boldsymbol{I}^T$ in $[T\mathcal{B}_R]_1{}^1$, respectively.

Note that *exclusively* in the case of "mixed" tensors isotropy is represented equivalently in the material and the spatial pictures, as the identities $\boldsymbol{I}$ and $\boldsymbol{i}$ (and their transposes) are directly related by a push-forward/pull-back operation.

4.2 Transversely Isotropic Second-Order Tensors

Transverse isotropy in $\mathcal{B}_R$ is the symmetry defined as the invariance under rotations about a given direction $\boldsymbol{M} \in \mathbb{S}^2\mathcal{B}_R$, called *axial direction.* These rotations are all material orthogonal tensors $\boldsymbol{Q} \in [T\mathcal{B}_R]^1{}_1$ such that $\boldsymbol{QM} = \boldsymbol{MQ}^T = \boldsymbol{M}$, i.e., with respect to which any vector parallel to $\boldsymbol{M}$ is invariant. The plane orthogonal (in the given metric) to the axial direction is called the *transverse plane.* Second-order tensors with transverse isotropy with respect to a given direction constitute a subspace of dimension two. Here we are interested in material tensors in $[T\mathcal{B}_R]^2_0$, i.e., "contravariant" tensors.

The natural basis of the subspace of symmetric second-order tensors in $[T\mathcal{B}_R]^2_0$, with transverse isotropy with respect to $\boldsymbol{M}$ is obtained by decomposing the inverse material metric tensor $\boldsymbol{G}^{-1}$ (the "contravariant identity") into (Walpole, 1981)

$$\boldsymbol{G}^{-1} = \boldsymbol{A} + \boldsymbol{T}, \tag{92}$$

where the basis tensors $\boldsymbol{A}$ and $\boldsymbol{T}$ are given by

$$\boldsymbol{A} = \boldsymbol{M} \otimes \boldsymbol{M}, \tag{93a}$$

$$\boldsymbol{T} = \boldsymbol{G}^{-1} - \boldsymbol{M} \otimes \boldsymbol{M} = \boldsymbol{G}^{-1} - \boldsymbol{A}. \tag{93b}$$

In the basis $\{\boldsymbol{A}, \boldsymbol{T}\}$, a symmetric tensor $\boldsymbol{Z} \in [T\mathcal{B}_R]^2_0$ with transverse isotropy with respect to $\boldsymbol{M}$ is expressed as the linear combination

$$\boldsymbol{Z} = Z_{\parallel}\, \boldsymbol{A} + Z_{\perp}\, \boldsymbol{T}, \tag{94}$$

where the axial and transverse components $Z_{\parallel}$ and $Z_{\perp}$ are obtained as the scalar products

$$Z_{\parallel} = \langle \boldsymbol{Z}, \boldsymbol{A} \rangle = Z^{AB}\, G_{AC}\, G_{BD}\, A^{CD}, \tag{95a}$$

$$Z_{\perp} = \langle \boldsymbol{Z}, \boldsymbol{T} \rangle = Z^{AB}\, G_{AC}\, G_{BD}\, T^{CD}. \tag{95b}$$

Tensors $\boldsymbol{A}$ and $\boldsymbol{T}$ have the geometrical meaning of projection operator on the axial direction $\boldsymbol{M}$ and of projection operator on the transverse plane (indeed, $\boldsymbol{T}$ is the complement of $\boldsymbol{A}$ to the "unit" tensor $\boldsymbol{G}^{-1}$), respectively. Thus, when contracted with a material vector $\boldsymbol{W} \in T\mathcal{B}_R$ by means of the metric tensor, they yield its axial and transverse components $W_{\parallel}$, $W_{\perp}$:

$$W_{\parallel} = \boldsymbol{A}.\boldsymbol{W} = \boldsymbol{A}\,\boldsymbol{G}\,\boldsymbol{W}, \tag{96a}$$

$$W_{\perp} = \boldsymbol{T}.\boldsymbol{W} = \boldsymbol{T}\,\boldsymbol{G}\,\boldsymbol{W} = \boldsymbol{W} - \boldsymbol{A}\,\boldsymbol{G}\,\boldsymbol{W}. \tag{96b}$$

Tensors $\boldsymbol{A}$ and $\boldsymbol{T}$ are called *axial projection operator* and *transverse projection operator*, respectively. In the context of fibre-reinforced materials, when vector $\boldsymbol{M}$ represents the fibre direction, tensor $\boldsymbol{A}$ is usually called *fabric tensor*, *texture tensor*, or *structure tensor*. In the following, we shall employ the latter nomenclature. Tensors $\boldsymbol{A}$ and $\boldsymbol{T}$ enjoy the properties of idempotence and orthogonality (see, e.g., Walpole, 1981), i.e.,

$$\boldsymbol{A}.\boldsymbol{A} = \boldsymbol{A}, \tag{97a}$$

$$\boldsymbol{T}.\boldsymbol{T} = \boldsymbol{T}, \tag{97b}$$

$$\boldsymbol{A}.\boldsymbol{T} = \boldsymbol{T}.\boldsymbol{A} = \mathbf{O}, \tag{97c}$$

the proof of which is an easy exercise ($\mathbf{O}$ is the null tensor in $[T\mathcal{S}]^2_0$).

Analogously to the material case, transverse isotropy in $\mathcal{S}$ is the symmetry defined as the invariance under rotations about a the axial direction $\boldsymbol{m} \in \mathbb{S}^2\mathcal{S}$, i.e., all spatial orthogonal tensors $\boldsymbol{q} \in [T\mathcal{S}]^1{}_1$ such that $\boldsymbol{q}\boldsymbol{m} = \boldsymbol{m}\boldsymbol{q}^T = \boldsymbol{m}$.

The natural basis of the subspace of symmetric second-order tensors in $[T\mathcal{S}]^2_0$ with transverse isotropy with respect to $\boldsymbol{m}$ is obtained by decomposing the inverse spatial metric tensor $\boldsymbol{g}^{-1}$ into (Walpole, 1981)

$$\boldsymbol{g}^{-1} = \boldsymbol{a} + \boldsymbol{t}, \tag{98}$$

where the basis tensors $\boldsymbol{a}$ and $\boldsymbol{t}$ given by

$$\boldsymbol{a} = \boldsymbol{m} \otimes \boldsymbol{m}, \tag{99a}$$

$$\boldsymbol{t} = \boldsymbol{g}^{-1} - \boldsymbol{m} \otimes \boldsymbol{m} = \boldsymbol{g}^{-1} - \boldsymbol{a}. \tag{99b}$$

The decomposition of a spatial symmetric tensor $\boldsymbol{z} \in [T\mathcal{S}]^2_0$, the projections of a spatial vector $\boldsymbol{w} \in T\mathcal{S}$ and the properties of idempotence and orthogonality of $\boldsymbol{a}$ and $\boldsymbol{t}$ are defined analogously to the material case.

The tensor bases $\{\boldsymbol{A}, \boldsymbol{T}\}$ and $\{\boldsymbol{a}, \boldsymbol{t}\}$ are *not* related by simple operations of push-forward/pull-back. Indeed, if one imposes that the spatial unit vector $\boldsymbol{m}$ is parallel to the push-forward $\boldsymbol{F}\boldsymbol{M}$ of $\boldsymbol{M}$, then $\boldsymbol{m}$ must coincide with the normalised push-forward of $\boldsymbol{M}$, i.e.,

$$\boldsymbol{m} = \|\boldsymbol{F}\boldsymbol{M}\|^{-1}\boldsymbol{F}\boldsymbol{M}. \tag{100}$$

Then, since the square of the norm $\|\boldsymbol{F}\boldsymbol{M}\|$ is given by

$$\|\boldsymbol{F}\boldsymbol{M}\|^2 = (\boldsymbol{F}\boldsymbol{M}).(\boldsymbol{F}\boldsymbol{M}) = (\boldsymbol{F}^T\!.\boldsymbol{F}) : (\boldsymbol{M} \otimes \boldsymbol{M}) = \boldsymbol{C} : \boldsymbol{A}, \tag{101}$$

the spatial structure tensor $\boldsymbol{a}$ can be expressed as the normalised push-forward of the material structure tensor $\boldsymbol{A}$, i.e.,

$$\begin{aligned}\boldsymbol{a} &= (\|\boldsymbol{F}\boldsymbol{M}\|^{-1}\boldsymbol{F}\boldsymbol{M}) \otimes (\|\boldsymbol{F}\boldsymbol{M}\|^{-1}\boldsymbol{F}\boldsymbol{M}) \\ &= (\boldsymbol{C} : \boldsymbol{A})^{-1}\boldsymbol{F}\boldsymbol{A}\boldsymbol{F}^T = (\boldsymbol{C} : \boldsymbol{A})^{-1}\chi_*[\boldsymbol{A}].\end{aligned} \tag{102}$$

Therefore, the push-forward of $\boldsymbol{A}$ does *not* coincide with $\boldsymbol{a}$, although it is proportional to it:

$$\chi_*[\boldsymbol{A}] = \boldsymbol{F}\boldsymbol{A}\boldsymbol{F}^T = (\boldsymbol{C} : \boldsymbol{A})\,\boldsymbol{a}. \tag{103}$$

For the case of the transverse operators $\boldsymbol{T}$ and $\boldsymbol{t}$, such a simple expression does *not* hold. Indeed, the push-forward of $\boldsymbol{T}$ is

$$\chi_*[\boldsymbol{T}] = \boldsymbol{b} - \boldsymbol{F}\boldsymbol{A}\boldsymbol{F}^T = \boldsymbol{b} - (\boldsymbol{C} : \boldsymbol{A})\,\boldsymbol{a}, \tag{104}$$

and is evidently *not* proportional to $\boldsymbol{t}$ ($\boldsymbol{b} = \chi_*[\boldsymbol{G}^{-1}] = \boldsymbol{F}\boldsymbol{G}^{-1}\boldsymbol{F}^T = \boldsymbol{F}.\boldsymbol{F}^T$ is the left Cauchy-Green deformation tensor).

Finally, we introduce the notation

$$\hat{\boldsymbol{A}}(\boldsymbol{M}) = \boldsymbol{A} = \boldsymbol{M} \otimes \boldsymbol{M}, \tag{105}$$

to denote the material structure tensor seen as an explicit function of the material direction $\boldsymbol{M}$.

Remark 4.1. In either the material or the spatial picture, the use of the structure tensor is appropriate whenever the sense of $\boldsymbol{M}$ or $\boldsymbol{m}$ is not relevant to the problem, e.g., for fibre-reinforcement. Indeed, $\boldsymbol{A}$ and $\boldsymbol{a}$ are invariant for reflection of $\boldsymbol{M}$ and $\boldsymbol{m}$ (i.e., transformations of the type $\boldsymbol{M} \mapsto -\boldsymbol{M}$ or $\boldsymbol{m} \mapsto -\boldsymbol{m}$), respectively.

Remark 4.2. Note that, in the literature, $\boldsymbol{m}$ and $\boldsymbol{a} = \boldsymbol{m} \otimes \boldsymbol{m}$ are often defined as the push-forwards of the unit vector $\boldsymbol{M}$ and the associated structure tensor $\boldsymbol{A} = \boldsymbol{M} \otimes \boldsymbol{M}$, respectively. With this convention, therefore, $\boldsymbol{m}$ is *not* a unit vector. Here, we prefer to keep both $\boldsymbol{m}$ and $\boldsymbol{a}$ normalised both in the material and the spatial pictures.

Remark 4.3. In the past (see Federico and Grillo, 2012, and previous works cited therein), we used to denote the unit sphere at point $X \in \mathcal{B}_R$ simply as $\mathbb{S}^2_X$, but we now (e.g., Tomic et al., 2014) prefer to explicitly indicate the space to which the base point belongs. So, for the material sphere, we have $\mathbb{S}^2_X\mathcal{B}_R$, and the notation is analogical to that of the tangent space $T_X\mathcal{B}_R$, of which the unit sphere is a subset. Also, this notation allows for smoothly defining the bundle unit sphere $\mathbb{S}^2\mathcal{B}_R$. The same naturally goes for the spatial unit sphere $\mathbb{S}^2_x\mathcal{S}$ at point $x \in \mathcal{S}$ and the spatial bundle unit sphere $\mathbb{S}^2\mathcal{S}$.

4.3 Transversely Isotropic Invariants

In the case of transverse isotropy, in addition to the three principal invariants of the deformation,

$$I_1(\boldsymbol{C}) = \mathrm{tr}\,(\boldsymbol{C}) = \boldsymbol{G}^{-1} : \boldsymbol{C}, \tag{106a}$$

$$I_2(\boldsymbol{C}) = \tfrac{1}{2}\,[(\mathrm{tr}\,(\boldsymbol{C}))^2 - \mathrm{tr}\,(\boldsymbol{C}^2)], \tag{106b}$$

$$I_3(\boldsymbol{C}) = \det(\boldsymbol{C}), \tag{106c}$$

other two invariants need to be considered (Spencer, 1984):

$$I_4(\boldsymbol{C}) = \boldsymbol{C} : \boldsymbol{A} = \boldsymbol{M}\boldsymbol{C}\boldsymbol{M} = \lambda^2_{\boldsymbol{M}}, \tag{107a}$$

$$I_5(\boldsymbol{C}) = \boldsymbol{C}^2 : \boldsymbol{A}. \tag{107b}$$

Note that the fourth invariant $I_4(\boldsymbol{C})$ has already appeared in Equations (101) and (102) as the square of the norm of $\boldsymbol{FM}$, which means that it represents the square of the stretch $\lambda_{\boldsymbol{M}}$ in direction $\boldsymbol{M}$. The fifth invariant $I_5(\boldsymbol{C})$ does not have a readily available physical interpretation. An alternative invariant that is more satisfying from the geometrical point of view has been presented by Holzapfel and Ogden (2010).

5 Review of Hyperelasticity

This section briefly reviews the general definition of internal (or deformation) power, hyperelasticity, and the derivation of stress and elasticity tensor, in the material and spatial pictures. Then, the use of the volumetric-distortional decomposition of the deformation in Elasticity is recalled, along with the basics of strictly incompressible and quasi incompressible bodies.

5.1 From the Internal Power to the Elastic Potential

In a continuum, the power of the external forces can be shown to cause, aside from a variation in the kinetic energy, measured by the kinetic power, also the *internal power* (or *stress power*, or *deformation power*), given by

$$\mathcal{P}_{\text{int}} = \int_{\chi(\mathcal{B}_R,\cdot)} \boldsymbol{\sigma} : \boldsymbol{d} = \int_{\mathcal{B}_R} \boldsymbol{\tau} : \boldsymbol{d} = \int_{\mathcal{B}_R} (\boldsymbol{gP}) : \dot{\boldsymbol{F}} = \int_{\mathcal{B}_R} \boldsymbol{S} : \dot{\boldsymbol{E}}, \tag{108}$$

where $\boldsymbol{\sigma}$ and $\boldsymbol{\tau}$ are the Cauchy stress and the Kirchhoff stress, conjugate to the deformation rate $\boldsymbol{d}$, $\boldsymbol{gP}$ is the first Piola-Kirchhoff stress with its first foot transformed into a covector (i.e., with its first index, the small one, lowered, i.e., $(\boldsymbol{gP})_a{}^B = g_{ab}P^{bB}$), conjugate to the time derivative $\dot{\boldsymbol{F}}$ of the deformation gradient, $\boldsymbol{S}$ is the second Piola-Kirchhoff stress, conjugate to the time derivative $\dot{\boldsymbol{E}}$ of the Green-Lagrange strain. Each of $\boldsymbol{\sigma}$-$\boldsymbol{d}$, $\boldsymbol{\tau}$-$\boldsymbol{d}$, $(\boldsymbol{gP})$-$\dot{\boldsymbol{F}}$, $\boldsymbol{S}$-$\dot{\boldsymbol{E}}$ is called a power-conjugate pair. The stress tensors in Equation (108) are related by

$$\boldsymbol{\sigma} = J^{-1}\boldsymbol{\tau} = J^{-1}\boldsymbol{PF}^T = J^{-1}\boldsymbol{FSF}^T. \tag{109}$$

Another conjugate stress-strain pair that is sometimes quite useful in Elasticity (Ogden, 1997) is given by the Biot stress $\boldsymbol{T}$ and the time derivative $\dot{\boldsymbol{E}}^{(1)} = \dot{\boldsymbol{U}}$ of the strain of order one $\boldsymbol{E}^{(1)} = \boldsymbol{U} - \boldsymbol{G}$, where $\boldsymbol{U}$ is the right stretch tensor of the polar decomposition (64), and $\boldsymbol{G}$ is, as usual, the material metric tensor. Note that, whereas the first and second Piola-Kirchhoff stresses $\boldsymbol{gP}$ and $\boldsymbol{S}$ and the Biot stress $\boldsymbol{T}$ are strictly the conjugates of $\boldsymbol{F}$, $\boldsymbol{E}$ and $\boldsymbol{E}^{(1)}$ via the rates $\dot{\boldsymbol{F}}$, $\dot{\boldsymbol{E}}$ and $\dot{\boldsymbol{E}}^{(1)}$, the Cauchy and Kirchhoff stresses

$\boldsymbol{\sigma}$ and $\boldsymbol{\tau}$ are conjugate with $\boldsymbol{d}$, which is *not* a rate, i.e., there is no tensor field such that its time derivative is $\boldsymbol{d}$.

A material is called hyperelastic if the internal power density, i.e., the integrand in one of the forms reported in Equation (108), is an exact differential in the strain. If one chooses the second Piola-Kirchhoff stress $\boldsymbol{S}$ and the Green-Lagrange strain $\boldsymbol{E}$ as the conjugate pair, it is possible to show, based on the Second Principle of Thermodynamics, that $\boldsymbol{S} : \dot{\boldsymbol{E}}$ is the exact differential in question, and equals the product of the referential mass by the time derivative of the Helmholtz free energy, i.e.,

$$\rho_R \, \dot{\Psi} = \boldsymbol{S} : \dot{\boldsymbol{E}} = \rho_R \, \frac{\partial \hat{\Psi}}{\partial \boldsymbol{E}}(\boldsymbol{E}) : \dot{\boldsymbol{E}}, \tag{110}$$

where $\rho_R = J \, \rho$ is the (constant, in the absence of mass sources or sinks) referential mass density and $\Psi = \hat{\Psi}(\boldsymbol{E})$ is the Helmholtz free energy, with constitutive function $\hat{\Psi}$. If we denote the *elastic potential* (often called *elastic strain energy*) by

$$W = \rho_R \, \Psi, \qquad \hat{W}(\boldsymbol{E}) = \rho_R \, \hat{\Psi}(\boldsymbol{E}), \tag{111}$$

then we can write

$$\boldsymbol{S} = \hat{\boldsymbol{S}}(\boldsymbol{E}) = \frac{\partial \hat{W}}{\partial \boldsymbol{E}}(\boldsymbol{E}), \tag{112}$$

where $\hat{\boldsymbol{S}}$ is the constitutive function associated with $\boldsymbol{S}$. The time derivative of the second Piola-Kirchhoff stress for a hyperelastic material is given by

$$\dot{\boldsymbol{S}} = \frac{\partial \hat{\boldsymbol{S}}}{\partial \boldsymbol{E}}(\boldsymbol{E}) : \dot{\boldsymbol{E}} = \frac{\partial^2 \hat{W}}{\partial \boldsymbol{E}^2}(\boldsymbol{E}) : \dot{\boldsymbol{E}} = \mathbb{C} : \dot{\boldsymbol{E}}, \tag{113}$$

where

$$\mathbb{C} = \hat{\mathbb{C}}(\boldsymbol{E}) = \frac{\partial^2 \hat{W}}{\partial \boldsymbol{E}^2}(\boldsymbol{E}), \tag{114}$$

is the *material elasticity tensor*, along with its constitutive function $\hat{\mathbb{C}}$.

Note that, if the elastic potential $\hat{W}$ and the constitutive functions $\hat{\boldsymbol{S}}$ and $\hat{\mathbb{C}}$ are written as a function of the right Cauchy-Green deformation tensor $\boldsymbol{C}$ rather than of the Green-Lagrange strain $\boldsymbol{E}$, which is often done, then Equations (112) and (114) must be modified considering that $\partial/\partial \boldsymbol{E} = 2 \, \partial/\partial \boldsymbol{C}$, so that

$$\boldsymbol{S} = \hat{\boldsymbol{S}}(\boldsymbol{C}) = 2 \, \frac{\partial \hat{W}}{\partial \boldsymbol{C}}(\boldsymbol{C}), \tag{115}$$

and

$$\mathbb{C} = \hat{\mathbb{C}}(\boldsymbol{C}) = 4 \, \frac{\partial^2 \hat{W}}{\partial \boldsymbol{C}^2}(\boldsymbol{C}). \tag{116}$$

The Cauchy stress $\boldsymbol{\sigma}$ and the spatial elasticity tensor c are then obtained as the inverse Piola transforms (loosely speaking, push-forwards "rescaled" by J^{-1}; (for the rigorous geometrical definition of the Piola transformation, see, e.g., Marsden and Hughes, 1983) of $\boldsymbol{S}$ and $\mathbb{C}$, respectively:

$$\boldsymbol{\sigma} = J^{-1}\chi_*[\boldsymbol{S}], \qquad \sigma^{ab} = J^{-1}F^a{}_A F^b{}_B\, S^{AB}, \tag{117a}$$

$$\mathrm{c} = J^{-1}\chi_*[\mathbb{C}], \qquad \mathrm{c}^{abcd} = J^{-1}F^a{}_A F^b{}_B F^c{}_C F^d{}_D\, \mathrm{C}^{ABCD}. \tag{117b}$$

Note that the left part of of Equation (117a) is equivalent to $\boldsymbol{\sigma} = J^{-1}\boldsymbol{F}\boldsymbol{S}\boldsymbol{F}^T$ in Equation (109).

Finally, we recall that $\mathbb{L}$ is the *material linear elasticity tensor*, obtained by evaluating the material elasticity tensor $\hat{\mathbb{C}}$ at $\boldsymbol{E} = \mathbf{O}$ (or $\boldsymbol{C} = \boldsymbol{G}$), and the L is the *spatial linear elasticity tensor*, obtained as the inverse Piola transform of $\mathbb{L}$ performed with $\boldsymbol{F} = \mathbf{1}$ and $J = 1$, with components $\mathrm{L}^{abcd} = \mathbf{1}^a{}_A \mathbf{1}^b{}_B \mathbf{1}^c{}_C \mathbf{1}^d{}_D\, \mathrm{L}^{ABCD}$.

5.2 Decomposition of $\boldsymbol{F}$ and Hyperelasticity

The deformation gradient $\boldsymbol{F}$ can be multiplicatively decomposed into a purely volumetric part, and a purely distortional part (Flory, 1961; Ogden, 1978) as

$$\boldsymbol{F} = J^{1/3}\bar{\boldsymbol{F}}, \quad \bar{\boldsymbol{F}} = J^{-1/3}\boldsymbol{F}, \tag{118}$$

where the distortional tensor $\bar{\boldsymbol{F}}$ is such that $\det(\bar{\boldsymbol{F}}) = 1$. The volumetric-distortional decomposition of $\boldsymbol{F}$ induces those of the right and left Cauchy-Green deformation tensors:

$$\boldsymbol{C} = J^{2/3}\bar{\boldsymbol{F}}^T.\bar{\boldsymbol{F}} = J^{2/3}\bar{\boldsymbol{C}}, \qquad \bar{\boldsymbol{C}} = J^{-2/3}\boldsymbol{C}, \tag{119}$$

$$\boldsymbol{b} = J^{2/3}\bar{\boldsymbol{F}}.\bar{\boldsymbol{F}}^T = J^{2/3}\bar{\boldsymbol{b}}, \qquad \bar{\boldsymbol{b}} = J^{-2/3}\boldsymbol{b}. \tag{120}$$

By expressing the elastic potential $\hat{W}$ as an explicit function of the volumetric deformation J and the distortional right Cauchy-Green deformation tensor $\bar{\boldsymbol{C}}$ (which are in turn both functions of the Cauchy-Green deformation tensor $\boldsymbol{C}$), it is possible to write the Cauchy stress $\boldsymbol{\sigma}$ as the sum of its hydrostatic and deviatoric components, i.e.,

$$\boldsymbol{\sigma} = -\pi\, \boldsymbol{g}^{-1} + \boldsymbol{\sigma}_{\mathrm{dev}}, \tag{121}$$

where we call the hydrostatic pressure in the spherical term $-\pi\, \boldsymbol{g}^{-1}$,

$$\pi = -\frac{\partial \hat{W}}{\partial J}(J, \bar{\boldsymbol{C}}), \tag{122}$$

constitutive pressure, and the deviatoric term is given by

$$\boldsymbol{\sigma}_{\text{dev}} = \text{dev}(\boldsymbol{\sigma}) = J^{-2/3}\,\text{dev}\left[J^{-1}\,\boldsymbol{F}\left[2\,\frac{\partial \hat{W}}{\partial \bar{\boldsymbol{C}}}(J,\bar{\boldsymbol{C}})\right]\boldsymbol{F}^T\right]$$
$$= J^{-1}\,\text{dev}\left[\bar{\boldsymbol{F}}\left[2\,\frac{\partial \hat{W}}{\partial \bar{\boldsymbol{C}}}(J,\bar{\boldsymbol{C}})\right]\bar{\boldsymbol{F}}^T\right]. \tag{123}$$

The derivation of the second Piola-Kirchhoff and the Cauchy stresses and of the material and spatial elasticity tensors written with the volumetric-distortional decomposition of the deformation require the definition of the spatial fourth-order symmetric identity, spherical operator and deviatoric operator, as well as their pull-backs (Federico, 2010, 2012). Note also that we use the symbol π for the hydrostatic pressure, and reserve the symbol p to the pore pressure of the biphasic case (Section 8).

5.3 Strict Incompressibility and Quasi Incompressibility

Here we limit ourselves to the case of monophasic materials, i.e., constituted by solid only, as opposed to the biphasic case (solid-fluid), which we shall study in Section 7.4. Those materials that exhibit much larger stiffness under volumetric deformation than under distortional deformation are usually called *incompressible.* Two possible approaches to describe their behaviour are strict incompressibility and quasi incompressibility.

Strict incompressibility rigorously means that the body is undergoing an isochoric (i.e., volume-preserving) motion, which is captured by the bilateral constraint (i.e., constraint expressed by an equality)

$$J - 1 = 0. \tag{124}$$

In this case, the hydrostatic pressure π of Equation (121) is not specified constitutively as the derivative of the potential $\hat{W}$ with respect to the volumetric deformation, but takes instead the meaning of Lagrange multiplier associated with the incompressibility constraint (124) (see, e.g., Ogden, 1997). Moreover, the application of Euler's theorem on homogeneous functions ensures that the potential $\hat{W}$ must be expressed as a function of $\bar{\boldsymbol{C}}$ alone, i.e., $\hat{W}(J,\bar{\boldsymbol{C}})$ of the general compressible case must be replaced by a function $\hat{W}_d(\bar{\boldsymbol{C}})$, which contributes a purely deviatoric stress. A very clear proof, with a slightly different notation, is given by Bonet and Wood (2008).

Quasi incompressibility is described by a decoupled elastic potential, given by the sum of a term function solely of the volumetric deformation J and a term function solely of a distortional measure of deformation, for example $\bar{\boldsymbol{C}}$, (Simo et al., 1985; Miehe, 1994):

$$\hat{W}(J, \bar{\boldsymbol{C}}) = \hat{U}(J) + \hat{W}_d(\bar{\boldsymbol{C}}). \tag{125}$$

Quasi incompressibility is enforced by writing the volumetric term $\hat{U}$ such that it attains large values (i.e., a large amount of energy is required) for small deviations of the volumetric deformation J from 1. The volumetric term $\hat{U}$ is often taken in the simple quadratic form

$$\hat{U}(J) = \tfrac{1}{2}\kappa\,(J-1)^2, \tag{126}$$

so that the constitutive pressure π is linear in J,

$$\pi = -\frac{\partial \hat{U}}{\partial J}(J) = -\kappa\,(J-1), \tag{127}$$

and the constant bulk modulus

$$\kappa = \frac{\partial^2 \hat{U}}{\partial J^2}(J) = \frac{\partial^2 \hat{U}}{\partial J^2}(1), \tag{128}$$

serves as the *penalty number* employed to enforce the incompressibility constraint $J - 1 = 0$. For this purpose, the bulk modulus is set to be 3 to 4 orders of magnitude larger than the shear modulus (Bonet and Wood, 2008), or some average of the shear moduli, for the case of anisotropy. Here, with "shear modulus" or "shear moduli" we mean, for example, those of the material linear elasticity tensor $\mathbb{L} = \hat{\mathbb{C}}(\boldsymbol{G})$, where $\hat{\mathbb{C}}$ is the constitutive function of the elasticity tensor associated with the potential (125) and $\boldsymbol{C} = \boldsymbol{G}$ represents the situation in which there is no deformation.

6 Monophasic Fibre-Reinforced Materials

In this section we shall focus on *monophasic* composites, i.e., composites comprised of solid only, as opposed to *biphasic* composites, which contain a solid and a fluid phase and shall be dealt with in Section 8. In order to avoid confusion, we use the term *constituent* to refer to each of the materials of which the solid composite is comprised. In the literature on composites, the term *phase* is very often employed in place of *constituent*, but in this case we prefer to reserve the term phase to "constituents with different physical state", i.e., solid and fluid. Furthermore, we shall restrict ourselves to the case of elasticity. We shall start with composites with a finite number of families of aligned fibres and then generalise to composites with a statistical distribution of fibre orientation. These materials represent well a wide class of soft biological tissues.

Tissues characterised by one family of aligned fibres are, e.g., tendons and ligaments (Weiss and Gardiner, 2001), whereas examples of tissues with two families are the annulus fibrosus of intervertebral discs (White and Panjabi, 1978) and the adventitial layer of blood vessels (Canham et al., 1989). Articular cartilage is instead a tissue characterised by a location-dependent fibre arrangement that cannot be represented with a finite number of families of aligned fibres. The fibres are roughly aligned in the deep zone, randomly oriented in the middle zone, and parallel to the surface in the superficial zone (Aspden and Hukins, 1981; Mollenhauer et al., 2003).

Biological tissues with statistical fibre arrangement have been first studied by Lanir (1983), whose approach has been widely used in subsequent works (e.g., Hurschler et al., 1997; Billiar and Sacks, 2000; Sacks, 2003; Gasser et al., 2006). A similar technique for accounting for the statistical fibre arrangement (with the same formalism employed here, but for the case of linear elasticity) was independently devised by Federico et al. (2004). Gasser et al. (2006) proposed to account for fibres with statistical orientation by calculating the directional average of the structure tensor, an approach that has been adopted in later works (e.g., Menzel et al., 2008). A microplane model based on stress directional averaging has been proposed by Caner and Carol (2006).

The methodology presented here stems from Mixture Theory. A conceptually similar approach, without the explicit use of volumetric fractions, has been adopted by Holzapfel et al. (2000) for the derivation of the elastic potential of a composite with a finite number of families of aligned fibres. This approach has then been generalised, within Mixture Theory, to the case of a continuous statistical distribution of fibres (Federico and Herzog, 2008a) and implemented numerically (Federico and Gasser, 2010). We first consider a generic physical quantity $\mathcal{F}$ for the case of constitutive dependence on deformation only, and then study the case of Elasticity. The evaluation of the permeability, another quantity that can be thought to depend on deformation only, is discussed in Section 8.

6.1 Volumetric Fractions and Rule of Mixtures

A solid-only fibre-reinforced material is a composite comprised of two constituents: the matrix and the fibres. Industrial composites often have a finite number of families of fibres, each aligned in a well-defined direction. If the the matrix is denoted by 0, each of the families of aligned fibres can be seen as as a constituent, denoted by 1, 2, etc. If the matrix is isotropic, anisotropy in the material properties of the composite arises because of the presence of the fibres and of their arrangement in space.

The simplest way to study any composite is to use the *rule of mixtures*, according to which the value of the physical quantity $\mathcal{F}$ relative to the whole composite is obtained as the weighted average of the values of the same physical quantity for the individual constituents, the weight of each constituent being its volumetric fraction. Moreover, the mixture is assumed to be *constrained*, i.e., all constituents are assumed to attain the *same* motion χ and, for the case of fibres, their reorientation is assumed to be governed by the *same* deformation gradient $\boldsymbol{F} = T\chi$ (Merodio and Ogden, 2002; Merodio, 2006; Gasser et al., 2006).

The volumetric fraction of constituent α is defined as the positive function ϕ_α (an *absolutely positive measure* in the language of the Theory of Integration) such that

$$\int_{\chi(\mathcal{B}_R,\,\cdot\,)} \phi_\alpha = V_\alpha, \tag{129}$$

where V_α is the portion of the current configuration $\chi(\mathcal{B}_R,\,\cdot\,)$ occupied by constituent α. In the material picture, Eq. (129) reads

$$\int_{\mathcal{B}_R} J\,\phi_\alpha = \int_{\mathcal{B}_R} \phi_{\alpha R} = V_\alpha, \tag{130}$$

where

$$\phi_{\alpha R} = J\,\phi_\alpha \tag{131}$$

is the material or referential volumetric fraction of constituent α. Note that Equation (131) is analogous to the material form of the equation of balance of mass,

$$\rho_R = J\,\rho, \tag{132}$$

of standard single-constituent Continuum Mechanics: the mass density ρ, indeed, is a measure as well. In the spatial and the material picture, the volumetric fractions must obey the conditions

$$\sum_{\alpha=1}^{N} \phi_\alpha = 1, \qquad \sum_{\alpha=1}^{N} \phi_{\alpha R} = 1. \tag{133}$$

The overall value of a given physical quantity $\mathcal{F}$ for a composite comprised of $N+1$ constituents, with constituent 0 being the matrix, is given, in the material picture, by

$$\mathcal{F} = \phi_{0R}\,\mathcal{F}_0 + \sum_{\alpha=1}^{N} \phi_{\alpha R}\,\mathcal{F}_\alpha. \tag{134}$$

In the following, we shall refer to this method as to the rule of mixture or the superposition method.

6.2 Constitutive Equations

In a general thermo-mechanical problem, any physical quantity $\mathcal{F}$ is associated with a constitutive function $\hat{\mathcal{F}}$ depending on motion χ, temperature θ, their gradients, their rates, etc, plus variables associated with the material symmetry, if necessary. We shall present the theory for the case in which $\mathcal{F}$ is a scalar quantity, as this can be easily generalised to a tensor of any order. We want to restrict ourselves to the case of elasticity, and therefore the dependence of the constitutive function $\hat{\mathcal{F}}$ on the motion χ reduces to the dependence on the deformation gradient $\boldsymbol{F} = T\chi$. However, by virtue of the Principle of Material Frame Indifference, according to which the constitutive equations must be invariant under change of observer, the dependence must reduce to that on a purely material measure of deformation, e.g., the right Cauchy-Green deformation tensor $\boldsymbol{C}$, or a purely material measure of strain, e.g., the Green-Lagrange strain $\boldsymbol{E} = \frac{1}{2}(\boldsymbol{C} - \boldsymbol{G})$.

If the matrix is isotropic, its constitutive function will exclusively depend on the deformation. If constituents $1, ..., N$ are fibre families each with a given referential orientation $\boldsymbol{M}_\alpha \in \mathbb{S}^2\mathcal{B}_R$, their constitutive functions shall explicitly depend on $\boldsymbol{M}_\alpha$. We recall that, in all those cases in which the system is invariant under reflection of each of the $\boldsymbol{M}_\alpha$, i.e., $\boldsymbol{M}_\alpha$ and $-\boldsymbol{M}_\alpha$ are equivalent, each constitutive function should be written as a function of the corresponding structure tensor $\boldsymbol{A}_\alpha = \boldsymbol{M}_\alpha \otimes \boldsymbol{M}_\alpha$. Under these hypotheses, Equation (134) reads, in terms of constitutive functions

$$\mathcal{F} = \hat{\mathcal{F}}(\boldsymbol{C}, \boldsymbol{A}_1, ..., \boldsymbol{A}_N) = \phi_{0R}\,\hat{\mathcal{F}}_0(\boldsymbol{C}) + \sum_{\alpha=1}^{N} \phi_{\alpha R}\,\hat{\mathcal{F}}_\alpha(\boldsymbol{C}, \boldsymbol{A}_\alpha). \tag{135}$$

6.3 Statistically Oriented Fibres

For a continuous distribution of statistically oriented fibres, the summation in (135) becomes an integral on the material unit sphere $\mathbb{S}^2\mathcal{B}_R$. In this integral, the weighing function is a probability distribution

$$\psi : \mathbb{S}^2\mathcal{B}_R \to \mathbb{R}_0^+ : \boldsymbol{M} \mapsto \psi(\boldsymbol{M}), \tag{136}$$

describing the probability to find, at point X, a fibre with orientation $\boldsymbol{M}$. Function ψ must be normalised over the material sphere, and must be invariant for reflection of $\boldsymbol{M}$, i.e.,

$$\int_{\mathbb{S}^2\mathcal{B}_R} \psi(\boldsymbol{M}) = 1, \qquad \psi(\boldsymbol{M}) = \psi(-\boldsymbol{M}). \tag{137}$$

Note that it is convenient to define the probability distribution (136) on the material sphere $\mathbb{S}^2\mathcal{B}_R$ because in principle the orientation of the fibres is known only in the reference configuration.

We can transform the variables in Equation (135) according to

$$\phi_{\alpha R} \mapsto \phi_{1R}\,\psi(\boldsymbol{M}), \tag{138a}$$

$$\hat{\mathcal{F}}_\alpha(\boldsymbol{C}, \boldsymbol{A}_\alpha) \mapsto \hat{\mathcal{F}}_1(\boldsymbol{C}, \hat{\boldsymbol{A}}(\boldsymbol{M})), \tag{138b}$$

$$\hat{\mathcal{F}}(\boldsymbol{C}, \boldsymbol{A}_1, \ldots, \boldsymbol{A}_N) \mapsto \hat{\mathcal{F}}(\boldsymbol{C}, \psi), \tag{138c}$$

i.e., we can call the collection of all fibres "constituent 1", and identify all orientations in $\mathbb{S}^2\mathcal{B}_R$ via the probability ψ. Note also that we are using the notation $\hat{\boldsymbol{A}}(\boldsymbol{M}) = \boldsymbol{A} = \boldsymbol{M} \otimes \boldsymbol{M}$ introduced in Equation (105). We therefore obtain

$$\mathcal{F} = \hat{\mathcal{F}}(\boldsymbol{C}, \psi) = \phi_{0R}\,\hat{\mathcal{F}}_0(\boldsymbol{C}) + \phi_{1R} \int_{\mathbb{S}^2\mathcal{B}_R} \psi(\boldsymbol{M})\,\hat{\mathcal{F}}_1(\boldsymbol{C}, \hat{\boldsymbol{A}}(\boldsymbol{M})). \tag{139}$$

Whereas in the discrete case of Equation (135) the overall constitutive function $\hat{\mathcal{F}}$ depended on all fibre directions, in Equation (139) it has a *functional dependence* on the probability distribution ψ (Federico and Grillo, 2012), i.e., it is a function

$$\hat{\mathcal{F}} : [T\mathcal{B}_R]^0_2 \times L^1_{\text{norm}}(\mathbb{S}^2\mathcal{B}_R, \mathbb{R}^+_0) \to \mathbb{R}, \tag{140}$$

where $[T\mathcal{B}_R]^0_2$ is the space of material second-order tensors with covariant components (in which the right Cauchy-Green deformation tensor $\boldsymbol{C}$ ist valued), and $L^1_{\text{norm}}(\mathbb{S}^2\mathcal{B}_R, \mathbb{R}^+_0)$ is the space of integrable functions normalised on the material unit sphere and valued in the positive real numbers. In a more general setting, the probability ψ may be regarded as an element in a space of *distributions* (i.e., *generalised functions*; see, e.g., Kolmogorov and Fomin, 1999), rather than in $L^1_{\text{norm}}(\mathbb{S}^2\mathcal{B}_R, \mathbb{R}^+_0)$. This functional dependence can be exploited to describe the evolution of the probability distribution when the composite undergoes *remodelling* (Grillo et al., 2012, 2014), i.e., the change of physico-mechanical properties in a material due to rearrangement of its internal structure. The most notable examples of remodelling occur in fact in biological tissues.

6.4 The Fibre Ensemble

The integral

$$\mathcal{F}_e = \hat{\mathcal{F}}_e(\boldsymbol{C}, \psi) = \int_{\mathbb{S}^2\mathcal{B}_R} \psi(\boldsymbol{M})\,\hat{\mathcal{F}}_1(\boldsymbol{C}, \hat{\boldsymbol{A}}(\boldsymbol{M})) \tag{141}$$

in Equation (139) is called the *fibre ensemble of* $\mathcal{F}$, and accounts for the effect of all fibres on the physical quantity $\mathcal{F}$ (Federico and Herzog, 2008a).

In general, the fibre ensemble of a physical quantity $\mathcal{F}$ cannot be calculated once and for all, because the deformation $\boldsymbol{C}$ cannot always be factorised out of the integral. Therefore, in the Finite Element implementation of this class of integrals, the integration needs to be performed numerically at each deformation increment (Federico and Gasser, 2010), which can be done using a spherical t-design (Hardin and Sloane, 1996; Hardin et al., 2008), i.e., a set $\{\boldsymbol{M}^{(1)}, \ldots, \boldsymbol{M}^{(N)}\} \in \mathbb{S}^2\mathcal{B}_R$ of N points on the unit material sphere such that, for polynomials $\mathcal{P}$ of degree $k \leq t$ we have

$$\int_{\mathbb{S}^2\mathcal{B}_R} \mathcal{P}(\boldsymbol{M}) = \frac{4\pi}{N} \sum_{j=1}^{N} \mathcal{P}(\boldsymbol{M}^{(j)}), \tag{142}$$

where 4π is the (surface) measure of the unit sphere $\mathbb{S}^2\mathcal{B}_R$.

Although in general integrals of the type of that in Equation (141) can only be evaluated at each increment of deformation, a once-and-for-all integration is possible when the constitutive function $\hat{\mathcal{F}}_1$ that is a polynomial in the structure tensor $\boldsymbol{A} = \hat{\boldsymbol{A}}(\boldsymbol{M})$, in which case the deformation $\boldsymbol{C}$ *can* be factorised out of the integral. Indeed, if $\hat{\mathcal{F}}_1$ is a polynomial of grade p, i.e., it has the form

$$\hat{\mathcal{F}}_1(\boldsymbol{C}, \hat{\boldsymbol{A}}(\boldsymbol{M})) = a \left[\sum_{k=1}^{p} \langle \mathbb{Q}_k(\boldsymbol{C}) | \underbrace{\hat{\boldsymbol{A}}(\boldsymbol{M}) \otimes \ldots \otimes \hat{\boldsymbol{A}}(\boldsymbol{M})}_{k \text{ times}} \rangle + q(\boldsymbol{C}) \right], \tag{143}$$

where a is a scalar constant, the $\mathbb{Q}_k$ are non-dimensional tensor functions of $\boldsymbol{C}$ of order $2k$, valued in $[T\mathcal{B}_R]^0_{2k}$, and q is a non-dimensional scalar function of $\boldsymbol{C}$, the integral (141) becomes

$$\mathcal{F}_e = \hat{\mathcal{F}}_e(\boldsymbol{C}, \psi) = a \left[\sum_{k=1}^{p} \langle \mathbb{Q}_k(\boldsymbol{C}) | \mathbb{H}_k \rangle + q(\boldsymbol{C}) \right], \tag{144}$$

where

$$\mathbb{H}_k = \int_{\mathbb{S}^2\mathcal{B}_R} \psi(\boldsymbol{M}) \; \underbrace{\hat{\boldsymbol{A}}(\boldsymbol{M}) \otimes \ldots \otimes \hat{\boldsymbol{A}}(\boldsymbol{M})}_{k \text{ times}} \tag{145}$$

are the $2k$-th order averaged structure tensors, elements of $[T\mathcal{B}_R]^{2k}_0$, and can be calculated independently of the deformation. Note that the tensors $\mathbb{Q}_1(\boldsymbol{C})$ and $\mathbb{H}_1$ are actually second-order tensors, and we shall write $\boldsymbol{Q}(\boldsymbol{C}) \equiv \mathbb{Q}_1(\boldsymbol{C})$ and $\boldsymbol{H} \equiv \mathbb{H}_1$. We also note that an expression in the form of Equation (144) could be though of as descending from the Taylor expansion of the fibre potential $\hat{W}_1$, with initial point $\boldsymbol{A}_0 = \boldsymbol{M}_0 \otimes \boldsymbol{M}_0$.

A very interesting particular case of the polynomial constitutive function (143) is the polynomial of order one, or *affine function*

$$\hat{\mathcal{F}}_1(\boldsymbol{C}, \hat{\boldsymbol{A}}(\boldsymbol{M})) = a\,[\boldsymbol{Q}(\boldsymbol{C}) : \hat{\boldsymbol{A}}(\boldsymbol{M}) + q(\boldsymbol{C})], \tag{146}$$

for which Equation (144) reduces to

$$\mathcal{F}_e = \hat{\mathcal{F}}_e(\boldsymbol{C}, \psi) = a\,[\boldsymbol{Q}(\boldsymbol{C}) : \boldsymbol{H} + q(\boldsymbol{C})] = \hat{\mathcal{F}}_1(\boldsymbol{C}, \boldsymbol{H}), \tag{147}$$

where we recall that the tensor

$$\boldsymbol{H} = \int_{\mathbb{S}^2\mathcal{B}_R} \psi(\boldsymbol{M})\,\hat{\boldsymbol{A}}(\boldsymbol{M}), \tag{148}$$

(which we had called $\boldsymbol{A}_{avg}$ in a previous work (Federico, 2012)) is the directional average of the structure tensor $\boldsymbol{A}$ over the material unit sphere $\mathbb{S}^2\mathcal{B}_R$, weighted by the probability ψ, and has been named *generalised structure tensor* by Gasser et al. (2006). Equation (147) also shows that if, and only if, a constitutive function is affine, then the averaging integral operator and the fibre constitutive function $\hat{\mathcal{F}}_1$ *commute.* This is the condition under which this approach (Federico and Herzog, 2008a) coincides with the approach proposed by Gasser et al. (2006) for the elastic potential, as it shall also be discussed in Section 6.7 and in Remark 6.1 below.

The fibre ensemble of $\hat{\mathcal{F}}_1$ is well approximated by that in Equation (147) when the probability ψ has a *weak* dispersion about a given direction $\boldsymbol{M}_0$, i.e., it is non-zero close to $\boldsymbol{M}_0$ and vanishes away from it. In this case, the fibre constitutive function $\hat{\mathcal{F}}_1$ can be Taylor-expanded in $\boldsymbol{A}$ to the first order, with initial point $\boldsymbol{A}_0 = \boldsymbol{M}_0 \otimes \boldsymbol{M}_0$, to yield

$$\hat{\mathcal{F}}_1(\boldsymbol{C}, \boldsymbol{A}) \simeq \hat{\mathcal{F}}_1(\boldsymbol{C}, \boldsymbol{A}_0) + \left[\frac{\partial \hat{\mathcal{F}}_1}{\partial \boldsymbol{A}}(\boldsymbol{C}, \boldsymbol{A}_0)\right] : [\boldsymbol{A} - \boldsymbol{A}_0], \tag{149}$$

which is an affine function of $\boldsymbol{A}$, of the type (146), and therefore generates an ensemble function of the form (147).

Remark 6.1. The fibre ensemble approach has been originally introduced in the context of non-linear elasticity in a previous work (Federico and Herzog, 2008a), in which we had erroneously mentioned that the integral defining the fibre ensemble $\hat{\mathcal{F}}_e$ of the fibre quantity $\hat{\mathcal{F}}_1$ (Equation (141)) is such that the deformation can be factorised out of the integral *if, and only if*, the constitutive function $\hat{\mathcal{F}}_1$ is affine in $\boldsymbol{A}$. As it has been shown above, this is a sufficient condition, but it is not necessary: indeed, the deformation can actually be factorised out of the integral for a polynomial of any order

(Equation (144)). M. Netušil and D. Steigmann are gratefully acknowledged for pointing out the incorrectness of this statement in our past work (Federico and Herzog, 2008a). What we really had in mind when we wrote that statement was the condition under which the proposed approach coincides with that presented by Gasser et al. (2006): indeed, the two approaches do coincide if, and only if, the fibre constitutive function $\hat{\mathcal{F}}_1$ is affine in $\boldsymbol{A}$, or can be reduced to an expression affine in $\boldsymbol{A}$ when the fibres have a weak dispersion about a given direction, as shown in Equation (149). This shall be further discussed in Section 6.7, which is specifically dedicated to the non-linear elasticity of fibre-reinforced composites.

6.5 Multiple Fibre Families with Statistical Orientation

Naturally, one can think that there are multiple families of statistically oriented fibres, and that the distribution of orientation in each fibre family is described by a probability distribution. In this case, the overall constitutive function of the physical quantity $\mathcal{F}$ reads

$$\mathcal{F} = \hat{\mathcal{F}}(\boldsymbol{C}, \psi_1, ..., \psi_N) = \phi_{0R}\, \hat{\mathcal{F}}_0(\boldsymbol{C}) + \sum_{\alpha=1}^{N} \phi_{\alpha R}\, \hat{\mathcal{F}}_{\alpha e}(\boldsymbol{C}, \psi_\alpha), \tag{150}$$

where the fibre ensemble $\hat{\mathcal{F}}_{\alpha e}$ of each family is given by

$$\hat{\mathcal{F}}_{\alpha e}(\boldsymbol{C}, \psi_\alpha) = \int_{\mathbb{S}^2\mathcal{B}_R} \psi_\alpha(\boldsymbol{M})\, \hat{\mathcal{F}}_\alpha(\boldsymbol{C}, \hat{\boldsymbol{A}}(\boldsymbol{M})). \tag{151}$$

The case of multiple families of statistically oriented fibres can always be retrieved by extension of the case of a single family (Equation (139)), to which we will thus restrict our analysis.

6.6 Forms of the Probability Distribution

In many applications, it is convenient to evaluate directional averaging integrals of the form (141) in polar coordinates. If $\{\boldsymbol{\mathcal{E}}_B\}_{B=1}^3$ is any orthonormal basis of $T\mathcal{B}_R$ (not necessarily coincident with the Cartesian basis $\{\boldsymbol{E}_A\}_{A=1}^3$), and $\boldsymbol{\mathcal{E}}_1$ is chosen as the polar axis, a given unit vector $\boldsymbol{M}$ is given as a function of the co-latitude Θ from the polar axis $\boldsymbol{\mathcal{E}}_1$ and longitude Φ from the $\boldsymbol{\mathcal{E}}_1$-$\boldsymbol{\mathcal{E}}_2$ plane (see box in Figure 1):

$$\boldsymbol{M}(\Theta, \Phi) = \cos\Theta\, \boldsymbol{\mathcal{E}}_1 + \sin\Theta \cos\Phi\, \boldsymbol{\mathcal{E}}_2 + \sin\Theta \sin\Phi\, \boldsymbol{\mathcal{E}}_3. \tag{152}$$

Hence, the probability distribution can be written as a function of Θ and Φ:

$$\varrho(\Theta, \Phi) = \psi(\boldsymbol{M}(\Theta, \Phi)) \tag{153}$$

and the fibre ensemble (directional average) of any function f of $\boldsymbol{M}$ becomes

$$f_e = \int_{\mathbb{S}^2} \psi(\boldsymbol{M}) f(\boldsymbol{M}) = \int_0^{2\pi} \int_0^{\pi} \varrho(\Theta, \Phi) f(\boldsymbol{M}(\Theta, \Phi)) \sin\Theta \, d\Theta \, d\Phi. \tag{154}$$

If the probability distribution is transversely isotropic with respect to the polar direction $\boldsymbol{\mathcal{E}}_1$, then the probability function ϱ depends only on the co-latitude angle Θ and not on the longitude Φ. In this case, the normalisation condition becomes

$$2\pi \int_0^{\pi} \varrho(\Theta) \sin\Theta \, d\Theta = 1. \tag{155}$$

If a transversely isotropic distribution tends to the Dirac delta distribution centred at $\Theta = \pi/2$ and $\Theta = 0$, then it represents the limit cases of fibres randomly oriented in the transverse plane (planar isotropy) and fibres all aligned in one direction, respectively.

All these cases can be represented well by the π-periodic von Mises distribution, which is the analog of the Gaussian distribution on the unit sphere. This distribution has been used by Gasser et al. (2006) for all cases ranging from isotropy to aligned fibres, and extended by Federico and Gasser (2010) to include also the case of planar isotropy. This has been done by means of two separate functions, one for the cases ranging from planar isotropy to isotropy and one for the cases ranging from isotropy to aligned fibres. The dispersion of the fibre orientation about the polar direction (i.e., about $\Theta = 0$) is modulated by the concentration parameter b (roughly the inverse of a dispersion; Gasser et al., 2006), which has value zero in the case of isotropy, and tends to infinity for the cases of planar isotropy and aligned fibres.

Pajerski (2010) used $b \in \mathbb{R}$, thereby including negative values, in order to include all possible cases in a single function (this approach has been then adopted by Tomic et al., 2014). The general von Mises distribution centred at $\Theta = \Theta_0 = 0$ is given by

$$\varrho_M(\Theta) = \frac{\exp[b\cos(2\Theta)]}{2\pi I_0(b)}, \tag{156}$$

where $I_0(b) = \frac{1}{\pi}\int_0^{\pi} \exp(b\cos\Theta)\, d\Theta$ denotes the modified Bessel function of the first kind of order zero. Normalisation of ϱ_M according to Equation (155) leads to the orientation probability distribution

$$\varrho(\Theta) = \frac{1}{\pi}\sqrt{\frac{b}{2\pi}} \frac{\exp[b(\cos(2\Theta) + 1)]}{\operatorname{erfi}(\sqrt{2b})}, \tag{157}$$

where erf(x) and erfi$(x) = -i\,\mathrm{erf}(i\,x)$ denote the error function at x and the imaginary error function at x, respectively (Weisstein, 2005). The limit $b \to -\infty$ describes fibres all lying on the plane, the limit $b \to 0$ represents isotropy, and the limit $b \to +\infty$ describes fibres all aligned in one direction. The transversely isotropic orientation probability distribution in Equation (157) is illustrated in Figure 1, by means of surface plots defined by the vector $\varrho(\Theta)\boldsymbol{M}(\Theta, \Phi)$.

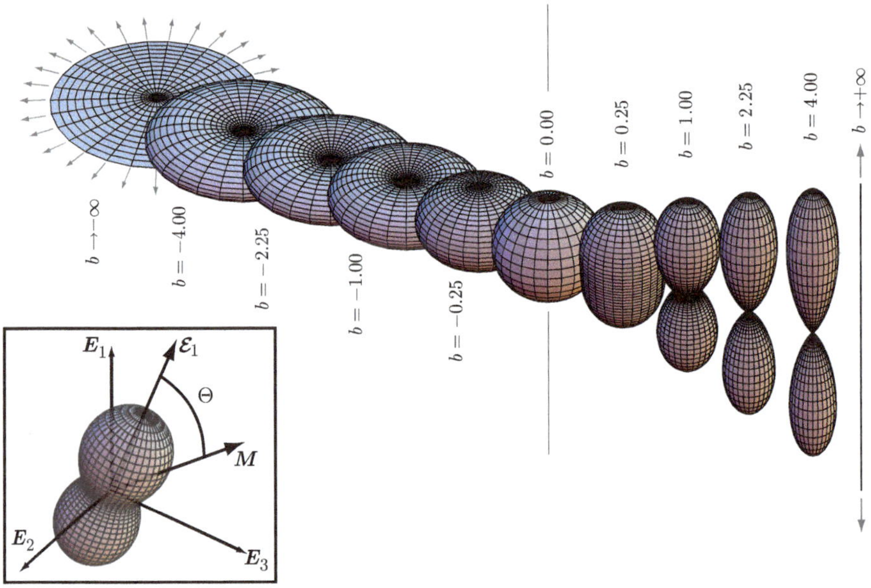

Figure 1: (Adapted from Federico and Gasser, 2010) Transversely isotropic probability distribution density functions describing fibre orientation, mapping points $\boldsymbol{M}$ on the material unit sphere $\mathbb{S}^2\mathcal{B}_R$ into the real numbers. Surface plots are defined by the vector $\varrho(\Theta)\boldsymbol{M}(\Theta, \Phi)$ and ϱ is based on the von Mises distribution. The limit $b \to -\infty$ describes planar isotropy, the limit $b \to 0$ describes isotropy, and the limit $b \to +\infty$ describes fibres all aligned in the polar direction $\boldsymbol{\mathcal{E}}_1$. The box in the left-bottom corner describes the general case in which the polar direction $\boldsymbol{\mathcal{E}}_1$ does not coincide with any of the axes of the Cartesian basis $\{\boldsymbol{E}_A\}_{A=1}^3$.

Another possible form is a standard Gaussian defined over the *hemisphere*, i.e., for $\Theta \in [0, \pi/2]$:

$$\varrho_G(\Theta) = \exp\left(-\frac{(\Theta - \Theta_0)^2}{\lambda^2}\right). \tag{158}$$

where Θ_0 is the mean value and λ is the variance. Normalisation leads to

$$\varrho(\Theta) = \frac{\varrho_G(\Theta)}{2\pi \int_0^{\pi/2} \varrho_G(t)\, \sin(t)\, dt}. \tag{159}$$

This probability distribution, with variance λ and mean value Θ_0 both parametric in the coordinate describing cartilage depth, has been used to represent the varying arrangement of the collagen fibres in articular cartilage, based on the X-ray diffraction experiments by Mollenhauer et al. (2003), to evaluate the permeability of the tissue (Federico and Herzog, 2008c), and is shown in Figure 4a in Section 8.4.

6.7 Elastic Potential

Using the rule of mixtures (or superposition method), the elastic potential of a fibre-reinforced material is obtained by simply replacing the generic constitutive function $\hat{\mathcal{F}}$ with the elastic potential $\hat{W}$.

In the case of a finite number of families of aligned fibres, Equation (135) gives

$$\hat{W}(\boldsymbol{C}) = \phi_{0R}\, \hat{W}_0(\boldsymbol{C}) + \sum_{\alpha=1}^{N} \phi_{\alpha R}\, \hat{W}_\alpha(\boldsymbol{C}, \boldsymbol{A}_\alpha), \tag{160}$$

where this time we do not explicitly write the dependence of the overall potential $\hat{W}$ on all the structure tensors $\boldsymbol{A}_\alpha$, for the sake of a lighter notation. This form of the elastic potential is essentially that used by Holzapfel et al. (2000) in their model of the two fibre families in arteries:

$$\hat{W}(\boldsymbol{C}) = \phi_{0R}\, \hat{W}_0(\boldsymbol{C}) + \phi_1\, \hat{W}_1(\boldsymbol{C}, \boldsymbol{A}_1) + \phi_2\, \hat{W}_2(\boldsymbol{C}, \boldsymbol{A}_2), \tag{161}$$

with the matrix potential being neo-Hookean

$$\hat{W}_0(\boldsymbol{C}) = \tfrac{c}{2}\, (\boldsymbol{C} : \boldsymbol{G}^{-1} - 3), \tag{162}$$

and the fibre potentials being exponential

$$\hat{W}_\alpha(\boldsymbol{C}, \boldsymbol{A}_\alpha) = \tfrac{k}{2h}\, \exp[h\, (\boldsymbol{C} : \boldsymbol{A}_\alpha - 1)], \tag{163}$$

where c and k have the dimension of an energy per unit volume, and h is non-dimensional (h and k were assumed to be the same for both fibre families).

Also, $\boldsymbol{C} : \boldsymbol{G}^{-1} = I_1(\boldsymbol{C})$ is the first invariant of $\boldsymbol{C}$, $\boldsymbol{C} : \boldsymbol{A}_1 = I_4(\boldsymbol{C})$ is the fourth invariant of $\boldsymbol{C}$ with respect to fibre family 1, and $\boldsymbol{C} : \boldsymbol{A}_2 = I_6(\boldsymbol{C})$ is the sixth invariant of $\boldsymbol{C}$, i.e., the "fourth invariant with respect to fibre family 2". Note that Holzapfel et al. (2000) did not explicitly use the volumetric fractions, which can be thought of to have been lumped together with the material constants in the potentials $\hat{W}_\alpha$ (with $\alpha \in \{0, 1, 2\}$).

In the case of one family of statistically oriented fibres, Equation (139) gives

$$\hat{W}(\boldsymbol{C}) = \phi_{0R}\, \hat{W}_0(\boldsymbol{C}) + \phi_{1R} \int_{\mathbb{S}^2 \mathcal{B}_R} \psi(\boldsymbol{M})\, \hat{W}_1(\boldsymbol{C}, \hat{\boldsymbol{A}}(\boldsymbol{M})). \tag{164}$$

where the integral

$$\hat{W}_e(\boldsymbol{C}) = \int_{\mathbb{S}^2 \mathcal{B}_R} \psi(\boldsymbol{M})\, \hat{W}_1(\boldsymbol{C}, \hat{\boldsymbol{A}}(\boldsymbol{M})) \tag{165}$$

is the *fibre ensemble potential* (Federico and Herzog, 2008a), and the functional dependence of both $\hat{W}$ and $\hat{W}_e$ on the probability distribution ψ has not been explicitly indicated, for the sake of a lighter notation. Again, we remark that, although for a $\hat{W}_1$ that is polynomial in $\boldsymbol{A}$ it is possible to evaluate the integral in Equation (165) independently of the deformation, in general $\hat{W}_e$ needs to be evaluated at each increment of deformation, within a computational framework.

As discussed in general in Section 6.4, when the fibre potential $\hat{W}_1$ is an affine function of $\boldsymbol{A}$, or when it can be reduced to such a function because the probability distribution ψ has a weak dispersion about a given direction $\boldsymbol{M}_0$, the overall potential reduces to

$$\hat{W}(\boldsymbol{C}) = \phi_{0R}\, \hat{W}_0(\boldsymbol{C}) + \phi_{1R}\, \hat{W}_1(\boldsymbol{C}, \boldsymbol{H}). \tag{166}$$

If we write this particular case for two families 1 and 2 of statistically oriented fibres (see Section 6.5, Equation (150)), associated with their probability distributions ψ_1 and ψ_2, respectively, we obtain

$$\hat{W}(\boldsymbol{C}) = \phi_{0R}\, \hat{W}_0(\boldsymbol{C}) + \phi_{1R}\, \hat{W}_1(\boldsymbol{C}, \boldsymbol{H}_1) + \phi_{2R}\, \hat{W}_2(\boldsymbol{C}, \boldsymbol{H}_2), \tag{167}$$

which coincides with the model by Gasser et al. (2006) for the arterial wall. The method by Gasser et al. (2006) was developed *directly* by using the averaged structure tensors $\boldsymbol{H}_1$ and $\boldsymbol{H}_2$, and therefore *does not* come from Mixture Theory. Rather, it is a different approach altogether, which coincides with the Mixture Theory approach of Equation (164) in the case of affine fibre potential or small orientation dispersion.

Remark 6.2. The potential (161) is in the form that been used by Holzapfel et al. (2000) and many other Authors for practical applications, and can be thought of as having been derived from the rule of mixtures because of its additive form. Note that the general form of potential reported by Holzapfel et al. (2000) does *not* come from Mixture Theory. Rather, it is expressed as the sum of an isotropic term, which accounts essentially for the matrix, and an anisotropic term, which accounts for the two fibre families, and in terms of the invariants of the deformation,

$$\hat{W} = \breve{W}_{iso}(I_1, I_2, I_3) + \breve{W}_{aniso}(I_1, I_2, I_3, I_4, I_5, I_6, I_7, I_8). \tag{168}$$

where the breve "$\breve{\ }$" denotes potentials written as a function of the invariants, I_4 and I_5 are defined as in Equations (107) with respect to fibre family 1, I_6 and I_7 are defined analogously with respect to fibre family 2, $I_8 = \operatorname{tr}(\boldsymbol{A}_1\, \boldsymbol{C}\, \boldsymbol{A}_2)$ (Spencer, 1984), and $I_3 = 1$ in the incompressible case. In particular, a model based on Mixture Theory *cannot* depend on I_8, which is a term coupling the structure tensors of the two families.

6.8 Effect of Fibres in Contraction

In a fibre-reinforced material, be it industrial or biological, the effect of the fibres on the elastic properties is dominant when the fibre are extended, but weak when they are contracted. A common modelling technique is to assign the fibres mechanical properties only in extension. Very often, in the literature, the fibres treated in this way are called "tension-only", although typically the criterion to "activate" or "deactivate" them is based on whether the squared stretch in the fibre direction,

$$\lambda_{\boldsymbol{M}}^2 = \boldsymbol{C} : \boldsymbol{A} = I_4(\boldsymbol{C}), \tag{169}$$

is greater or smaller than one. Therefore, in order to avoid confusion, we shall avoid the expressions "fibres in tension" and "fibres in compression" in favour of "fibres in extension" and "fibres in contraction", respectively, as we shall always adopt a stretch-based criterion for the activation of the fibres. Moreover, we shall also give the fibre potential an isotropic term, always active, regardless of whether the fibre is in extension or contraction, in order to account for the however small contribution of the fibres when they do undergo contraction.

The simplest possible fibre potential is given by the sum of an always active isotropic contribution, and an anisotropic contribution active only in extension (Federico and Herzog, 2008a; Federico and Gasser, 2010),

$$\hat{W}_1 = \breve{W}_{1i}(I_1, I_2, I_3) + \mathcal{H}(I_4 - 1)\, \breve{W}_{1a}(I_4, I_5), \tag{170}$$

where

$$\mathcal{H}(I_4 - 1) = \begin{cases} 1 & \text{for } I_4 > 1, \quad \text{fibre in extension,} \\ 0 & \text{for } I_4 \leq 1, \quad \text{fibre in contraction,} \end{cases} \tag{171}$$

is the Heaviside step function at $I_4 - 1$, which is used to suppress the contribution of fibres undergoing contraction. More general forms of the fibre potential can be devised, not necessarily additive, which still contain the Heaviside step to allow any anisotropic term to be active in extension only. One such form shall be shown in the biphasic (solid-fluid) case (Section 8.1), as an example of applicability to articular cartilage.

It is important to emphasise that, once the Heaviside step is introduced, direct, "once-and-for-all" evaluation of the integral defining the ensemble potential (165) is impossible. Indeed, even for an anisotropic fibre potential $\hat{W}_{1a}$ given by

$$\hat{W}_{1a}(\boldsymbol{C}, \hat{\boldsymbol{A}}(\boldsymbol{M})) = \mathcal{H}(I_4(\boldsymbol{C}) - 1)\, \hat{W}_{1b}(\boldsymbol{C}, \hat{\boldsymbol{A}}(\boldsymbol{M})), \tag{172}$$

where $\hat{W}_{1b}$ is a "base" potential, with polynomial form of the type of Equation (143),

$$\hat{W}_{1b}(\boldsymbol{C}, \hat{\boldsymbol{A}}(\boldsymbol{M})) = a \left[\sum_{k=1}^{p} \langle \mathbb{Q}_k(\boldsymbol{C}) | \underbrace{\hat{\boldsymbol{A}}(\boldsymbol{M}) \otimes \ldots \otimes \hat{\boldsymbol{A}}(\boldsymbol{M})}_{k \text{ times}} \rangle + q(\boldsymbol{C}) \right], \tag{173}$$

the presence of the Heaviside step makes the integrand of Equation (165) highly non-linear, and therefore impedes the factorisation of the deformation outside the integral sign. Thus, in all cases in which the fibres undergo contraction, a numerical integration at each step of deformation is as unavoidable as in the most general case.

Remark 6.3. A criterion based on stress is certainly a possibility, and could be based, e.g., on the second Piola-Kirchhoff stress in the fibre direction, $S_{\boldsymbol{M}} = \langle \boldsymbol{S}, \boldsymbol{A} \rangle$, and be such that the fibres are active whenever $S_{\boldsymbol{M}} > 0$ and inactive otherwise. Note the use of the scalar product $\langle \boldsymbol{S}, \boldsymbol{A} \rangle = S^{AB} G_{AC} G_{BD} A^{CD}$, necessary to contract the two tensors as both are contravariant, as opposed to the case of a covariant and a contravariant tensor, such as $\boldsymbol{C} : \boldsymbol{A} = C_{AB} A^{AB} = \langle \boldsymbol{C} | \boldsymbol{A} \rangle$, in which the double contraction must be used.

7 Mixtures of Incompressible Solid and Fluid Phases

A saturated porous material consists of a solid porous matrix whose pores are assumed to be completely filled by a fluid. In the spirit of Mixture Theory, such a system is regarded as biphasic, with the solid phase and the fluid phase assumed to *coexist* at each point. In these biphasic solid-fluid systems, the configuration map χ, and therefore the deformation gradient $\boldsymbol{F} = T\chi$ and all other derived measures of deformation and strain, always refer to the solid phase (Quiligotti, 2002; Quiligotti et al., 2003; Grillo et al., 2007; Tomic et al., 2014).

We shall restrict our analysis to the case of intrinsically incompressible solid and fluid phases. This hypothesis, which allows for great simplification of the equations governing the mixture (a discussion on this topic can be found in Federico and Grillo, 2012, Remark 3.1 and references therein), does not prevent the mixture from undergoing volumetric deformations. Indeed, locally, a point can undergo volumetric deformations if the fluid moves away from it or into it (local volumetric contraction and dilatation, respectively) and, globally, the mixture can undergo volumetric deformations if the fluid can escape or enter through the boundaries of the system (global volumetric contraction and dilatation, respectively). We shall use the terms "(saturated) porous material" and "biphasic material" interchangeably.

7.1 Balance of Mass

In Mixture Theory, it is of fundamental importance to distinguish between the *true* mass density and the *apparent* mass density: for a given phase of the mixture, the former is calculated with respect to the volume occupied by each phase in mixture, the latter is calculated with respect to the whole volume occupied by the whole mixture. If both the solid and the fluid phase are assumed to be intrinsically incompressible, their *true* mass densities ρ_{sT} and ρ_{fT} remain constant throughout the motion. Therefore, their *apparent* mass densities ρ_s and ρ_f vary only in response to the change of volumetric fractions

$$\phi_s = \frac{\rho_s}{\rho_{sT}}, \qquad \phi_f = \frac{\rho_f}{\rho_{fT}}. \tag{174}$$

The volumetric fractions are defined as in Equation (129), which has been written for the case of a single phase with multiple constituents (see Equation (133)), and always obey the saturation condition

$$\phi_s + \phi_f = 1. \tag{175}$$

A biphasic mixture comprised of intrinsically incompressible solid and fluid phases admits global volumetric deformations if the fluid can escape or enter through the boundary, and admits local volumetric deformations if the fluid moved away from or into the point. We say that the mixture is globally and locally compressible, respectively, and that it exhibits an apparent compressibility. The determinant J of the deformation gradient $\boldsymbol{F} = T\chi$ of the solid is a measure of the variation of the volumetric fractions. Indeed, at a given material point X and time t, $J(X,t) > 1$ means that fluid has moved into X, whereas $J(X,t) < 1$ means that fluid has moved away from X. This is rigorously expressed by the material form of the equation of balance of mass for the solid phase (see Equation (132))

$$\rho_{sR} = J\,\rho_s, \tag{176}$$

where ρ_{sR} is the apparent mass density of the solid in the reference configuration, which is a constant, in the absence of sources/sinks and fluxes of mass. The counterpart of Equation (176) in terms of volumetric fractions is obtained by dividing by the constant true mass density ρ_{sT} (see Equation (174)), as

$$\phi_{sR} = J\,\phi_s. \tag{177}$$

where ϕ_{sR} is the referential solid volumetric fraction. In order to obtain the material form of the equation of balance of mass for the fluid phase, we substitute the saturation condition (175) and its counterpart in the reference configuration,

$$\phi_{sR} + \phi_{fR} = 1, \tag{178}$$

where ϕ_{fR} is the referential fluid volumetric fraction, into Equation (177), which leads to (see, e.g., Morland, 1972):

$$1 - \phi_{fR} = J\,(1 - \phi_f). \tag{179}$$

It is important to note that, if all the fluid moves away from a given point X at time t, then the point is comprised solely of incompressible solid, and thus cannot undergo further compression. This is mathematically expressed by the *compaction condition*

$$J(X,t) = \phi_{sR}(X). \tag{180}$$

Therefore, the system is subjected, at all points $X \in \mathcal{B}_R$ and all times t, to the *unilateral* (i.e., expressed by an inequality) compaction constraint

$$J \geq \phi_{sR}, \tag{181}$$

rather than to the unilateral constraint $J > 0$ that holds for a general compressible material in the standard monophasic theory of Continuum Mechanics, or the bilateral constraint (i.e., constraint expressed by an equality) $J = 1$ of monophasic strictly incompressible materials.

7.2 Stress and Balance of Linear Momentum

The fluid phase is assumed to be macroscopically *inviscid*, i.e., the fluid can only bear hydrostatic stresses, but has a non-zero viscosity causing frictional drag forces and therefore resistance to the flow of the fluid within the solid matrix (e.g., Hassanizadeh, 1986). Moreover, the relative motion of the phases in the mixture induces a momentum flux contribution to the total stress in the mixture, given by $-\sum_j \rho_j \boldsymbol{u}_j \otimes \boldsymbol{u}_j$, where $\boldsymbol{u}_j = \boldsymbol{v}_j - \boldsymbol{v}$ is the relative velocity of the j-th constituent with respect to the velocity of the centre of mass of the mixture, defined by $\rho \boldsymbol{v} = \sum_j \rho_j \boldsymbol{v}_j$ (e.g., Hassanizadeh, 1986), where $\rho = \sum_j \rho_j$ is the mass density of the mixture and ρ_j is the apparent mass density of phase j. However, this contribution can be neglected when the relative velocities are small, an assumption that usually holds true in modelling biological tissues.

Under these hypotheses, the total Cauchy stress in the mixture and the Cauchy stresses in the solid and in the fluid are given by

$$\boldsymbol{\sigma} = -p\boldsymbol{g}^{-1} + \boldsymbol{\sigma}_{const}, \tag{182a}$$

$$\boldsymbol{\sigma}_s = -\phi_s p\boldsymbol{g}^{-1} + \boldsymbol{\sigma}_{const}, \tag{182b}$$

$$\boldsymbol{\sigma}_f = -\phi_f p\boldsymbol{g}^{-1}, \tag{182c}$$

where p is the *pore pressure*, Lagrange multiplier arising from the constraint of intrinsic incompressibility of the solid and fluid phases, and $\boldsymbol{\sigma}_{const}$ is the *constitutive contribution* to the solid stress (usually called "effective stress" in soil mechanics, see. e.g., Terzaghi, 1943). A formulation for the constitutive contribution $\boldsymbol{\sigma}_{const}$ shall be discussed in detail in Section 7.4.

In all cases in which the external volume forces and the effects of inertia can be neglected, the equations balance of linear momentum for the mixture, the solid and the fluid (see, e.g., the first term in Eq. (5.19) in Grillo et al., 2012) read

$$\mathrm{o} = \operatorname{div} \boldsymbol{\sigma}, \tag{183a}$$

$$\mathrm{o} = \operatorname{div} \boldsymbol{\sigma}_s - \boldsymbol{f}_d - (p\boldsymbol{g}^{-1} \operatorname{grad} \phi_f), \tag{183b}$$

$$\mathrm{o} = \operatorname{div} \boldsymbol{\sigma}_f + \boldsymbol{f}_d + (p\boldsymbol{g}^{-1} \operatorname{grad} \phi_f), \tag{183c}$$

where the right-hand sides of Equations (183b) and (183c) are the internal drag forces that the fluid and the solid exert on each other. These are given by the equilibrium contribution $-p\boldsymbol{g}^{-1}\operatorname{grad} \phi_f$ and the dissipative contribution $\boldsymbol{f}_d$, which is usually required to vanish identically in the absence of non-equilibrium processes.

7.3 Darcy's Law and Permeability

By substituting Equation (182c) into Equation (183c), we obtain

$$\boldsymbol{f}_d = \phi_f\, \boldsymbol{g}^{-1} \mathrm{grad}\, p. \tag{184}$$

The simplest admissible constitutive choice is that $\boldsymbol{f}_d$ is a linear function of its power-conjugated variable, i.e., the filtration velocity of the fluid, given by the relative velocity between fluid and solid, weighted by the volumetric fraction of the fluid,

$$\boldsymbol{w} = \phi_f(\boldsymbol{v}_f - \boldsymbol{v}_s). \tag{185}$$

In this case we have to assume that $\boldsymbol{f}_d$ has the form

$$\boldsymbol{f}_d = -\phi_f\, \boldsymbol{g}^{-1}\, \boldsymbol{r}\, \boldsymbol{w}, \tag{186}$$

where $\boldsymbol{r}$ a tensor field valued in $[T\mathcal{S}]^0_2$, called hydraulic resistivity. Comparing Equations (184) and (186), and provided that the resistivity tensor $\boldsymbol{r}$ is invertible, with inverse $\boldsymbol{k} = \boldsymbol{r}^{-1}$, leads to Darcy's law

$$\boldsymbol{w} = -\boldsymbol{k}\, \mathrm{grad}\, p = \boldsymbol{k}\, \mathfrak{h}, \tag{187}$$

where the tensor field $\boldsymbol{k}$, valued in $[T\mathcal{S}]^2_0$ (i.e., "contravariant": see, e.g., Biot, 1972), is the *hydraulic permeability* or simply *permeability* (generally referred to as hydraulic conductivity, in the literature of Porous Media), and the covector field $\mathfrak{h} = -\mathrm{grad}\, p$ is the *hydraulic gradient.*

Performing a full Piola transformation allows for deriving the material counterpart of Darcy's law (Federico et al., 2013) as

$$\boldsymbol{W} = -\boldsymbol{K}\ \mathrm{Grad} p = \boldsymbol{K}\, \boldsymbol{\mathcal{H}}, \tag{188}$$

where the Piola transform

$$\boldsymbol{W} = J\, \chi^*[\boldsymbol{w}] = J\, \boldsymbol{w}\, \boldsymbol{F}^{-T} \tag{189}$$

is the material filtration velocity, the full Piola transform (i.e., obtained by a Piola transformation on the second foot and a pull-back on the first foot)

$$\boldsymbol{K} = J\, \chi^*[\boldsymbol{k}] = J\, \boldsymbol{F}^{-1} \boldsymbol{k}\, \boldsymbol{F}^{-T} \tag{190}$$

is the material permeability, and the pull-back

$$\boldsymbol{\mathcal{H}} = \chi^*[\mathfrak{h}] = \boldsymbol{F}^T \mathfrak{h} = -\boldsymbol{F}^T(\mathrm{grad}\, p) = -\mathrm{Grad}\, p \tag{191}$$

is the material hydraulic gradient. Note the abuse of notation in the use of the same symbol p for the pressure seen as a function of x (on which grad operates) and as a function of X (on which Grad operates).

For the Finite Element implementation of a permeability whose constitutive function depends on the whole deformation via, e.g., the right Cauchy-Green deformation tensor, i.e., $\boldsymbol{K} = \hat{\boldsymbol{K}}(\boldsymbol{C})$, it is necessary (Ateshian and Weiss, 2010) to define the fourth-order material *tangent permeability* tensor

$$\mathbb{D} = \hat{\mathbb{D}}(\boldsymbol{C}) = 2\frac{\partial \hat{\boldsymbol{K}}}{\partial \boldsymbol{C}}(\boldsymbol{C}), \tag{192}$$

which has to be provided along with $\hat{\boldsymbol{K}}$ in the definition of the material properties. Its inverse Piola transform is the spatial tangent permeability tensor

$$\mathbb{d} = J^{-1}\chi_*[\mathbb{D}], \qquad \mathrm{d}^{abcd} = J^{-1}F^a{}_A F^b{}_B F^c{}_C F^d{}_D \, \mathrm{D}^{ABCD}. \tag{193}$$

Remark 7.1. In the work on which a good part of the material presented here is based (Federico and Grillo, 2012), we followed the convention according to which the gradient of a scalar field p is the vector field associated with the differential of p, which is a covector field by definition (e.g., Marsden and Hughes, 1983; De Marco, 1999). Consequently, we wrote Darcy's law as $\boldsymbol{w} = -\boldsymbol{k}.\mathrm{grad}\, p = \boldsymbol{k}.\mathfrak{h}$, i.e., considering $\mathfrak{h} = -\mathrm{grad}\, p$ and as a vector field and using the metric $\boldsymbol{g}$ (represented by the low dot) to contract it with the permeability. Here, however, we prefer to consider the gradient of a scalar field as a covector (e.g., Talpaert, 2003; Epstein, 2010), which is a more natural choice, as it can be defined independently on whether or not a metric is available. Thus, we express Darcy's law as in Equation (187).

7.4 Hyperelastic Biphasic Mixtures

In establishing the hyperelastic constitutive equations for the solid phase of a mixture of incompressible solid and fluid phases, we must take into account the apparent compressibility caused globally by the fluid escaping or entering through the boundaries of the system, and locally by the fluid moving away from or into a given point (see Section 7), as well as the fact that, at the compaction limit $J \to \phi_{sR}$, when all the fluid has escaped from a given point, that point is made solely of strictly incompressible solid, and therefore further volumetric deformation is not allowed. One may be tempted to use the same *decoupled* formulation as in Equation (125), with the term $\hat{U}$ diverging at the compaction point. However, this would be a gross error. Indeed, the use of the decoupled potential (125) is appropriate only for incompressible or quasi incompressible materials, as it would cause the elasticity tensor to be incompatible with the linear elasticity tensor obtainable from small-strain experiments (Federico, 2010; Federico et al., 2014). An effect of this would be the fact that, for the case of anisotropic materials, the potential would be unable to predict distortional distortional deformations under hydrostatic states of stress and, viceversa, of volumetric dilatations under shear stress (Sansour, 2008; Guo et al., 2008; Federico, 2010; Vergori et al., 2013). Therefore, the potential for biphasic mixtures of incompressible solid and fluid phases *cannot* be decoupled in terms of volumetric and distortional deformation.

Several formulations have been proposed for mixtures of incompressible solid and fluid phases. For instance, Chapelle et al. (2010) proposed a theory for quasi incompressible constituents by considering the discrepancy between the volume changes of the mixture as a whole and the fluid contribution to the volume changes. Wilson et al. (2007), in order to account for the volume variations due to the change in fluid fraction, modified the compressible isotropic neo-Hookean material, by expressing the Poisson's ratio ν as a function of J, such that $\nu \to 1/2$ for $J \to \phi_{sR}$.

The theory presented here (Federico and Grillo, 2012) generally follows the lines of the approach proposed by Ehlers and Eipper (1997, 1999), in which a term diverging at the compaction limit was employed. However, we prefer to keep the analogy with the theory of monophasic, compressible hyperelastic solids, which was recalled in Sections 5.1 and 5.2, so that the biphasic theory converges to the monophasic one in the limit $\phi_{sR} \to 1$. The theory is independent of material symmetry, and the possible dependence on structural parameters related to the material symmetry is not indicated explicitly. We propose a potential in the form

$$\hat{W}(J, \bar{\boldsymbol{C}}) = \phi_{sR} \left[\hat{U}(J) + \hat{V}(J, \bar{\boldsymbol{C}}) \right], \tag{194}$$

which is *fully coupled* in J and $\bar{\boldsymbol{C}}$, since $\hat{V}$ is assumed to be so (as discussed above, this is in contrast with the decoupled potential of Equation (125)). The purely volumetric term $\hat{U}$ serves to enforce incompressibility at compaction, exactly as in the potential proposed by Ehlers and Eipper (1997, 1999). Therefore, we impose that $\hat{U}$ diverges at compaction, so that also the associated constitutive pressure (the negative of its first derivative) and the associated non-linear bulk modulus (its second derivative) diverge as well:

$$\hat{U}(J) \to +\infty, \qquad \text{for } J \to \phi_{sR}, \tag{195a}$$

$$-\frac{\partial \hat{U}}{\partial J}(J) \to +\infty, \qquad \text{for } J \to \phi_{sR}, \tag{195b}$$

$$\frac{\partial^2 \hat{U}}{\partial J^2}(J) \to +\infty, \qquad \text{for } J \to \phi_{sR}. \tag{195c}$$

Note that conditions (195a) and (195b) are identical to equations (27) in the work by Ehlers and Eipper (1999), and that we prefer to also have the non-linear bulk modules diverge (Equation (195c)). The potential in Equation (194), considered along with the conditions in Equation (195), is very similar to that proposed by Ehlers and Eipper (1997, 1999) but there are some important differences. Firstly, we prefer, for the time being, to indicate the dependence of $\hat{W}$ and $\hat{V}$ from $\boldsymbol{C}$ trough J and $\bar{\boldsymbol{C}}$, which are in turns functions of $\boldsymbol{C}$; this facilitates some subsequent passages. Secondly, we explicitly have the referential solid fraction ϕ_{sR} as a multiplicative factor for the potential; if on the one hand this means just rescaling the material constants, on the other hand it has the advantage to allow for a seamless passage to the monophasic limit $\phi_{sR} \to 1$, describing a strictly incompressible solid. Indeed, for $\phi_{sR} \to 1$, the unilateral constraint $J \geq \phi_{sR}$ tends to the bilateral constraint $J = 1$ because of the incompressibility of the solid. Therefore, we also have $J \to 1$. Under this condition, from Equation (194) we obtain

$$\hat{U}(J) \to 0, \qquad \hat{V}(J, \bar{\boldsymbol{C}}) \to \hat{V}_d(\bar{\boldsymbol{C}}), \tag{196}$$

and therefore the overall potential becomes that of the the case of an incompressible monophasic solid (e.g., Bonet and Wood, 2008):

$$\hat{W}(J, \bar{\boldsymbol{C}}) = \hat{W}_d(\bar{\boldsymbol{C}}) = \hat{V}_d(\bar{\boldsymbol{C}}). \tag{197}$$

Furthermore, the pore pressure p of the biphasic theory becomes the regular pressure π, Lagrange multiplier arising from the incompressibility of the pure solid.

The most important difference with the model proposed by Ehlers and Eipper (1997, 1999), however, is in the further conditions that we are now going impose on the volumetric term $\hat{U}$, in addition to those in Equation (195). We deem this necessary because the potential (194) with the conditions (195) yields constitutive pressure and large-strain bulk modulus

$$\pi = -\phi_{sR}\left[\frac{\partial \hat{U}}{\partial J}(J) + \frac{\partial \hat{V}}{\partial J}(J, \bar{\boldsymbol{C}})\right], \tag{198a}$$

$$\mathrm{K} = \phi_{sR}\left[\frac{\partial^2 \hat{U}}{\partial J^2}(J) + \frac{\partial^2 \hat{V}}{\partial J^2}(J, \bar{\boldsymbol{C}})\right]. \tag{198b}$$

From these expressions, we note that the potential $\hat{U}$ plays an important role also in the characterisation of the linearised material: indeed, the small-strain bulk modulus, which is given by Equation (198b) evaluated in the undeformed configuration, i.e., when $J = 1$ and $\bar{\boldsymbol{C}} = \boldsymbol{G}$, reads

$$\kappa = \phi_{sR} \left[\frac{\partial^2 \hat{U}}{\partial J^2}(1) + \frac{\partial^2 \hat{V}}{\partial J^2}(1, \boldsymbol{G}) \right] . \tag{199}$$

However, it would be desirable that the purely volumetric potential $\hat{U}$ be such that it exclusively accounts for the compaction limit, and the largest part of the characterisation of the material is left to the potential $\hat{V}$. Following these considerations, we assume $\hat{U}$ to be active only in compression (more rigorously speaking, under volumetric contractions, i.e., for $J < 1$) and with zero stiffness at zero strain, a condition that we call "zero-zero" (Federico et al., 2008). Therefore, function $\hat{U}$ and its derivatives should be such that

$$\hat{U}(J) = 0, \qquad \forall J \geq 1, \tag{200a}$$

$$\frac{\partial \hat{U}}{\partial J}(1) = 0, \tag{200b}$$

$$\frac{\partial^2 \hat{U}}{\partial J^2}(1) = 0. \tag{200c}$$

Equation (200a) sets to zero the potential for volumetric dilatations, Equation (200b) imposes zero pressure at zero deformation ($J = 1$), so that the state of stress due to $\hat{U}$ is natural, and Equation (200c) expresses the "zero-zero" condition. Furthermore, we require convexity of $\hat{U}$ in J under volumetric contractions, which is expressed by the positiveness of the second derivative:

$$\frac{\partial^2 \hat{U}}{\partial J^2}(J) > 0, \qquad \forall J < 1. \tag{201}$$

Substituting the the "zero-zero" condition (200c) into Equation (198b), we obtain that the linear bulk modulus depends *only* on potential $\hat{V}$, and is *independent* of the particular choice of $\hat{U}$:

$$\kappa = \phi_{sR} \frac{\partial^2 \hat{V}}{\partial J^2}(1, \boldsymbol{G}). \tag{202}$$

With the conditions (200), $\hat{U}$ takes it the physical meaning of a *correction* to potential $\hat{V}$, with the purpose of accounting for the singularity at compaction, without contributing for volumetric dilatations ($J \geq 1$) and in the undeformed configuration ($J = 1$).

A relatively simple form of $\hat{U}$ that satisfies all requirements (195), (200) and (201) could be

$$\hat{U}(J) = \mathcal{H}(J_{crit} - J)\,(J - J_{crit})^{2q}\,(J - \phi_{sR})^{-r}, \tag{203}$$

where $\mathcal{H}$ is the Heaviside step function, $J_{crit} \in]\phi_{sR}, 1]$ is a certain critical value of the volumetric deformation under which the correction $\hat{U}$ is active, $q \geq 2$ is a positive integer (2 is the smallest possible value that guarantees the "zero-zero condition" (200c) for this particular function), and $r \in]0, 1]$. The two parameters J_{crit} and ϕ_{sR} are essentially geometrical, and the exponents q and r could be related to intrinsic material properties. With the choice (203) of the correction potential $\hat{U}$, the potential $\hat{V}$ is the only "active" potential for most of the interval $]\phi_{sR}, +\infty[$ in which J can range (precisely, $\forall J \in [J_{crit}, +\infty[$).

Figure 2 shows an example of the the behaviour of $\hat{U}$, $-\partial\hat{U}/\partial J$, and $\partial^2\hat{U}/\partial J^2$, for $\phi_{sR} = 0.2$, $J_{crit} = 0.6$, $q = 2$, and $r = 1/2$. Note that $\hat{U}$ is convex in J, and therefore its pressure contribution $-\partial\hat{U}/\partial J$ is monotonic in J and its bulk modulus contribution $\partial^2 U/\partial J^2$ is strictly positive. Moreover, all three functions, and the second derivative in particular ("zero-zero" condition (200c)), take the value zero for $J = J_{crit}$, and therefore also for $J \geq J_{crit}$, by virtue of the Heaviside step in (203).

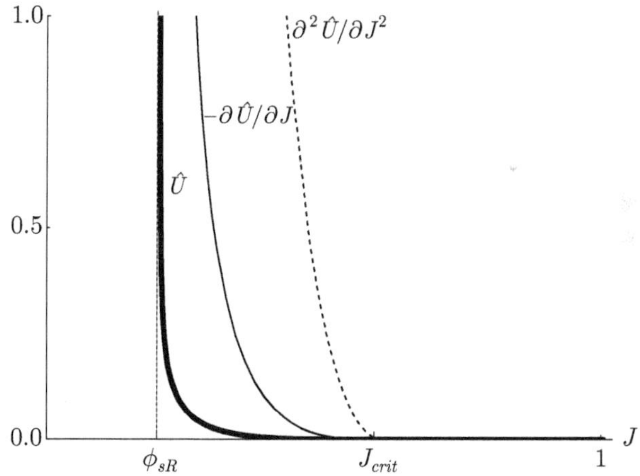

Figure 2: (Adapted from Federico and Grillo, 2012) Behaviour of function $\hat{U}$ (thick solid line), the negative of its first derivative (thin solid line) and its second derivative (dashed line), for the form of $\hat{U}$ in Equation (203), with a value of the referential solid fraction $\phi_{sR} = 0.2$, a critical volumetric deformation $J_{crit} = 0.6$, and exponents $q = 2$ and $r = 1/2$.

8 Biphasic Fibre-Reinforced Materials

In this section, we synthesise the the results on monophasic fibre-reinforced materials (Section 6) and on and biphasic materials (Section 7) in order to describe biphasic fibre-reinforced materials. In this case, the solid phase s is comprised of two constituents: the matrix (index 0) and the fibres (index 1). All constituents of the solid phase are assumed to attain the same motion χ and therefore the same deformation gradient $\boldsymbol{F} = T\chi$ (see Section 6.1).

Great care must be taken in the definition of the volumetric fractions because, in principle, not only the matrix but also the fibres can be assumed to be porous. Below, we shall refer to *inclusions* in general, rather than to fibres in particular, because the reasoning holds regardless of the geometry of the inclusion (Federico and Grillo, 2012; Tomic et al., 2014).

Let us consider a representative element of volume (REV) Ω, with volume measure equal to $|\Omega|$, within which we identify two complementary subsets: Ω_0, representing the region occupied by the matrix together with the fluid in its pores, and Ω_1, representing the region occupied by the inclusion together with the fluid in its pores. Let us also assume that the interface between matrix and inclusions is a sufficiently smooth surface, which can be thought to belong to Ω_1. Therefore, if *disjoint* union is denoted by the symbol $\sqcup$, we have $\Omega = \Omega_0 \sqcup \Omega_1$. Since disjoint volumes are additive, we also have $|\Omega| = |\Omega_0| + |\Omega_1|$. By dividing this expression by the volume $|\Omega|$ of the REV, we find

$$1 = \frac{|\Omega_0|}{|\Omega|} + \frac{|\Omega_1|}{|\Omega|} \quad \Rightarrow \quad 1 = \phi_{0A} + \phi_{1A}, \tag{204}$$

which, since both matrix and inclusions are assumed to be porous, defines the *apparent* volumetric fractions of matrix and inclusion, ϕ_{0A} and ϕ_{1A}, respectively. Let us now distinguish the portions of Ω_0 and Ω_1 occupied by solid and fluid, respectively, and denote them Ω_{0s}, Ω_{0f}, Ω_{1s}, Ω_{1f}, with obvious meaning of the symbols. By enforcing again the additivity of disjoint volumes, the portions of the REV Ω occupied by solid and fluid are obtained as $\Omega_s = \Omega_{0s} \sqcup \Omega_{1s}$ and $\Omega_f = \Omega_{0f} \sqcup \Omega_{1f}$, respectively, and the corresponding volumes are $|\Omega_s| = |\Omega_{0s}| + |\Omega_{1s}|$ and $|\Omega_f| = |\Omega_{0f}| + |\Omega_{1f}|$. By dividing these two expressions by the volume $|\Omega|$ of the REV, we find the solid and fluid volumetric fractions, as

$$\frac{|\Omega_s|}{|\Omega|} = \frac{|\Omega_{0s}|}{|\Omega|} + \frac{|\Omega_{1s}|}{|\Omega|} \quad \Rightarrow \quad \phi_s = \phi_0 + \phi_1, \tag{205a}$$

$$\frac{|\Omega_f|}{|\Omega|} = \frac{|\Omega_{0f}|}{|\Omega|} + \frac{|\Omega_{1f}|}{|\Omega|} \quad \Rightarrow \quad \phi_f = \phi_{0f} + \phi_{1f}, \tag{205b}$$

where ϕ_0 and ϕ_1 are the *true solid fractions* of matrix and inclusion, and ϕ_{0f} and ϕ_{1f} are the *true fluid fractions* of matrix and inclusion. In the material picture, these read

$$\phi_{sR} = \phi_{0R} + \phi_{1R}, \tag{206a}$$

$$\phi_{fR} = \phi_{0fR} + \phi_{1fR}, \tag{206b}$$

where, following Equation (177), the referential solid fractions are given by

$$\phi_{0R} = J\,\phi_0, \qquad \phi_{1R} = J\,\phi_1. \tag{207}$$

Furthermore, the saturation conditions (175) and (178) continue to hold:

$$\phi_s + \phi_f = 1, \qquad \phi_{sR} + \phi_{fR} = 1. \tag{208}$$

The few modifications needed to obtain the elastic potential of biphasic (solid-fluid) fibre-reinforced materials from the monophasic (solid-only) case of Section 6 are discussed in Section 8.1, whereas the remainder of the Section is dedicated to the several steps needed to evaluate the permeability, and to an example of application to articular cartilage. We recall that we use the terminologies "saturated porous material" and "biphasic material" interchangeably.

8.1 Biphasic Hyperelastic Fibre-Reinforced Materials

In the biphasic case, the elastic potential has the same form of Equation (160) for a finite number of families of aligned fibres,

$$\hat{W}(\boldsymbol{C}) = \phi_{0R}\,\hat{W}_0(\boldsymbol{C}) + \sum_{\alpha=1}^{N} \phi_{\alpha R}\,\hat{W}_\alpha(\boldsymbol{C}, \boldsymbol{A}_\alpha). \tag{209}$$

and of Equation (164) for one family of statistically oriented fibres,

$$\hat{W}(\boldsymbol{C}) = \phi_{0R}\,\hat{W}_0(\boldsymbol{C}) + \phi_{1R} \int_{\mathbb{S}^2\mathcal{B}_R} \psi(\boldsymbol{M})\,\hat{W}_1(\boldsymbol{C}, \hat{\boldsymbol{A}}(\boldsymbol{M})), \tag{210}$$

Again, we are not indicating the dependence on the probability distribution ψ for the sake of a lighter notation.

The apparent compressibility due to the presence of the fluid phase is accounted for as elucidated in Section 7.4, with a term depending exclusively on J that enforces incompressibility at compaction, and a term depending on the whole right Cauchy-Green deformation tensor $\boldsymbol{C}$ that characterises most of the material behaviour. Here we discuss the case of one family of statistically oriented fibres: the case of multiple families of aligned fibres (as well as that of multiple families of statistically oriented fibres, Section 6.5) can be easily derived by analogy.

By comparing Equations (194) and (164), and imposing

$$\phi_{sR}\,\hat{U}(J(\boldsymbol{C})) = \phi_{0R}\,\hat{U}(J(\boldsymbol{C})) + \phi_{1R}\,\hat{U}(J(\boldsymbol{C})), \tag{211a}$$

$$\phi_{sR}\,\hat{V}(\boldsymbol{C}) \quad = \phi_{0R}\,\hat{V}_0(\boldsymbol{C}) + \phi_{1R}\,\hat{V}_e(\boldsymbol{C}), \tag{211b}$$

so that the volumetric term $\hat{U}$ is the same for matrix and fibres, we obtain matrix, fibre, and ensemble potentials as

$$\phi_{0R}\,\hat{W}_0(\boldsymbol{C}) \quad = \phi_{0R}\left[\hat{U}(J(\boldsymbol{C})) + \hat{V}_0(\boldsymbol{C})\right], \tag{212a}$$

$$\phi_{1R}\,\hat{W}_1(\boldsymbol{C},\boldsymbol{A}) = \phi_{1R}\left[\hat{U}(J(\boldsymbol{C})) + \hat{V}_1(\boldsymbol{C},\boldsymbol{A})\right], \tag{212b}$$

$$\phi_{1R}\,\hat{W}_e(\boldsymbol{C}) \quad = \phi_{1R}\left[\hat{U}(J(\boldsymbol{C})) + \hat{V}_e(\boldsymbol{C})\right], \tag{212c}$$

where we are forgiving ourselves for the abuse of notation in writing $J = J(\boldsymbol{C})$. By use of Equation (206a), we obtain the overall potential as

$$\hat{W}(\boldsymbol{C}) = \phi_{sR}\,\hat{U}(J(\boldsymbol{C})) + \phi_{0R}\,\hat{V}_0(\boldsymbol{C}) + \phi_{1R}\,\hat{V}_e(\boldsymbol{C}), \tag{213}$$

where

$$\hat{V}_e(\boldsymbol{C}) = \int_{\mathbb{S}^2\mathcal{B}_R} \psi(\boldsymbol{M})\,\hat{V}_1(\boldsymbol{C},\hat{\boldsymbol{A}}(\boldsymbol{M})), \tag{214}$$

is the ensemble potential relative to the fibre potential $\hat{V}_1$.

The matrix is assumed to be isotropic, and therefore the term $\hat{V}_0$ can be written as a function $\check{V}_0$ of the principal invariants (106) I_1, I_2, I_3 of the Cauchy-Green deformation tensor $\boldsymbol{C}$. As mentioned in Section 6.7 and shown in Equation (170), the simplest possible form of fibre potential $\hat{V}_1$ is additive, and it is given by the sum of an isotropic term representing the contribution of the fibres regardless of whether they are in extension or contraction, function of I_1, I_2, I_3, and an anisotropic term representing the contribution of the fibres in extension only, function of the additional transversely isotropic invariants (Equation (107)) I_4 and I_5, i.e.,

$$\hat{V}_1 = \check{V}_{1i}(I_1, I_2, I_3) + \mathcal{H}(I_4 - 1)\,\check{V}_{1a}(I_4, I_5), \tag{215}$$

where $\mathcal{H}(I_4 - 1)$ is again the Heaviside step function at $I_4 - 1$, used to suppress the contribution of fibres in contraction. An example of non-additive fibre potential could be derived from the anisotropic version proposed by

Almeida and Spilker (1998) of the isotropic potential of Holmes and Mow (1990), i.e.,

$$\hat{V}_1 = \check{V}_1(I_1, I_2, I_3, I_4, I_5) = \alpha \exp\left[f(I_1, I_2, I_4, I_5) - \beta_1 \ln(I_3)\right], \qquad (216)$$

with

$$f(I_1, I_2, I_4, I_5) = \sum_{i,j=1}^{n} \left[a_{ij}(I_1 - 3)^i (I_2 - 3)^j + \right.$$
$$\left. + \mathcal{H}(I_4 - 1)\left(b_{ij}(I_4 - 1)^i (I_5 - 1)^j + c_{ij}(I_1 - 3)^i (I_4 - 1)^j\right)\right]. \qquad (217)$$

where α is a material constant with the physical units of an energy per unit volume, a_{ij}, b_{ij} and c_{ij} are non-dimensional material constants, and f is a polynomial modified with the introduction of the Heaviside step, which suppresses all terms containing the anisotropic invariants I_4 and I_5 when the fibres are in contraction.

8.2 Permeability: REV with an Infinitesimal Fraction of Fibres

In order to evaluate the permeability of a fibre-reinforced porous material, we consider, at each spatial point x, a representative element of volume (REV) comprised of a cylinder centred at x, containing a cylindric and rectilinear (or linearised) segment of a fibre oriented in direction $\boldsymbol{m} \in \mathbb{S}^2_x\mathcal{S}$, and the surrounding matrix (Figure 3). Under the hypothesis of *scale separation*, at the fibre level, the matrix is a porous continuum, i.e., the *microscopic pores* in the matrix are much smaller than the *mesoscopic pores* constituted by the interstices between the fibres (Federico and Herzog, 2008b). At the REV level, the aim is to find an expression for the local permeability (or REV permeability) $\boldsymbol{k}_{\mathrm{REV}}$. At the global level, the aim is to find the overall permeability $\boldsymbol{k}$ by means of a directional averaging integral, as outlined in Section 6.3.

This is achieved in three steps: *a*) a REV in the case of infinitesimal volumetric fraction of fibres, which also covers the case of an infinitesimal volumetric fraction of aligned fibres (this section); *b*) a REV in the case of large volumetric fraction of fibres, which also covers the case of a finite volumetric fraction of aligned fibres (Section 8.3); *c*) the case of a finite volumetric fraction of statistically oriented fibres, both in the large deformation and the small deformation settings (Section 8.4).

The permeability of a system such as the REV in Figure 3 or, equivalently, the permeability of a composite with aligned fibres, can be evaluated by exploiting known solutions for formally identical problems. Indeed, Darcy's law reported in Equation (187) and the equation of linear dielectric

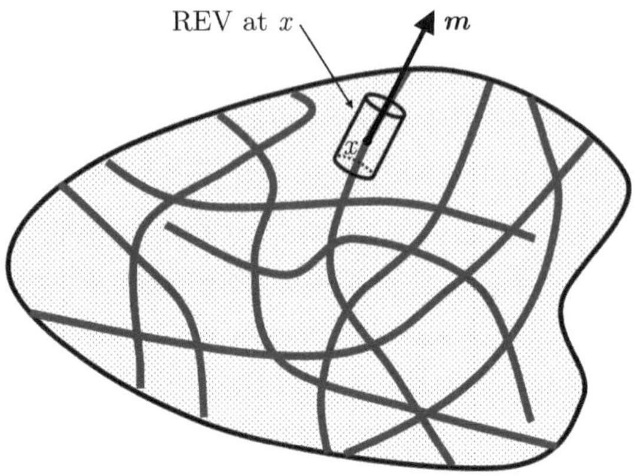

Figure 3: (Adapted from Federico and Grillo, 2012) A porous fibre-reinforced material, with the representative element of volume (REV) at spatial point x. The REV contains a small portion of an impermeable fibre (which can be considered as rectilinear), surrounded by the fluid-saturated matrix.

materials are formally identical, as they are the first-order approximation describing the interaction of a field with matter. These two laws read

$$\boldsymbol{w} = -\boldsymbol{k}\,\mathrm{grad}\,p = \boldsymbol{k}\,\mathfrak{h}, \tag{218a}$$

$$\boldsymbol{d} = -\boldsymbol{\epsilon}\,\mathrm{grad}\,\varphi = \boldsymbol{\epsilon}\,\boldsymbol{e}, \tag{218b}$$

where it is clear that the dielectric displacement $\boldsymbol{d}$, the dielectric permittivity tensor $\boldsymbol{\epsilon}$, the scalar potential φ and the electric field $\boldsymbol{e}$ correspond to the filtration velocity $\boldsymbol{w}$, the permeability tensor $\boldsymbol{k}$, the pore pressure p and the hydraulic gradient $\mathfrak{h}$, respectively. Based on this perfect analogy, and the solutions found by Landau and Lifshitz (1960) for the dielectric case, Pozdniakov and Tsang (2004) solved the case of ellipsoidal inclusions, and Federico and Herzog (2008b) that of fibre-like inclusions.

Let us impose an external hydraulic gradient $\mathfrak{h} = -\mathrm{grad}\,p$ on the REV and consider the balance of the flow $\boldsymbol{w}_{\mathrm{REV}} = -\boldsymbol{k}_{\mathrm{REV}}\,\mathrm{grad}\,p = \boldsymbol{k}_{\mathrm{REV}}\,\mathfrak{h}$ and of the hydraulic gradient $\mathfrak{h}_{\mathrm{REV}} = \mathfrak{h}$. These are expressed following the usual rule of mixtures, by weighing the contribution of each constituent with its apparent volumetric fraction (Pozdniakov and Tsang, 2004):

$$\boldsymbol{k}_{\mathrm{REV}}\,\mathfrak{h} = \phi_{0A}\,\boldsymbol{k}_0\,\mathfrak{h}_0 + \phi_{1A}\,\boldsymbol{k}_1\,\mathfrak{h}_1, \tag{219a}$$

$$\mathfrak{h} = \phi_{0A}\,\mathfrak{h}_0 + \phi_{1A}\,\mathfrak{h}_1. \tag{219b}$$

The balance equations (219) are valid in the general case in which both matrix and inclusion are porous. However, let us now restrict our analysis to the case of impermeable inclusions, for which we have that $\boldsymbol{k}_1 = \mathbf{o}$, and that $\phi_{1A} = \phi_1$, i.e., since the inclusions contain no fluid, their apparent and true volumetric fractions coincide. This also implies that the apparent matrix volumetric fraction can be conveniently expressed as $\phi_{0A} = 1 - \phi_{1A} = 1 - \phi_1$. Thus, the balance equations (219) become

$$\boldsymbol{k}_{\mathrm{REV}}\, \mathfrak{h} = (1 - \phi_1)\, \boldsymbol{k}_0\, \mathfrak{h}_0, \tag{220a}$$

$$\mathfrak{h} = (1 - \phi_1)\, \mathfrak{h}_0 + \phi_1\, \mathfrak{h}_1, \tag{220b}$$

from which, by eliminating $(1 - \phi_1)\, \mathfrak{h}_0$, we obtain

$$\boldsymbol{k}_{\mathrm{REV}}\, \mathfrak{h} = \boldsymbol{k}_0\, \mathfrak{h} - \phi_1\, \boldsymbol{k}_0\, \mathfrak{h}_1. \tag{221}$$

In Equation (221), the unknown is the hydraulic gradient in the inclusion, $\mathfrak{h}_1$. This can be found by exploiting the analogy between porous and dielectric materials (Equation (218)) and the solution for an infinite dielectric cylinder (fibre inclusion) in an infinite dielectric matrix reported by Landau and Lifshitz (1960), for the case of a cylinder with zero dielectric permittivity, from which we obtain

$$\mathfrak{h}_1 = \boldsymbol{g}\ [\boldsymbol{a} + 2\boldsymbol{t}]\ \mathfrak{h}, \tag{222}$$

where tensors $\boldsymbol{a}$ and $\boldsymbol{t}$ are the basis of the space of second-order tensors in $[T\mathcal{S}]^2_0$ with transversely isotropy in direction $\boldsymbol{m}$ (Equations (99)), and the metric tensor $\boldsymbol{g}$ is used to transform the vector $[\boldsymbol{a} + 2\boldsymbol{t}]\, \mathfrak{h}$ into a covector. Equation (222) states that the hydraulic gradient in the fibre, $\mathfrak{h}_1$, is transversely isotropic in the direction $\boldsymbol{m}$ of the fibre, that the axial component of $\mathfrak{h}_1$ is equal to that of the external hydraulic gradient $\mathfrak{h}$, and that the transverse component of $\mathfrak{h}_1$ is twice as large as that of $\mathfrak{h}$.

Substituting Equation (222) into Equation (221), and considering that the external hydraulic gradient $\mathfrak{h}$ is arbitrary, we obtain

$$\boldsymbol{k}_{\mathrm{REV}} = \boldsymbol{k}_0 - \phi_1\, \boldsymbol{k}_0.[\boldsymbol{a} + 2\boldsymbol{t}], \tag{223}$$

which, by using the identities $\boldsymbol{g}^{-1} = \boldsymbol{a} + \boldsymbol{t}$ and $\boldsymbol{i}^T = \boldsymbol{g}\,\boldsymbol{g}^{-1}$, can be written as (Federico and Herzog, 2008b)

$$\boldsymbol{k}_{\mathrm{REV}} = \boldsymbol{k}_0.\left[(1 - \phi_1)\boldsymbol{a} + (1 - 2\phi_1)\boldsymbol{t}\right]. \tag{224}$$

Therefore, the axial permeability is larger than the transverse permeability, as one would intuitively expect: the fibre creates more obstruction to the fluid flow in the transverse direction than it does in the axial direction. This result, however, is valid only for infinitesimal volumetric fraction of fibres, as it is based on the solution by Landau and Lifshitz (1960), which has been found for the case of a cylinder in an *infinite* matrix.

8.3 Permeability: REV with a Finite Fraction of Fibres

The case of finite volumetric fraction of fibres can be approached by means of the differential methods described by McLaughlin (1977) and Norris (1985), which here we employ in the form reported by Zimmermann (1991).

Let us perform a thought experiment or what we could call, perhaps more appropriately, a thought manufacturing process, and imagine to construct a fibre-reinforced composite by adding fibres to a homogeneous matrix with an incremental procedure. At the beginning of the procedure, the fibre volumetric fraction is $\phi_1^{(0)} = 0$, i.e., the composite is made by pure homogeneous matrix. At the first increment, we replace a small fraction $\Delta\Gamma$ of the composite (which is still made of pure matrix) by fibres, and obtain a composite with fibre fraction $\phi_1^{(1)} = \Delta\Gamma$. At the second increment, we remove another portion $\Delta\Gamma$ of the composite, thus removing a fraction $(1 - \phi_1^{(1)})\,\Delta\Gamma$ of matrix, and a fraction $\phi_1^{(1)}\,\Delta\Gamma$ of fibres, and we add another fraction $\Delta\Gamma$ of fibres. After the second increment, the fibre fraction is then $\phi_1^{(2)} = \phi_1^{(1)} - \phi_1^{(1)}\,\Delta\Gamma + \Delta\Gamma$. By induction, we conclude that, after the $n+1$-th increment, the fibre fraction is

$$\phi_1^{(n+1)} = \phi_1^{(n)} - \phi_1^{(n)}\,\Delta\Gamma + \Delta\Gamma. \tag{225}$$

We can now redefine ϕ_1 as a continuous function of $\Gamma = n\,\Delta\Gamma$, i.e.,

$$\phi_1(\Gamma) = \phi_1(n\,\Delta\Gamma) = \phi_1^{(n)}, \tag{226}$$

so that

$$\phi_1(\Gamma + \Delta\Gamma) = \phi_1((n+1)\,\Delta\Gamma) = \phi_1^{(n+1)}. \tag{227}$$

By substituting Equations (226) and (227) into Equation (225), and rearranging, we obtain

$$\frac{\phi_1(\Gamma + \Delta\Gamma) - \phi_1(\Gamma)}{\Delta\Gamma} = 1 - \phi_1(\Gamma). \tag{228}$$

By performing the limit for $\Delta\Gamma \to 0$, Equation (228) becomes a separable differential equation, which can be considered together with the initial condition stating that the fibre fraction ϕ_1 is zero at $\Gamma = 0$:

$$\phi_1'(\Gamma) = 1 - \phi_1(\Gamma), \tag{229a}$$

$$\phi_1(0) = \phi_1^{(0)} = 0. \tag{229b}$$

The solution of this differential equation describes the incremental process with which fibres are added:

$$\phi_1(\Gamma) = 1 - \exp(-\Gamma). \tag{230}$$

Note that, for $\Gamma \to \infty$, $\phi_1 \to 1$, and that, when Γ is small, MacLaurin expansion of ϕ_1 in Equation (230) yields

$$\phi_1(\Gamma) \simeq \Gamma. \tag{231}$$

i.e., a small Γ approximates ϕ_1 to the first order.

Rearranging Equation (223), and considering that it is written for the case of infinitesimal fibre fraction, we can use Equation (231) to approximate the fibre fraction, and obtain

$$\frac{\boldsymbol{k}_{\mathrm{REV}} - \boldsymbol{k}_0}{\Gamma} = -\boldsymbol{k}_0.[\boldsymbol{a} + 2\boldsymbol{t}], \tag{232}$$

which, by means of the variable transformations

$$\boldsymbol{k}_{\mathrm{REV}} \mapsto \boldsymbol{k}_{\mathrm{REV}}(\Gamma + \Delta\Gamma), \tag{233a}$$
$$\boldsymbol{k}_0 \mapsto \boldsymbol{k}_{\mathrm{REV}}(\Gamma), \tag{233b}$$
$$\Gamma \mapsto \Delta\Gamma, \tag{233c}$$

allows for describing $\boldsymbol{k}_{\mathrm{REV}}$ as a function of Γ, and the composite as a homogenised mixture with finite fibre fraction characterised by Γ, to which an additional infinitesimal fraction $\Delta\Gamma$ of fibres has been added, and obtain

$$\frac{\boldsymbol{k}_{\mathrm{REV}}(\Gamma + \Delta\Gamma) - \boldsymbol{k}_{\mathrm{REV}}(\Gamma)}{\Delta\Gamma} = -\boldsymbol{k}_{\mathrm{REV}}(\Gamma).[\boldsymbol{a} + 2\boldsymbol{t}]. \tag{234}$$

Performing the limit $\Delta\Gamma \to 0$, we obtain again a separable differential equation, which we consider together with the initial condition prescribing that, for $\Gamma = 0$ (no fibres), the permeability of the REV equals that of the matrix:

$$\boldsymbol{k}'_{\mathrm{REV}}(\Gamma) = -\boldsymbol{k}_{\mathrm{REV}}(\Gamma).[\boldsymbol{a} + 2\boldsymbol{t}], \tag{235a}$$
$$\boldsymbol{k}_{\mathrm{REV}}(0) = \boldsymbol{k}_0. \tag{235b}$$

The solution,

$$\boldsymbol{k}_{\mathrm{REV}}(\Gamma) = \boldsymbol{k}_0.\exp[-\Gamma\ (\boldsymbol{a} + 2\boldsymbol{t})], \tag{236}$$

needs to be brought back to an expression in terms of $\phi_1 = 1 - \exp(-\Gamma)$. We achieve this in a few steps. First, we write the exponential in MacLaurin series, separate the term of order zero, which is equal to the inverse metric (the "contravariant unit tensor"), i.e., $(-\Gamma)^0(\boldsymbol{a} + 2\,\boldsymbol{t})^0/0! = \boldsymbol{g}^{-1}$, from the others, and then decompose it into $\boldsymbol{g}^{-1} = \boldsymbol{a} + \boldsymbol{t}$,

$$
\begin{aligned}
\exp[-\Gamma\,(\boldsymbol{a} + 2\,\boldsymbol{t})] &= \sum_{n=0}^{\infty} \frac{(-\Gamma)^n\,(\boldsymbol{a} + 2\,\boldsymbol{t})^n}{n!} = \\
&= \boldsymbol{g}^{-1} + \sum_{n=1}^{\infty} \frac{(-\Gamma)^n\,(\boldsymbol{a} + 2\,\boldsymbol{t})^n}{n!} = \\
&= \boldsymbol{a} + \boldsymbol{t} + \sum_{n=1}^{\infty} \frac{(-\Gamma)^n\,(\boldsymbol{a} + 2\,\boldsymbol{t})^n}{n!} = \\
&= \frac{(-\Gamma)^0\,\boldsymbol{a}}{0!} + \frac{(-2\,\Gamma)^0\,\boldsymbol{t}}{0!} + \sum_{n=1}^{\infty} \frac{(-\Gamma)^n\,(\boldsymbol{a} + 2\,\boldsymbol{t})^n}{n!},
\end{aligned} \tag{237}
$$

where the identities $(-\Gamma)^0/0! = 1$ and $(-2\,\Gamma)^0/0! = 1$ have been used. Then, we exploit the idempotence and the orthogonality of $\boldsymbol{a}$ and $\boldsymbol{t}$ (see Equations (97), which report the case of the material $\boldsymbol{A}$ and $\boldsymbol{T}$) so that, in each term of order $n \geq 1$, with $n, k \neq 0$, we have $\boldsymbol{a}^{n-k}.\boldsymbol{t}^k = \boldsymbol{a}.\boldsymbol{t} = \mathbf{o}$, and only the terms in $\boldsymbol{a}^n = \boldsymbol{a}$ and $\boldsymbol{t}^n = \boldsymbol{t}$ survive. Thus, we obtain

$$
\begin{aligned}
\exp[-\Gamma\,(\boldsymbol{a} + 2\,\boldsymbol{t})] &= \frac{(-\Gamma)^0\,\boldsymbol{a}}{0!} + \frac{(-2\,\Gamma)^0\,\boldsymbol{t}}{0!} + \sum_{n=1}^{\infty} \frac{(-\Gamma)^n\,(\boldsymbol{a} + 2\,\boldsymbol{t})^n}{n!} = \\
&= \frac{(-\Gamma)^0\,\boldsymbol{a}}{0!} + \frac{(-2\,\Gamma)^0\,\boldsymbol{t}}{0!} + \sum_{n=1}^{\infty} \frac{(-\Gamma)^n\,(\boldsymbol{a} + 2^n\,\boldsymbol{t})}{n!} = \\
&= \left[\sum_{n=0}^{\infty} \frac{(-\Gamma)^n}{n!}\right] \boldsymbol{a} + \left[\sum_{n=0}^{\infty} \frac{(-2\,\Gamma)^n}{n!}\right] \boldsymbol{t},
\end{aligned} \tag{238}
$$

where the terms of order zero have been put together with the remaining terms. Equation (238) provides an expression of the exponential that is a linear combination of $\boldsymbol{a}$ and $\boldsymbol{t}$, and in which the coefficients are indeed MacLaurin expansions of exponentials, i.e.,

$$
\begin{aligned}
\exp[-\Gamma\,(\boldsymbol{a} + 2\,\boldsymbol{t})] &= \left[\sum_{n=0}^{\infty} \frac{(-\Gamma)^n}{n!}\right] \boldsymbol{a} + \left[\sum_{n=0}^{\infty} \frac{(-2\,\Gamma)^n}{n!}\right] \boldsymbol{t}, \\
&= \exp(-\Gamma)\,\boldsymbol{a} + \exp(-2\,\Gamma)\,\boldsymbol{t} = \\
&= \exp(-\Gamma)\,\boldsymbol{a} + (\exp(-\Gamma))^2\,\boldsymbol{t}.
\end{aligned} \tag{239}
$$

Substituting this result in Equation (236) and using the relationship (230) between ϕ_1 and Γ, we obtain the final expression

$$
\boldsymbol{k}_{\mathrm{REV}} = \boldsymbol{k}_0.\left[(1 - \phi_1)\boldsymbol{a} + (1 - \phi_1)^2\boldsymbol{t}\right]. \tag{240}
$$

By means of the definition of $\boldsymbol{t}$ (Equation (99b)) and some straightforward algebraic manipulation, $\boldsymbol{k}_{\mathrm{REV}}$ can be put into the alternative form

$$\boldsymbol{k}_{\mathrm{REV}} = \boldsymbol{k}_0 . \left[(1-\phi_1)\phi_1 \boldsymbol{a} + (1-\phi_1)^2 \boldsymbol{g}^{-1}\right] . \tag{241}$$

8.4 Permeability: Statistically Oriented Fibres

In order to evaluate the overall permeability of a porous material reinforced by statistically oriented fibres, we need a way to perform a weighted average of the permeability of the REV over the material unit sphere $\mathbb{S}^2\mathcal{B}_R$, over which the probability distribution ψ describing fibre orientation is defined (Section 6.3). However, although the upscaling procedure that we employed to find the permeability of the REV in Equation (241) is entirely based on the spatial picture, the current structure tensor $\boldsymbol{a}$ and the current fibre volumetric fraction ϕ_1 are in fact known *only* in terms of their material counterparts, through Equation (102) and the second of Euqations (207), respectively. We could proceed by expressing $\boldsymbol{a}$ and ϕ_1 in terms of $\boldsymbol{A}$ and ϕ_{1R}, substitute into Equation (241) and then perform the averaging integral (Federico and Grillo, 2012). Equivalently, we could work in terms of material permeability. Here, we choose the latter method, as it simplifies discussing the material symmetries.

Following the definition of material permeability in Equation (190), the material permeability of the REV is obtained by performing a full Piola transformation on Equation (241), as

$$\begin{aligned} \boldsymbol{K}_{\mathrm{REV}} &= \hat{\boldsymbol{K}}_{\mathrm{REV}}(\boldsymbol{C}, \boldsymbol{A}) = J\,\boldsymbol{F}^{-1}\boldsymbol{k}_{\mathrm{REV}}\boldsymbol{F}^{-T} \\ &= \boldsymbol{K}_0 . \left[J^{-2}(J-\phi_{1R})\phi_{1R}\,(\boldsymbol{C}:\boldsymbol{A})^{-1}\boldsymbol{A} + J^{-2}(J-\phi_{1R})^2\boldsymbol{B}\right], \end{aligned} \tag{242}$$

where $\boldsymbol{K}_0 = J\,\boldsymbol{F}^{-1}\boldsymbol{k}_0\boldsymbol{F}^{-T}$ is the material permeability of the matrix and Equation (207) has been used. The tensor $\hat{\boldsymbol{K}}_{\mathrm{REV}}$ can be directly integrated over the material unit sphere, to yield the overall material permeability

$$\begin{aligned} \boldsymbol{K} = \hat{\boldsymbol{K}}(\boldsymbol{C}) &= \int_{\mathbb{S}^2\mathcal{B}_R} \psi(\boldsymbol{M})\,\hat{\boldsymbol{K}}_{\mathrm{REV}}(\boldsymbol{C}, \hat{\boldsymbol{A}}(\boldsymbol{M})) \\ &= \boldsymbol{K}_0 . \left[J^{-2}(J-\phi_{1R})\phi_{1R}\,\hat{\boldsymbol{Z}}(\boldsymbol{C}) + J^{-2}(J-\phi_{1R})^2\boldsymbol{B}\right], \end{aligned} \tag{243}$$

where the linearity of the integral operation has been used to split the two terms of the sum in the expression of the material permeability $\hat{\boldsymbol{K}}_{\mathrm{REV}}$ in Equation (242), the normalisation condition (136) has been applied to the term in $\boldsymbol{B} = \boldsymbol{C}^{-1}$, and the symmetric tensor $\boldsymbol{Z} = \hat{\boldsymbol{Z}}(\boldsymbol{C})$ is given by

$$\boldsymbol{Z} = \hat{\boldsymbol{Z}}(\boldsymbol{C}) = \int_{\mathbb{S}^2\mathcal{B}_R} \psi(\boldsymbol{M})\,(\boldsymbol{C}:\hat{\boldsymbol{A}}(\boldsymbol{M}))^{-1}\hat{\boldsymbol{A}}(\boldsymbol{M}). \tag{244}$$

Also in this case, the probability distribution ψ has not been explicitly indicated as an argument of $\hat{\boldsymbol{K}}$ and $\hat{\boldsymbol{Z}}$, for the sake of a lighter notation. It is important to note that the integral in the definition (244) of the tensor $\boldsymbol{Z} = \hat{\boldsymbol{Z}}(\boldsymbol{C})$ cannot be reduced to a polynomial in the structure tensor $\boldsymbol{A}$. Therefore, we are in the general case described by Equation (141), in which a once-and-for-all integration is not possible, and the integral must be calculated, within a Finite Element implementation, at every increment of deformation, e.g., by means of the method of the spherical designs. This has been recently done by Tomic et al. (2014), in the open source Finite Element package FEBio (Ateshian and Weiss, 2010; Maas et al., 2012), which allows for the introduction of a permeability dependent on the whole Cauchy-Green deformation $\boldsymbol{C}$. The implementation requires the evaluation of the material permeability $\boldsymbol{K} = \hat{\boldsymbol{K}}(\boldsymbol{C})$ and the fourth-order material tangent permeability $\mathbb{D} = \hat{\mathbb{D}}(\boldsymbol{C})$ defined in Equation (192), which are then inverse Piola transformed to obtain the spatial permeability

$$\boldsymbol{k} = \boldsymbol{k}_0 . \left[(1-\phi_1)\phi_1 \, \boldsymbol{z} + (1-\phi_1)^2 \boldsymbol{g}^{-1}\right] , \tag{245}$$

where the symmetric tensor $\boldsymbol{z} = \boldsymbol{F}\boldsymbol{Z}\boldsymbol{F}^T$ is the push-forward of $\boldsymbol{Z}$, and the spatial tangent permeability $\mathbb{d}$ (Equation (193)).

We also note that, in the undeformed configuration, when $\boldsymbol{F} = \mathbf{1}$, $\boldsymbol{C} = \boldsymbol{G}$, $\boldsymbol{B} = \boldsymbol{G}^{-1}$, and $J = 1$, it can be easily verified that $\hat{\boldsymbol{Z}}(\boldsymbol{G}) = \boldsymbol{H}$, where $\boldsymbol{H}$ is the simple average of the material structure tensor $\boldsymbol{A}$, and therefore the overall material permeability reduces to

$$\hat{\boldsymbol{K}}(\boldsymbol{G}) = \boldsymbol{K}_0 . \left[(1-\phi_{1R})\phi_{1R} \, \boldsymbol{H} + (1-\phi_{1R})^2 \boldsymbol{G}^{-1}\right] , \tag{246}$$

which is equivalent to the final expression found by Federico and Herzog (2008b) directly in the undeformed configuration.

8.5 Permeability: Material Symmetries, Physical Admissibility

We emphasise that, while the spatial permeability of the REV, $\boldsymbol{k}_{\mathrm{REV}}$, is transversely isotropic in the basis $\{\boldsymbol{a}, \boldsymbol{t}\}$, its full Piola transform, the material permeability of the REV, $\boldsymbol{K}_{\mathrm{REV}}$, is *not* transversely isotropic in the basis $\{\boldsymbol{A}, \boldsymbol{T}\}$. This is due to the particular upscaling procedure that we employed, based on considering, at each current configuration, a REV made of a fibre segment and the surrounding fluid-saturated matrix (Figure 3). However, both the spatial $\boldsymbol{k}_{\mathrm{REV}}$ and the material $\boldsymbol{K}_{\mathrm{REV}}$ are particular cases of transversely isotropic tensor-valued functions constructed within general tensor representation theorems: an ample discussion on this can be found in the work by Tomic et al. (2014). A permeability model based on the representation theorems for tensor-valued functions (see the work by Ateshian and Weiss, 2010) is certainly the most general way to approach the problem. On the other hand, the upscaling procedure presented here (Federico and Herzog, 2008b; Federico and Grillo, 2012) allows for an analytically and numerically tractable representation of the effect of the fibres.

Similar considerations on symmetry can be made for the matrix permeability $\boldsymbol{k}_0$. We assume that $\boldsymbol{k}_0$ is spherical for under *any* deformation, i.e., that it has the form $\boldsymbol{k}_0 = k_0\, \boldsymbol{g}^{-1}$ (Federico and Grillo, 2012). Again, within a theory based on the representation theorems, this is a particular case of isotropic tensor-valued function. The advantage is that, if $\boldsymbol{k}_0$ is spherical, it is coaxial to any tensor, and thus also to the symmetric tensor $\boldsymbol{z}$ of Equation (245), so that the spatial overall permeability $\boldsymbol{k}$ remains symmetric.

Finally, the permeability of a deformable porous medium is physically admissible if it tends to zero at compaction. Since the fibres are impermeable, the onus of satisfying this physical requirement is on the matrix permeability. Therefore, if the matrix permeability is assumed to be always isotropic, i.e., $\boldsymbol{k}_0 = k_0\, \boldsymbol{g}^{-1}$, it suffices that the scalar matrix permeability $k_0 = \hat{k}_0(J)$ vanishes for $J \to \phi_{sR}$. For the case of articular cartilage, such a constitutive law has been proposed by Holmes and Mow (1990), with

$$\hat{k}_0(J) = k_{0R} \left(\frac{J - \phi_{sR}}{1 - \phi_{sR}} \right)^{\kappa} \exp\left[\tfrac{1}{2} M (J^2 - 1) \right], \tag{247}$$

where k_{0R}, κ, and M are constants to be determined experimentally.

8.6 The Permeability of Articular Cartilage

The expression of the permeability in the undeformed configuration reported in Equation (246) (Federico and Herzog, 2008b) has been applied to the evaluation of the axial permeability in articular cartilage (Federico and Herzog, 2008c). Although the resistance to fluid flow in articular cartilage has been deemed to be mainly due to the electrically charged proteoglycan matrix, Maroudas and Bullough (1968) found a controversial experimental result. Since the fixed charge density (which is proportional to the proteoglycan volumetric fraction) decreases almost linearly from the deep to the superficial zone (from the bone-cartilage interface to the articular surface), they would have expected to find a corresponding almost linear increase of the permeability: less charges cause less resistance to fluid flow, and therefore higher permeability. This was true up to the middle zone, but then they found that the permeability drastically decreased in the superficial zone (open squares in Figure 4b). Therefore, they inferred that the collagen fibres in the superficial zone, which are parallel to the surface, and thus orthogonal to the fluid flow that they were measuring, create a further obstacle, causing the permeability to decrease.

In order to give a theoretical explanation to this phenomenon, and to verify whether the intuitive explanation given by Maroudas and Bullough (1968) was correct, the permeability was calculated at several points along the tissue depth (solid diamonds in Figure 4b) by means of Equation (246), using a Gaussian fibre orientation probability of the type of Equation (159), parametric in the tissue depth. The mean co-latitude Θ_0 was taken from the X-ray diffraction experiments performed by Mollenhauer et al. (2003) on human ankle cartilage, and an expression of the variance λ as a function of the depth was assumed in order to represent the typical orientation of fibres in cartilage (Figure 4a): fibres nearly orthogonal to the bone-cartilage interface in the deep zone (Θ_0 close to 0, small λ), nearly randomly oriented fibres in the middle zone (large λ), and fibres nearly parallel to the articular surface in the superficial zone (Θ_0 close to $\pi/2$, small λ). The results gave a good fit with the experimental data obtained by Maroudas and Bullough (1968) (Figure 4b), thus confirming the idea that the collagen fibres do play a role in determining the local anisotropy and global inhomogeneity of permeability in articular cartilage (Federico and Herzog, 2008c).

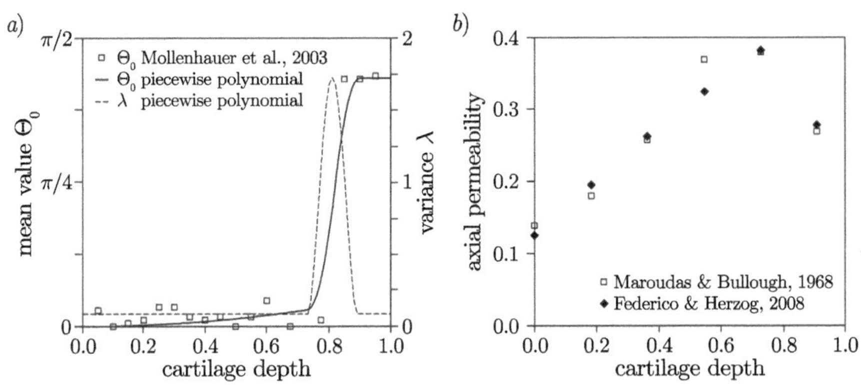

Figure 4: (Adapted from Federico and Herzog, 2008c) In both plots, the horizontal axis is cartilage depth, with 0 and 1 denoting the bone-cartilage interface and the articular surface, respectively. *a*) Mean value of the fibre co-latitude Θ_0, measured with respect to the direction of cartilage depth, from the experimental data by Mollenhauer et al. (2003) (open squares), interpolating piecewise polynomial for Θ_0 (solid line), and assumed piecewise polynomial for the variance λ (dashed line). *b*) Comparison of the axial permeability (units of $10^{-3}\text{mm}^4\text{N}^{-1}\text{s}^{-1}$) measured experimentally (open squares) by Maroudas and Bullough (1968) with the model's predictions, as a function of tissue depth.

Acknowledgements

I would like to gratefully acknowledge Dr. Alfio Grillo (Politecnico di Torino, Italy), Dr. T. Christian Gasser (KTH Stockholm, Sweden) and Dr. Walter Herzog (The University of Calgary, Canada), for their contributions in the publications from which this chapter takes inspiration. I would also like to thank Mr. Marek Netušil (Univerzita Karlova, Praha, Czech Republic) and Dr. David Steigmann (University of California Berkeley, USA), who suggested the existence of an exact expression of Equation (141) for the case of polynomials in the structure tensor, which had completely escaped me. Furthermore, I would like to thank my Group Members and the students in my graduate course ENME 653 Continuum Mechanics in Engineering, for the help in spotting several mistypings and unclear points. Finally, I would like to thank Dr. Ray W. Ogden (University of Glasgow, United Kingdom) and Dr. Marcelo Epstein (The University of Calgary, Canada) for the many and very useful suggestions they have given me through the past years.

Bibliography

E. S. Almeida and R. L. Spilker. Finite element formulations for hyperelastic transversely isotropic biphasic soft tissues. *Comput. Meth. Appl. Mech. Eng.*, 151:513–538, 1998.

R. M. Aspden and D. W. L. Hukins. Collagen organization in articular cartilage, determined by X-ray diffraction, and its relationship to tissue function. *Proc. Roy. Soc. Lond. B*, 212:299–304, 1981.

G. A. Ateshian and J. A. Weiss. Anisotropic hydraulic permeability under finite deformation. *J. Biomech. Eng.*, 132:111004, 2010.

C. Bellini and S. Federico. Green-Naghdi rate of the Kirchhoff stress and deformation rate: the elasticity tensor. *Z. Angew. Math. Phys.*, 2014.

K. L. Billiar and M. S. Sacks. Biaxial mechanical properties of the natural and glutaraldehyde treated aortic valve cusp - Part I: Experimental results. *J. Biomech.*, 122:23–30, 2000.

M. A. Biot. Theory of finite deformations of porous solids. *Indiana Univ. Math. J.*, 21:597–620, 1972.

R. L. Bishop and S. I. Goldberg. *Tensor Analysis on Manifolds.* Prentice-Hall, NJ, USA, 1968.

J. Bonet and R. D. Wood. *Nonlinear Continuum Mechanics for Finite Element Analysis (Second Edition).* Cambridge University Press, Cambridge, UK, 2008.

F.C. Caner and I. Carol. Microplane constitutive model and computational framework for blood vessel tissue. *J. Biomech. Eng.*, 128:419–427, 2006.

P. B. Canham, H. M. Finlay, J. G. Dixon, D. R. Boughner, and A. Chen. Measurements from light and polarised light microscopy of human coronary arteries fixed at distending pressure. *Cardiovasc. Res.*, 23:973–982, 1989.

P. Chadwick. *Continuum Mechanics, Concise Theory and Problems.* George Allen & Unwin Ltd., London, UK, 1976.

D. Chapelle, J.-F. Gerbeau, J. Sainte-Marie, and I. E. Vignon-Clementel. A poroelastic model valid in large strains with applications to perfusion in cardiac modeling. *Comput. Mech.*, 46:91–101, 2010.

W. Ehlers and G. Eipper. Finite elastizität bei fluidgesättigten hochporösen festkörpern. 77:S79–S80, 1997.

W. Ehlers and G. Eipper. Finite elastic deformations in liquid-saturated and empty porous solids. *Transport Porous Med.*, 34:179–191, 1999.

M. Epstein. *The Geometrical Language of Continuum Mechanics.* Cambridge University Press, Cambridge, UK, 2010.

A. C. Eringen. *Mechanics of Continua.* Robert E. Krieger Publishing Company, Huntington, NY, USA, 1980.

S. Federico. Volumetric-distortional decomposition of deformation and elasticity tensor. *Math. Mech. Solids*, 15:672–690, 2010.

S. Federico. Covariant formulation of the tensor algebra of non-linear elasticity. *Int. J. Non-Linear Mech.*, 47:273–284, 2012.

S. Federico. Some remarks on metric and deformation. *Math. Mech. Solids*, in press, DOI: 10.1177/1081286513506432, 2014.

S. Federico and T. G. Gasser. Non-linear elasticity of biological tissues with statistical fibre orientation. *J. R. Soc. Interface*, 7:955–966, 2010.

S. Federico and A. Grillo. Elasticity and permeability of porous fibre-reinforced materials under large deformations. *Mech. Mat.*, 44:58–71, 2012.

S. Federico and W. Herzog. Towards an analytical model of soft tissues. *J. Biomech.*, 41:3309–3313, 2008a.

S. Federico and W. Herzog. On the permeability of fibre-reinforced porous materials. *Int. J. Solids Struct.*, 45:2160–2172, 2008b.

S. Federico and W. Herzog. On the anisotropy and inhomogeneity of permeability in articular cartilage. *Biomech. Model. Mechanobiol.*, 7:367–378, 2008c.

S. Federico, A. Grillo, and W. Herzog. A transversely isotropic composite with a statistical distribution of spheroidal inclusions: a geometrical approach to overall properties. *J. Mech. Phys. Solids*, 52:2309–2327, 2004.

S. Federico, A. Grillo, S. Imatani, G. Giaquinta, and W. Herzog. An energetic approach to the analysis of anisotropic hyperelastic materials. *Int. J. Eng. Sci.*, 46:164–181, 2008.

S. Federico, A. Grillo, and R. Segev. Material counterpart of Darcy's law in terms of differential forms. In *Proceedings of CanCNSM 2013, 4th Canadian Conference on Nonlinear Solid Mechanics*, 2013.

S. Federico, A. Grillo, and S. Imatani. The linear elasticity tensor of incompressible materials. *Math. Mech. Solids*, in press, 2014, DOI: 10.1177/1081286514550576.

P. J. Flory. Thermodynamic relations for high elastic materials. *Trans. Faraday Soc.*, 57:829–838, 1961.

Y. C. Fung. *Biomechanics. Mechanical Properties of Living Tissues.* Springer-Verlag, New York, 2nd edition, 1993.

T. C. Gasser, R. W. Ogden, and G. A. Holzapfel. Hyperelastic modelling of arterial layers with distributed collagen fibre orientations. *J. R. Soc. Interface*, 3:15–35, 2006.

A. Grillo, G. Zingali, G. Borrello, S. Federico, W. Herzog, and G. Giaquinta. A multiscale description of growth and transport in biological tissues. *Theor. Appl. Mech.*, 34:51–87, 2007.

A. Grillo, S. Federico, and G. Wittum. Growth, mass transfer, and remodeling in fiber-reinforced, multi-constituent materials. *Int. J. Non-Linear Mech.*, 47:388–401, 2012.

A. Grillo, G. Wittum, A. Tomic, and S. Federico. Remodelling in statistically oriented fibre-reinforced composites and biological tissues. *Math. Mech. Solids*, in press, DOI: 10.1177/1081286513515265, 2014.

Z. Y. Guo, F. Caner, X. Q. Peng, and B. Moran. On constitutive modelling of porous neo-Hookean composites. *J. Mech. Phys. Solids*, 56:2338–2357, 2008.

R. H. Hardin and N. J. A. Sloane. McLaren's improved snub cube and other new spherical designs in three dimensions. *Discrete Comput. Geom.*, 15: 429–441, 1996.

R. H. Hardin, N. J. A. Sloane, and W. D. Smith. *Spherical Codes.* 2008. http://www.research.att.com/~njas/packings/.

S. M. Hassanizadeh. Derivation of basic equations of mass transport in porous media, part ii. generalized Darcy's and Fick's laws. *Adv. Water Resour.*, 9:208–222, 1986.

M. H. Holmes and V. C. Mow. The nonlinear characteristics of soft gels and hydrated connective tissues in ultrafiltration. *J. Biomech.*, 23:1145–1156, 1990.

G. A. Holzapfel and R. W. Ogden. Constitutive modelling of arteries. *Proc. Roy. Soc. Lond. A*, 466:1551–1597, 2010.

G. A. Holzapfel, T. C. Gasser, and R. W. Ogden. A new constitutive framework for arterial wall mechanics and a comparative study of material models. *J. Elasticity*, 61:1–48, 2000.

C. Hurschler, B. Loitz-Ramage, and R. Vanderby Jr. A structurally based stress-stretch relationship for tendon and ligament. *J. Biomech. Eng.*, 119:392–399, 1997.

A. N. Kolmogorov and S. V. Fomin. *Elements of the Theory of Functions and Functional Analysis.* Dover, 1999.

L. D. Landau and E. M. Lifshitz. *Electrodynamics of Continuous Media.* Pergamon Press, Oxford, UK., 1960.

Y. Lanir. Constitutive equations for fibrous connective tissues. *J. Biomech.*, 16:1–12, 1983.

S. A. Maas, B. J. Ellis, G. A. Ateshian, and J. A. Weiss. Febio: Finite elements for biomechanics. *J. Biomech. Eng.*, 134:011005–1–10, 2012.

A. Maroudas and P. Bullough. Permeability of articular cartilage. *Nature*, 219:1260–1261, 1968.

J. E. Marsden and T. J. R. Hughes. *Mathematical Foundations of Elasticity.* Prentice-Hall, Englewood Cliff, NJ, USA, 1983.

G. De Marco. *Analisi Due.* Zanichelli, Bologna, Italy, 2nd edition, 1999.

R. McLaughlin. A study of the differential scheme for composite materials. *Int. J. Eng. Sci.*, 15:237–244, 1977.

A. Menzel, M. Harrysson, and M. Ristinmaa. Towards an orientation-distribution-based multi-scale approach for remodelling biological tissues. *Comput. Meth. Biomech. Biomed. Eng.*, 11:505–524, 2008.

J. Merodio. On constitutive equations for fiber-reinforced nonlinearly viscoelastic solids. *Mech. Res. Commun.*, 33:764–770, 2006.

J. Merodio and R. W. Ogden. Material instabilities in fiber-reinforced nonlinearly elastic solids under plane deformation. *Arch. Mech.*, 54:525–552, 2002.

C. Miehe. Aspects of the formulation and finite element implementation of large strain isotropic elasticity. *Int. J. Numer. Meth. Eng.*, 37:1981–2004, 1994.

J. Mollenhauer, M. Aurich, C. Muehleman, G. Khelashvilli, and T. C. Irving. X-ray diffraction of the molecular substructure of human articular cartilage. *Conn. Tiss. Res.*, 44:201–207, 2003.

L. W. Morland. A simple constitutive theory for a fluid-saturated porous solid. *J. Geophys. Res.*, 77:890–900, 1972.

A. N. Norris. A differential scheme for the effective moduli of composites. *Mech. Mat.*, 4:1–16, 1985.

R. W. Ogden. Nearly isochoric elastic deformations: Application to rubber-like solids. *J. Mech. Phys. Solids*, 26:37–57, 1978.

R. W. Ogden. *Non-linear Elastic Deformations.* Dover, New York, USA, 1997.

J. Pajerski. *Nonlinear Biphasic Microstructural Numerical Analysis of Articular Cartilage and Chondrocytes.* MSc Thesis, The University of Calgary, Canada, 2010.

S. Pozdniakov and C.-F. Tsang. A self-consistent approach for calculating the effective hydraulic conductivity of a binary, heterogeneous medium. *Water Resour. Res.*, 40:1–13, W05105, 2004.

S. Quiligotti. On bulk growth mechanics of solid-fluid mixtures. *Theor. Appl. Mech.*, 28/29:277–288, 2002.

S. Quiligotti, G. A. Maugin, and F. dell'Isola. An eshelbian approach to the nonlinear mechanics of contrained solid-fluid mixtures. *Acta Mech.*, 160:45–60, 2003.

M. S. Sacks. Incorporation of experimentally-derived fiber orientation into a structural constitutive model for planar collagenous tissues. *J. Biomech. Eng.*, 125:280–287, 2003.

C. Sansour. On the physical assumptions underlying the volumetric-isochoric split and the case of anisotropy. *Eur. J. Mech. A/Solids*, 27: 28–39, 2008.

J. C. Simo, R. L. Taylor, and K. S. Pister. Variational and projection methods for the volume constraint in finite deformation elasto-plasticity. *Comput. Meth. Appl. Mech. Eng.*, 51:177–208, 1985.

A. J. M. Spencer. Constitutive theory for strongly anisotropic solids. In A. J. M. Spencer, editor, *Continuum Theory of the Mechanics of Fibre-Reinforced Compositcs*, pages 1–32. Springer-Verlag, Wien, Austria, 1984. CISM Courses and Lectures No. 282, International Centre for Mechanical Sciences.

Y. R. Talpaert. *Tensor Analysis and Continuum Mechanics.* Springer (Kluwer), Heidelberg, Germany, 2003.

K. Terzaghi. *Theoretical Soil Mechanics.* Wiley, New York, USA, 1943.

A. Tomic, A. Grillo, and S. Federico. Poroelastic materials reinforced by statistically oriented fibres - numerical implementation and application to articular cartilage. *IMA J. Appl. Math.*, 79:1027–1059, 2014.

L. Vergori, M. Destrade, P. McGarry, and R. W. Ogden. On anisotropic elasticity and questions concerning its Finite Element implementation. *Comput. Mech.*, 52:1185–1197, 2013.

L. J. Walpole. Elastic behavior of composite materials: Theoretical foundations. *Adv. in Appl. Mech.*, 21:169–242, 1981.

J. A. Weiss and J. C. Gardiner. Computational modeling of ligament mechanics. *Crit. Rev. Biomed. Eng.*, 29:303–371, 2001.

E. W. Weisstein. Erfi. From MathWorld-A Wolfram Web Resource. 2005. `http://mathworld.wolfram.com/Erfi.html`.

A. A. White and M. M. Panjabi. *Clinical Biomechanics of the Spine.* Lippincott, Philadelphia, USA, 1978.

W. Wilson, J. M. Huyghe, and C. C. van Donkelaar. Depth-dependent compressive equilibrium properties of articular cartilage explained by its composition. *Biomech. Model. Mechanobiol.*, 6:43–53, 2007.

R. W. Zimmermann. Elastic moduli of a solid containing spherical inclusions. *Mech. Mat.*, 12:17–24, 1991.

Nonlinear Elasticity for Soft Fibrous Materials

Giuseppe Saccomandi
Dipartimento di Ingegneria, University of Perugia, Perugia, Italy
Email: giuseppe.saccomandi@unipg.it

Abstract. This chapter is devoted to a review of several aspects of the nonlinear theory of elasticity that may be useful to deduce in a rational way mathematical models that may be useful for describing the mechanical behaviour of soft fibrous materials. We review some general aspects of the nonlinear theory and a list of conditions that seems to be plausible to impose on mechanical models of soft fibrous materials. We consider also some experimental data. Then we study the possibility of using continuum mechanics models to study single fibre materials.

1 Introduction

In a purely mechanical theory a uniform homogeneous elastic solid has the properties that its material response is rate-independent and non dissipative, it can support deviatoric stress without flowing and it returns to its original unique volume and shape when unloaded to zero stress.

Whereas the linear theory of infinitesimal elasticity is well and uniquely established, in the finite strain theory the elastic material response is modelled using three classes of models: Cauchy elasticity, hyperelasticity and hypoelasticity. There is no general agreement on which of these three ways is the right one for describing nonlinear elasticity. Famous is the Rivlin–Truesdell clash[1] about hyperelasticity and Cauchy elasticity based on the work done in closed cycle of deformation.[2]

[1] A trace of this discussion may be found in the proceedings edited by M. Reiner, *Second-order Effects in Elasticity, Plasticity and Fluid Dynamics* (Macmillan, New York, 1964).

[2] Recently Rajagopal (2007) and Rubin (2009) have reexamined, in a original and interesting way, the notion of rate-independent elastic response. The ideas considered by Rajagopal and Rubin are extremely important because they place elastic materials in a more general perspective.

L. Dorfmann, R. W. Ogden (Eds.), *Nonlinear Mechanics of Soft Fibrous Materials*,
CISM International Centre for Mechanical Sciences
DOI 10.1007/978-3-7091-1838-2_3 © CISM Udine 2015

Here, to restrict the discussion and for the sake of the simplicity we consider only *hyperelasticity* where the existence of a strain-energy function ensures that for all deformations and for all specific choices of model conservation of energy holds (Carroll, 2009). It is a common believe that hyperelasticity may be a very good mathematical model for describing the mechanical behaviour of natural and synthetic rubber.

Rubber-like mechanics has had one of the longest and most distinguished histories in all polymer science. As a first approximation hyperelasticity is the right tool for obtaining reliable mathematical models of elastomeric materials and this is mainly for two reasons. First, any simple material (Noll, 1972; Truesdell and Noll, 2004) is an elastic material when static deformations are considered. Therefore, going through hyperelasticity is a natural step in the understanding of the mechanical response of elastomeric materials. Second, the statistical approach to rubber mechanics based on molecular concepts is fully compatible with the phenomenological theory of hyperelasticity. For these reasons after the World War II there has been a huge research activity aimed at finding a specific strain-energy density able to describe the experimental data obtained for natural and synthetic rubber.[3]

In spite of decades of researches regarding this subject until today there is no effective model for describing the experimental data of rubber-like material in a careful and universal way (i.e. for any kind of deformation field and for the full range of stretches in which we are interested). This state of affairs has not to be confused with the fact that it may happen that some simulations based on some special models may be successful in describing some special phenomena. These simulations are concretely descriptive, but apply only to particular cases. On the other hand, mathematical models are abstract and universal and it must be clear that such **universal** models for rubber-like materials seem not to exist.[4] Here, a *satisfactory* model is defined as a model able to describe the experimental data first of all from a qualitative point of view and then from a quantitative point with an acceptable relative error.

In the literature it is not rare to read announcements of strain-energy functions able to describe in a wonderful and accurate way any kind of

[3]Clearly hyperelasticity is only a first step in the understanding of rubber-like materials. Fundamental effects in the mechanics of rubber such as the Mullins or Payne effects cannot be modelled using a theory as simple as elasticity.

[4]I am not speaking of a single model able to describe the mechanical behaviour of any elastomeric material. I am claiming that when if you consider just a single material (e.g., natural rubber) there are no models able to describe in a satisfactory way for all deformations fields and for all stretch ranges the mechanical response for this material.

experimental data and for any elastomeric material. Unfortunately, a careful analysis of these models reveals that all the usual problems that we have been faced with for decades are always under the sand and the proposed models have in reality a quite limited descriptive and predictive ability. The fact that, in the literature, there is a continuous flow of new models is a clear indicator that we are far away from the solution of the problem. In my opinion, what is very disappointing about all this history is not only the scarce intellectual integrity we have to face in reading the results of some researches hiding the real state of affairs, but the fact that often the reasons explaining the failure of the various models to describe the experimental data may be found in simple theoretical arguments that still are not common knowledge. We do not need complicated experimental testing machines and sophisticated sets of data to uncover many of the *bugs* in the various models. What we need is just a firm application of the classical hypothetical-deductive method and basic tools of continuum mechanics.

In our opinion the reason for this state of affairs is due to the fact that still today many people working in rubber-like materials and soft-tissue mechanics believe in statements like (Chui et al., 2004):

> *A constitutive equation describes a physical property of a material. Its derivation should begin with empirical measurements.*

This is an evidently childish statement. The only measurement possible without a theoretical framework is *count.*[5] This point of view is clearly stated in Truesdell and Noll's (2004) monumental work *The Non-Linear Field Theories of Mechanics*:

> *While laymen and philosophers of science often believe, contend, or at least hope, that physical theories are directly inferred from experiments, anyone who has faced the problem of discovering a good constitutive equation or anyone who has sought and found the historical origin of the successful field theories knows how childish is such a prejudice. The task of the theorist to bring order into the chaos of the phenomena can be described efficiently and simply. Here is the place for* intuition, *and here the old preconception, common among natural philosophers, that nature is simple and elegant, has led to many great successes. Of course, physical theory must be based on experience, but experiment comes after, not before, theory. Without theoretical con-*

[5]For the sake of precision, to count we need numbers and therefore a theoretical construction such as Peano's axioms.

cepts one would neither know what experiments to perform nor be able to interpret their outcome.

These problems are still actual because in recent times there has been a renewed interest in the nonlinear theory of elasticity. The reason can be found in the field of biomechanics and in our effort to understand the mechanical behaviour of *soft tissues*, i.e. the biomaterials that connect, support, or surround any structure and/or organ of the body, and which are not to be considered bones. This includes a long list of organs and structures such as tendons, ligaments, fascia, skin, fibrous tissues, fat, synovial membranes, muscles, nerves, blood vessels and many others.

In animals connective tissues are based on the ECM, the *extracellular matrix*. The ECM is the scaffold part of the animal tissue, i.e. the part that provides structural support to the cells. The characteristic substances inside the ECM of soft tissues are two important proteins, *collagen* and the *elastin*, plus a certain number of ground substances including many proteinaceous components (e.g., proteoglycans) and water. In most situations collagen and elastin are organized in *fibres* and for this reason we speak of soft fibrous materials.

The complicating factor in modelling such materials is mainly due to the fact that they self-organize into mesoscopic physical structures that are much larger than the microscopic scale of atoms and molecules, and yet are much smaller than the macroscopic scale of the material. The properties and interactions of these mesoscopic structures may determine the macroscopic behaviour of the material, and a challenge for continuum models is in accounting for this situation. For this reason, it is often necessary to adopt a *multi-scale* approach to the problem, i.e. we need to establish a hierarchical link between the various spatial and temporal scales, to exhibit the output at the larger scale, passing through a finely detailed representation at lower scales.

There are several *homogenization* techniques for deriving a macroscopic constitutive law starting from the meso/microscopic network structure of a material. Despite the complexity of the mesoscopic and microscopic structures of the materials, it must be clear that at the macroscopic level we cannot escape from the general framework of continuum mechanics. This fact suggests that there is another approach for tackling these kinds of problems, an approach fully compatible and organic with the axiomatic theory of thermomechanics of continua: it is possible to introduce mesoscopic information directly into the constitutive equations. At first sight it seems that this last pathway is less valuable with respect the first one because it is purely phenomenological, but we may show that this is not the case.

The mechanical behaviour of soft fibrous materials is so complex that we need to attack the problem from any possible perspective. Therefore, any point of view and general coherent framework is valuable and it may help to grasp the problem. Moreover, the approach we are proposing not only may help to make important advances in the framework of materials but may give some interesting ideas also in the framework of the mechanical modelling of single molecules and/or filaments.

The aim of my lecture notes is to review some of the basic mistakes and urban myths with respect to the derivation of specific forms of the strain-energy function and to try to overcome the various shortcomings. The results of this exercise are in some sense discouraging. At the end of this essay I am obliged to leave the reader with the idea that the problem of finding a mathematically reliable model of a strain energy is a Chimera, the fire-breathing female monster with a lion's head, a goat's body, and a serpent's tail that Greek heroes thought illusory and impossible to fight. Our only solace are the lines of Homer (Iliad VI, 180-184, translation by Ian Johnston):

> *But Bellerophon killed the Chimera, putting his trust in omens from the gods.*

Acknowledgements. There is a long list of collaborators and friends that I have to thank for several reasons and it is correct to do this just at the starting point of my lecture notes. Because it would take too long to explain the reasons in detail I am just mentioning their names in a simple alphabetical order: Michel Destrade (Galway), Alain Goriely (Oxford), Mike Hayes (Dublin), Jerry Murphy (Dublin), Ray Ogden (Glasgow), Geppe Puglisi (Bari), Kumbakonam R. Rajagopal (College Station), Ivonne Sgura (Lecce). I am grateful to all these people not only because they have shared with me their knowledge, but also because they have been always very good friends. The section of this chapter on experimental data is deeply based on joint work with Ivonne Sgura, the section on anisotropy on joint work with Jerry Murphy and the section on DNA mechanics on joint work with Ivonne Sgura and Ray Ogden.

2 Ut Tensio, Sic Vis

The first constitutive model of interest in elasticity (which is the simplest theory we may encounter) is the so-called Hooke's law. This law involves the kind of force that in the approximation of classical mechanics and within a certain range of deformations we may regard as exerted by a spring on a particle: the elastic force. Analytically

$$\boldsymbol{F} = -k(\boldsymbol{x} - \boldsymbol{x}_0), \tag{1}$$

where $\boldsymbol{x}$ is the distance that the spring has been stretched or compressed away from the equilibrium position $\boldsymbol{x}_0$, $\boldsymbol{F}$ is the restoring force and $k\,(> 0)$ is the spring constant.

The cartoon of the mass attached to a spring is misleading. What in classical mechanics is called an elastic force is an example of a central force, i.e. a force whose magnitude is a function of only the scalar distance r of the particle $\boldsymbol{x}$ from $\boldsymbol{x}_0$ and is directed along the unit vector $\boldsymbol{u}$ in the direction $\boldsymbol{x} - \boldsymbol{x}_0$. Hooke's law is the linear central force, i.e. $\boldsymbol{F} = -kr\boldsymbol{u}$ (usually, $\boldsymbol{x}_0$ does not depend on time). All central forces are conservative forces and they are derived from a potential U. For the linear central force we have $U = -kr^2/2$.

Several generalizations of (1) are possible. It is possible to imagine nonlinear versions of Hooke's law, for example $\boldsymbol{F} = -(k_l r + k_{nl} r^3)\boldsymbol{u}$. We point out that in the linear case the only equilibrium position is for $r = r_{eq} = 0$ and the same is true in the nonlinear case if $k_{nl} > 0$. When $k_{nl} < 0$ we have $r_{eq} = 0$ and additional equilibrium positions with $r_{eq}^{\pm} = \pm\sqrt{k_l/\left|k_{nl}\right|}$.

Hooke's law, as just considered, is a purely phenomenological zero-order model. It is clear that any *linear* phenomenological model is a quite powerful tool because, modulo some constants, linear functions are easily and uniquely characterized. Phenomenological nonlinear models are special choices in the interminable prairies of nonlinear functions. There is no *a priori* reason why we have to consider nonlinear polynomial functions instead of other function classes. The usual idea that polynomial functions are significant because the laws of nature are *smooth* and it is therefore natural to consider Taylor expansions of such laws, although they may not be realistic.

Another possible generalization of (1) is to consider the *evolution* of the natural (i.e. equilibrium) configuration. In (1) the natural configuration is given by $\boldsymbol{x}_0$ and clearly there is no *a priori* reason to exclude $\boldsymbol{x}_0 = \boldsymbol{x}_0(t)$. This simple modification in the constitutive equation (1) opens a full range of mechanical behaviours. A clear example is elasto-plastic behaviour that is well described by adding to (1) a *flow* rule for $\boldsymbol{x}_0$.

A last possible generalization of (1) is to consider an *implicit* theory. There is no need (and no good reason) to give force explicitly in terms of kinematical quantities (here we are just considering displacement). Why not consider general laws in the form $\mathcal{F}(\boldsymbol{F}, \boldsymbol{x}) = \boldsymbol{0}$? Once again the linear theory is misleading because in this case constitutive equations are always invertible, but in general nonlinear theories invertibility cannot in general be the case and interesting branching phenomena may occur.

For rubber-like materials it is usual to consider models based on molecular considerations. Because rubber-like materials and soft tissues are *amorphous* materials it is not easy to develop, in a rigorous way, a model from molecular considerations, but molecular cartoons may be a source of very powerful insights into the mesoscopic structure of the material.

The first step is to understand, from elementary thermodynamics, what is usually described as *entropic elasticity*. We start by considering the following experiment with a rubber band. One end of the band is held fixed and on the other end a tensile force, F, is applied. The experiment is performed under quasi-static and isothermal conditions: the temperature is θ everywhere and at all times in the band. The reference undistorted length is L and the current deformed length under F is l. The first law of thermodynamics says that

$$dE = dQ + dW,$$

where W is the total work. Consider, as usual, that $dQ = \theta dS$ or

$$dE = \theta dS + dW,$$

where $S = \int \eta dx$, η being the entropy per unit reference volume. Under *isothermal* conditions, we have for the Helmholtz free energy Ψ

$$d\Psi = dE - \theta dS = dW,$$

with $\Psi = \int \psi dx$, ψ being the Helmholtz free energy per unit reference volume.

The work of the external tensile force is $dW = Fdl$ (we are considering that the volume of our rubber band does not change because the material is incompressible) and we obtain that under *isothermal* conditions $F = d\Psi/dl$ or, using the usual thermodynamical notation,

$$F = \left(\frac{\partial \Psi}{\partial l}\right)_\theta . \tag{2}$$

On the other hand by differentiating the free energy, at constant θ, we obtain

$$\left(\frac{\partial \Psi}{\partial l}\right)_\theta = \left(\frac{\partial E}{\partial l}\right)_\theta - \theta \left(\frac{\partial S}{\partial l}\right)_\theta .$$

Since $dE = Fdl + \theta dS$ and $d\Psi = dE - \theta dS - Sd\theta$, we have

$$d\Psi = Fdl - Sd\theta,$$

and since Ψ is an exact differential then by the Maxwell relation $(\partial S/\partial l)_\theta = -(\partial F/\partial \theta)_l$, i.e.

$$F = \left(\frac{\partial E}{\partial l}\right)_\theta - \theta\left(\frac{\partial S}{\partial l}\right)_\theta = \left(\frac{\partial E}{\partial l}\right)_\theta + \theta\left(\frac{\partial F}{\partial \theta}\right)_l . \tag{3}$$

Equation (3) says that the force is composed of two components, one due to the change in the elastomer's internal energy as a result of the elongation (or compression), and the other due to the entropy change accompanying the deformation. Because, in the classical approach, F and θ are dependent and independent experimental variables, respectively, equation (3) allows us to obtain $(\partial E/\partial l)_\theta$ from a plot of F versus θ. These relations are valid for all materials (if volume variations are negligible).

When ideal elastomers are extended, the change in length (and energy) comes almost entirely from a change in conformation, and there is negligible stretching of bonds composing the macromolecular chain. Therefore, at constant temperature, it can be approximated that the internal energy of the bonds does not change and we assume

$$F = -\theta\left(\frac{\partial S}{\partial l}\right)_\theta . \tag{4}$$

This is what we call an *entropic spring.*

The key ingredient in the molecular modelling of polymeric materials is the end-to-end average (over time) distance r of the extrema of the macromolecular chain. To compute this distance we consider the highly idealized case of a macromolecular chain (a sort of zero-order model) composed of n equal rigid rods connected by hinges that may lie only in a straight line. Therefore the hinges can only be either fully open or fully closed. If the link originally to the right is directed to the right we shall use the symbol $+$, otherwise the symbol $-$.

In this oversimplified case if n^+ is the number of rods directed to the right and n^- the number of rods directed to the left and l is the length of a rod, we have that

$$r = (n^+ - n^-)l, \tag{5}$$

and considering that $n = n^+ + n^-$, we obtain

$$n^+ = \frac{1}{2}\left(n + \frac{r}{l}\right), \; n^- = \frac{1}{2}\left(n - \frac{r}{l}\right). \tag{6}$$

Therefore, the probability of realizing a given length r is simply given by

$$P(r) = \frac{n!}{n^+!n^-!}. \tag{7}$$

We suppose n, n^+ and n^- to be large numbers so that we may use Stirling's approximation

$$\log(n!) \approx \left(n + \frac{1}{2}\right)\log(n) - n + \frac{1}{2}\log(2\pi). \tag{8}$$

In writing the statistical entropy associated with the distribution (7) we obtain

$$\log(P) = \log\left(\frac{n!}{n^+!n^-!}\right) = \log(n!) - \log(n^+!) - \log(n^-!). \tag{9}$$

This last relation may be approximated as

$$\begin{aligned}\log(P) = &\left(n + \frac{1}{2}\right)\log(n) - \left(n^+ + \frac{1}{2}\right)\log(n^+)\\ &- \left(n^- + \frac{1}{2}\right)\log(n^-) - \frac{1}{2}\log(2\pi).\end{aligned} \tag{10}$$

The assumption of the purely entropic theory means that

$$\Psi(r) = -kT\log\left(P(r)\right),$$

and therefore the tensile force is given by

$$F = -kT\frac{d}{dr}\log\left(P(r)\right).$$

Using the previous computations, we obtain

$$\begin{aligned}F(r) = -kT&\left\{\frac{1}{2l}\left[\log\left(\frac{nl-r}{l}\right) - \log\left(\frac{nl+r}{l}\right)\right]\right.\\ &\left.+\frac{r}{(nl+r)(nl-r)}\right\},\end{aligned} \tag{11}$$

or if n is large

$$F(r) = -kT\frac{1}{2l}\left[\log\left(\frac{nl-r}{l}\right) - \log\left(\frac{nl+r}{l}\right)\right]. \tag{12}$$

When $r/nl << 1$ a further simplification gives

$$F(r) = -kT\frac{r}{nl^2}. \tag{13}$$

In this limit we recover the linear phenomenological equation for the linear elastic force, but here we are able to relate the spring constant to microscopic quantities.

We rewrite (11) by introducing the notations $\tilde{F} = Fl/kT$ and $\xi = r/nl$ and we obtain

$$\tilde{F} = -\frac{1}{2}\log\left(\frac{1-\xi}{1+\xi}\right) - \frac{\xi}{n(1+\xi)(1-\xi)}.$$

Now it is clear that as we approach the contour length nl, i.e. $\xi \to 1$, we have that $F \to \infty$: the *finite chain length* effect.

In a three-dimensional framework the computations we have outlined are much more complex. Just to give some idea of the various models that may be deduced in a three-dimensional framework we review some of them. First of all, we consider an ideal chain composed by a collection of rigid rods hinged together, namely the *Freely Jointed Chain* (FJC). In the non-Gaussian limit for this model the force is expressed implicitly via

$$r = l_c\left(\coth\left(\frac{lF}{k\theta}\right) - \frac{k\theta}{lF}\right), \tag{14}$$

where $l_c = nl$ is the contour length. A formal inversion of this law is possible using a special function, the inverse Langevin function. It is important to note that the asymptotic behaviour of the FJC model at $\xi \to 1$ is as $1/(1-\xi)$.

Another very important model for a single macromolecular chain is the *Worm-Like Chain* (WLC). In this case the rigid segments between bonds of the FJC are substituted by Kirchhoff rods. These are rods that may bend but they cannot change their length, they are inextensible. In such a way it is possible to obtain via a numerical procedure the force versus extension formula. To make operative such an idea various *interpolation formulas* have been proposed. The first and commonest of such formulas is given by

$$F = \frac{k\theta}{l_p}\frac{1}{4}\left(\frac{1}{(1-r/l_c)^2} - 1 + \frac{4r}{l_c}\right). \tag{15}$$

Here, not only do we have the contour length l_c, but also a new parameter, the *persistence length* l_p. This is an important additional parameter measuring the distance over which correlations among neighbouring links in the direction of the tangent are lost. The main difference between the FJC and WLC models is the asymptotic behaviour as the contour length l_c is approached. For the WLC we have $F \to \infty$ as $1/x^2$ and for the FJC as $1/x$.

The basic idea is that macromolecular chains relevant in soft fibrous materials are *semiflexible* filaments, i.e. that they have a finite rigidity associated with bending of the chain. Flexible filaments are well modelled as Gaussian chains at *small* elongations, i.e. they exhibit linear force–extension relations when pulled for *small* displacements. These filaments begin to behave nonlinearly only when subject to *large* strains. The problem is that at the microscopic level things are much more complex than in a macroscopic world because we are in the presence of two competing effects. First, the thermal motion which promotes bent configurations. Second, the rigidity which promotes straight ones. The interplay between these two effects produces a very complex elastic response, in particular nonlinear elasticity is manifest already at small strains. What happens when we start to pull a filament by increasing the tensile force F?

In the smallest force regime, the extension increases linearly with the force. This feature is predicted by any model in the small force regime, but as we increase the tensile force a little bit several ideas have been proposed. By just increasing the force a little bit what we observe is that many polymeric chains become organized in a series of *globular structures*. These structures are for a certain range of tensile force more important that the single elements (rigid segment, rods or beams) composing the idealized chain and for this reason we observe the so called Pincus regime (Pincus, 1976). In this range we have $r \sim F^{1/\nu - 1}$, where ν is the scaling exponent such that $\nu \sim [1/2, 3/5]$ depending on the quality of the solvent in which the polymer is immersed. With further increases in the tensile force this globular structure is less and less relevant for the mechanical behaviour of the chain and then we observe the WLC regime $r/l_c \sim 1 - F^{-1/2}$ and then, because a very high tensile force helps the rigidity of the chain, we have the FJC regime $r/l_c \sim 1 - F^{-1}$.

To show how it is possible to formulate phenomenological constitutive equations by considering molecular information and maintaining mathematical feasibility, we consider a bar[6] that can only stretch in one dimension. In this case the surface area of each cross section of the bar is A in the reference configuration and a in the current configuration. All we need is a coordinate $x \in [0, L]$ such that we are able to locate a given cross section in the reference configuration. Therefore, the motion is described by an equation of the form

$$y = y(x, t), \tag{16}$$

where $y \in [0, l]$ determines the position of the cross section x in the current configuration. The geometry and the kinematics of our problem are described by two simple quantities: y and the stretch $\lambda = \partial y / \partial x$. We require $\lambda > 0$ because the length of the bar cannot be reduced to zero.

[6]The term bar is here used only to describe a one-dimensional continuum.

Our aim is to develop a theory considering only *elastic local interactions* at any point, x_0, of the reference domain, and at any instant of time, t_0. Then

$$y(x,t) \approx y(x_0,t_0) + (x - x_0) \left.\frac{\partial y}{\partial x}\right|_{x=x_0,t=t_0} + \dots,$$

and therefore for this kind of interaction it is sufficient to consider that the necessary constitutive equation for the stress is a function of the quantities y and λ, the motion and the stretch. Requiring that the values of the constitutive functions are unaffected by superposing rigid motions[7] rules out y itself. Indeed, this requirement means that if two motions y and $\bar{y}$ are related by

$$\bar{y}(x,t) = y(x,t) + g(t),$$

where g is an arbitrary smooth function, the constitutive equations must not change their value. Moreover, because we are interested only in hyperelastic materials, we postulate the existence of a strain-energy density function $\psi = \psi(\lambda)$, and from the balance of energy we obtain $s = \partial\psi/\partial\lambda$, where s is the one-dimensional Piola–Kirchhoff stress.

Here the main problem is to choose $s = s(\lambda)$. For example, in some experiments on soft tissues coming from the mesentery, Fung (1967) identified the existence of a linear relation between the nonlinear version of Young's modulus (or stiffness) and s. As a first approximation Fung fitted the experimental curve by using a linear relationship according to

$$\frac{ds}{d\lambda} = k_1 + k_2\lambda,$$

where k_1 and k_2 are two constitutive parameters. Integrating this simple differential equation and requiring $s(1) = 0$, we obtain the celebrated law

$$s(\lambda) = \frac{k_1}{k_2}\left[\exp(k_2(\lambda - 1)) - 1\right]. \tag{17}$$

This is a good example of how a phenomenological law may be obtained. With respect to the laws obtained in the previous subsection we have no idea how to relate *a priori* the constitutive coefficients to physical quantities. Only *a posteriori* is it possible to give a mechanical meaning to the various constitutive parameters. For example, here is clear that k_1 is exactly the Young's modulus of the linear limit of the model (17), but it is quite hard to propose a *meso/micro* interpretation of the parameter k_2.

[7]This is because in a rigid motion internal forces do no work.

Another way of generating constitutive equations is to *presume* that natural laws are always smooth laws and therefore they may be approximated by Taylor expansions. This means considering $\psi = \sum_{i=1}^{n} \alpha_i(\lambda^2 - 1)^i$. The stress associated with this choice is always a polynomial function. Clearly, the use of Taylor series is not always a good choice, for several reasons. For example, to model the limiting chain extensibility effect we have pointed out in the previous section that a Taylor expansion is problematic because this expansion cannot take into account the presence of a singularity. In this case, a more general expansion has to be considered. In particular, it is possible to consider

$$\psi = -\frac{\mu J^*}{2} \ln \left(1 - \frac{e^2}{J^*}\right),$$

where μ and J^* are two parameters and $e = \lambda - 1$. In so doing we deduce that

$$s = \mu \frac{e}{1 - e^2/J^*},$$

and when the parameter $J^* \to \infty$ we have the classical linear law $s = \mu e$. The positive parameter J^* is an upper bound for e^2, so it is related to a contour length and when the contour length is infinite we recover a stress function which does not blow up for a finite value of the stretch. We remark that in this case s is not polynomial; we are approximating the stress by considering a rational function such as a Padé approximant.

3 Three-dimensional Theory

The aim of this section is not to give a crash course in nonlinear elasticity, but only to fix the notation. For a detailed account of the subject we refer to, for example, Ogden (1984).

We call $\boldsymbol{x}(\boldsymbol{X})$ the current position of a particle which was located at $\boldsymbol{X}$ in the reference configuration. Two kinematic quantities associated with this motion are the deformation gradient and the left Cauchy–Green deformation tensor,

$$\boldsymbol{F} = \partial \boldsymbol{x}/\partial \boldsymbol{X}, \quad \boldsymbol{B} = \boldsymbol{F}\boldsymbol{F}^T, \tag{18}$$

respectively. The principal values of $\boldsymbol{B}$ are $\lambda_1^2, \lambda_2^2, \lambda_3^2$, where λ_1, λ_2 and λ_3 are the principal stretches. We consider hyperelastic, isotropic materials, and so we introduce the strain-energy density $W = W(I_1, I_2, I_3)$, where I_1 and I_2 are the first and second principal invariants of $\boldsymbol{B}$, respectively, given by

$$I_1 = \operatorname{tr} \boldsymbol{B}, \quad I_2 = \tfrac{1}{2}\left[I_1^2 - \operatorname{tr}(\boldsymbol{B}^2)\right], \tag{19}$$

and $I_3 = J^2$, where $J = \det \boldsymbol{F}$. Equivalently, it is possible to regard the strain energy as a symmetric function, say $\hat{W}$, of $\lambda_1, \lambda_2, \lambda_3$. In particular,

$$\hat{W}(\lambda_1, \lambda_2, \lambda_3) = \hat{W}(\lambda_1, \lambda_3, \lambda_2) = \hat{W}(\lambda_3, \lambda_1, \lambda_2), \tag{20}$$

for all $\lambda_1, \lambda_2, \lambda_3 \in (0, \infty)$.

Using the principal invariants, the general representation formula for the Cauchy stress tensor $\boldsymbol{T}$ reads

$$\boldsymbol{T} = \beta_0 \boldsymbol{I} + \beta_1 \boldsymbol{B} + \beta_{-1} \boldsymbol{B}^{-1}, \tag{21}$$

where

$$\beta_0 = 2J^{-1}\left[I_2 W_2 + I_3 W_3\right], \quad \beta_1 = 2J^{-1} W_1, \quad \beta_{-1} = -2JW_2, \tag{22}$$

and $W_1 \equiv \partial W / \partial I_1$, $W_2 \equiv \partial W / \partial I_2$, $W_3 \equiv \partial W / \partial I_3$.

If the material is incompressible then the only admissible deformations are isochoric, i.e. $J = 1$ at all times, so that $I_3 \equiv 1$, and $W = W(I_1, I_2)$ only. In this case we have that

$$\boldsymbol{T} = -p\boldsymbol{I} + 2W_1 \boldsymbol{B} - 2W_2 \boldsymbol{B}^{-1}, \tag{23}$$

where p is the as yet undetermined Lagrange multiplier introduced by the constraint of incompressibility.

So far no restrictions have been imposed on the nature of the strain-energy function besides those arising from frame indifference and material symmetry. Here we list some general requirements.

- *Normalization* conditions. We require W (or $\hat{W}$) to vanish in the reference configuration:

$$W(3,3,1) = 0, \quad \text{or} \quad \hat{W}(1,1,1) = 0. \tag{24}$$

  Moreover, if the reference configuration is stress free then, in terms of derivatives with respect the stretches, we must have

$$\frac{\partial \hat{W}}{\partial \lambda_i}(1,1,1) = 0, \quad i = 1,2,3. \tag{25}$$

- *Compatibility with the classical linear theory of elasticity.* If μ is the infinitesimal shear modulus and κ the bulk modulus, then we have the connections

$$\begin{aligned} \mu &= 2(W_1 + W_2)|_{I_1 = I_2 = 3,\, I_3 = 1}, \\ \frac{\kappa}{4} + \frac{\mu}{3} &= (W_{11} + 2W_{13} + W_{33})\,|_{I_1 = I_2 = 3,\, I_3 = 1}. \end{aligned} \tag{26}$$

- *Constitutive inequalities.* The question of what reasonable restrictions on the form of the strain energy must be imposed is a delicate and central matter sometimes referred to as Truesdell's Hauptproblem and is related to the interplay between, stability, thermodynamics and the mathematical conditions which guarantee the existence and uniqueness of solutions in nonlinear elasticity. It is usual to impose the so-called *empirical inequalities.* These inequalities are, for compressible materials,

$$\beta_0 \leq 0, \quad \beta_1 \geq 0, \quad \beta_{-1} \leq 0$$

  and, for incompressible materials,

$$\beta_1 \geq 0, \quad \beta_{-1} \leq 0.$$

  It is clear that these inequalities are not a definitive solution to Truesdell's Hauptproblem or the determination of conditions on the strain-energy function that guarantee reasonable mechanical behaviour. It is only known that empirical inequalities are useful in some physical situations, because they reflect the intuitive behaviour of materials. Other sets of analytical restrictions have been postulate by many authors in nonlinear elasticity. Among the various possibilities it is possible to cite as examples the Baker– Ericksen (B–E) inequalities and the Coleman–Noll inequality (the C–N condition). At first sight, some of these inequalities are connected with some reasonable expectations, but at a deeper level of investigation it is often found that these requirements may be in contradiction with experimental findings.

- *Existence of solutions.* In hyperelasticity this question is reduced to the existence of a minimum of a suitable functional: the potential energy (strain energy minus work of exterior loads). We point out that in general the strain energy over the set of admissible functions is not convex (usually we require polyconvexity). Moreover, to ensure existence we must have a coerciveness inequality, which means that we require not only an infinite amount of strain energy in order to expand a continuum body to the infinite range, but also that the growth condition must be sufficiently strong. Obviously this is a very hard condition to test experimentally because the elastic range of any material is finite. On the other hand, the strain energy must also tend to infinity as the volume ratio vanishes. When all these conditions are satisfied it is possible to show that for the functional potential there exists a minimizing sequence that converges weakly to a minimum that belongs to the set of admissible functions. From some mathematical results by L.C. Young we know that there are examples of reasonable functionals for which a minimizing sequence does attain a minimum and that this situation may also have a physical interpretation in the appearance of micro-structures. Therefore even non-existence may be an admissible, physically acceptable and desirable behaviour.

All these questions are reported in detail in Beatty (1987), Antman (2005) and Truesdell and Noll (2004), for example. They are not strictly essential to our discussion, because our main criticisms of the literature seem to be independent and robust with respect such inequalities or qualitative properties.

3.1 Weakly Nonlinear Theory

Before considering the full nonlinear theory of elasticity we shall review briefly what it is referred to as the *weakly nonlinear* theory of elasticity.

It is well known that the linear theory of elasticity is based on two assumptions. First a linear relationship between stress and strain (or a quadratic strain-energy density function), then a linear strain measure. To see this in detail let us consider

$$\boldsymbol{x} = \boldsymbol{X} + \varepsilon \boldsymbol{u}(\boldsymbol{X}, t),$$

where $\varepsilon \boldsymbol{u}$ is the displacement and ε is a *small* parameter. At order $\mathcal{O}(\varepsilon)$

$$\boldsymbol{B} \approx \boldsymbol{C} \approx \boldsymbol{I} + \varepsilon \left[\nabla_{\boldsymbol{X}} \boldsymbol{u} + (\nabla_{\boldsymbol{X}} \boldsymbol{u})^T\right] + \dots,$$

and therefore we introduce the linear strain measure

$$\epsilon = \frac{1}{2}\varepsilon \left[\nabla_{\boldsymbol{X}} \boldsymbol{u} + (\nabla_{\boldsymbol{X}} \boldsymbol{u})^T\right].$$

Then,

$$\boldsymbol{B}^{-1} \approx \boldsymbol{I} - \varepsilon \left[\nabla_{\boldsymbol{X}} \boldsymbol{u} + (\nabla_{\boldsymbol{X}} \boldsymbol{u})^T\right] + \dots,$$

and from (21) we obtain

$$\boldsymbol{T} \approx (\beta_0 + \beta_1 + \beta_{-1})\boldsymbol{I} + (\beta_1 - \beta_{-1})\varepsilon \left[\nabla_{\boldsymbol{X}} \boldsymbol{u} + (\nabla_{\boldsymbol{X}} \boldsymbol{u})^T\right] + \dots, \qquad (27)$$

where

$$\beta_0 + \beta_1 + \beta_{-1} = 2J^{-1}\left[I_2 W_2 + I_3 W_3\right] + 2J^{-1}W_1 - 2JW_2$$

must be approximated at order $\mathcal{O}(\varepsilon)$, whereas

$$\beta_1 - \beta_{-1} = 2J^{-1}W_1 + 2JW_2$$

must be approximated at the order zero.

In so doing we obtain the classical formula

$$\boldsymbol{T} = \lambda(\mathrm{tr}\,\boldsymbol{\epsilon})\boldsymbol{I} + 2\mu\boldsymbol{\epsilon},$$

where λ and μ are the Lamé coefficients[8]. This constitutive equation is associated with the quadratic strain-energy density

$$W = \mu \mathrm{tr}\,(\boldsymbol{\epsilon}^2) + \frac{1}{2}\lambda[\mathrm{tr}\,(\boldsymbol{\epsilon})]^2 + \dots .$$

It is clear that it is possible to push forward this approximation, for example to the third order, to obtain

$$W = \frac{\lambda + 2\mu}{8} J_1^2 + \frac{\mu}{2} J_2 + \frac{l + 2m}{24} J_1^3 + \frac{m}{4} J_1 J_2 + \frac{n}{8} J_3, \tag{28}$$

where l, m, n are the Murnaghan third-order constants (Murnaghan, 1937). Here, J_1, J_2, J_3 constitute another set of independent invariants, related to the principal deformation invariants through

$$J_1 = I_1 - 3, \quad J_2 = 2I_1 - I_2 - 3, \quad J_3 = I_3 - I_2 + I_1 - 1. \tag{29}$$

It is clear that in (28) we consider only the terms up to $\mathcal{O}(\varepsilon^3)$.

In the next pages we shall focus our interest on the incompressible case. In this framework it is meaningful to consider the fourth-order weakly nonlinear theory of elasticity characterized by the strain-energy density

$$W = \mu \mathrm{tr}\,(\boldsymbol{E}^2) + \frac{\mathcal{A}}{3}\mathrm{tr}\,(\boldsymbol{E}^3) + \mathcal{D}[\mathrm{tr}\,(\boldsymbol{E})]^2, \tag{30}$$

where $\boldsymbol{E} = (\boldsymbol{F}^T\boldsymbol{F} - \boldsymbol{I})/2$ is the Green strain tensor and we are using another set of invariants. Here μ is the infinitesimal shear modulus (the only constitutive parameter in the incompressible linear theory) and $\mathcal{A}$, $\mathcal{D}$ are the third-, and fourth-order elasticity constants.

Equation (30) may be related to the full nonlinear strain-energy density

$$W(I_1, I_2) = C_{10}(I_1 - 3) + C_{20}(I_1 - 3)^2 + E(I_2 - 3), \tag{31}$$

which is equivalent, up to the fourth order in the amplitude of the displacement, to (30) if

$$\mu = 2(C_{10} + E), \quad \mathcal{A} = -8(C_{10} + 2E), \quad \mathcal{D} = 2(C_{10} + 3E + 2C_{20}).$$

[8]We notice that $\kappa = \lambda + 2\mu/3$.

Once again in (30) only terms up to $\mathcal{O}(\varepsilon^4)$ are included.

Let us consider the class of deformations[9]

$$x = (1+\varepsilon\lambda_1)X + \varepsilon k_1 Z, \quad y = (1+\varepsilon\lambda_2)Y + \varepsilon k_2 Z, \quad z = (1+\varepsilon\lambda_3)Z,$$

where to ensure isochoricity we requires $(1+\varepsilon\lambda_1)(1+\varepsilon\lambda_2)(1+\varepsilon\lambda_3) = 1$. If we restrict our attention to (31) we have to consider

$$\boldsymbol{T} = -p\boldsymbol{I} + 2[C_{10} + 2C_{20}(I_3 - 1)]\boldsymbol{B} - 2E\boldsymbol{B}^{-1}.$$

With some simple computations it is possible to write down, for example for the shear stress T_{13}, the following expression

$$\begin{aligned} T_{13} = {} & \mu k_1\varepsilon - 2[2\mu\lambda_1 + (6\mu + \mathcal{A})\lambda_2 +]k_1\varepsilon^2 \\ & + [4(5\lambda_1^2 + 7\lambda_1\lambda_2 + 6\lambda_2^2 + k_1^2 + k_2^2)\mu \\ & + (8\lambda_1^2 + 10\lambda_1\lambda_2 + 9\lambda_2^2 + 2k_1^2 + 2k_2^2)\mathcal{A} \\ & + 4(4\lambda_1^2 + 4\lambda_1\lambda_2 + 4\lambda_2^2 + k_1^2 + k_2^2)\mathcal{D}]k_1\varepsilon^3 + \dots . \end{aligned} \tag{32}$$

This formula is important because by using (32) we show that in the linear theory (order $\mathcal{O}(\epsilon)$) we have no interaction at all between various components of the displacement, as we know very well. Using third-order elasticity (i.e. $\mathcal{O}(\epsilon^2)$) we have only the interaction between the shear component k_1 and the axial components. To have interactions between k_1 and k_2 we have to go to order $\mathcal{O}(\epsilon^3)$, i.e. at fourth-order elasticity. Therefore, to have a weakly nonlinear theory which considers all the possible interactions we have to consider (30) and this means that at least *three* constitutive parameters are needed to have full generality. Indeed, if we consider a full nonlinear incompressible constitutive equation $W(I_1, I_2)$ with only two constitutive parameters when we recover from this general model the fourth-order approximation. We introduce a link between $\mu, \mathcal{A}$ and $\mathcal{D}$, but there is no *a priori* reason for such a link.

3.2 The Kinetic Theory

In the previous sections we were able to derive some models for the elastic free energy of a single (idealized) chain. To obtain the free energy for a macroscopic material we have to consider that amorphous materials are indeed composed of a network of ideal chains.

A network is obtained by permanently linking together some chains in a three-dimensional structure. This structure may be described in several

---

[9] The balance equation of linear momentum, *in the absence of body forces*, is $\operatorname{div}\boldsymbol{T} = \boldsymbol{0}$. This equation is always satisfied if we consider homogeneous deformations.

ways. A simple way is to consider the number of junctions, say n_1, i.e. the points where the different chains are chemically and permanently joined together, the functionality, f_u, of the junction, i.e. the number of chains that meet at a junction (this is an average) and the number of chains, n_2, that should be cut to reduce a network to a tree (a tree is a structure that contains no closed cycles).

In the elementary molecular theory of networks the basic postulate is that the free energy of the network is just equal to the sum of the elastic free energies of the individual chains. Then it is assumed that the free energy of a single chain (in the isothermal case) is just given by

$$\psi = \frac{3kT}{2 < r^2 >_0} r^2,$$

where $< r^2 >_0$ is the mean-square end-to-end distance. This is just a Gaussian approximation for the free energy of the single chain. Then we suppose that

$$\Psi = \frac{3kT}{2 < r^2 >_0} \sum_{i=1}^{N} r_i^2,$$

for a network composed by N equal chains. Considering the normalization

$$\Psi - \Psi_0 = \frac{3kT}{2 < r^2 >_0} \sum_{i=1}^{N} \left(r_i^2 - r_{0,i}^2 \right),$$

and noting that $< r^2 >= \sum_{i=1}^{N} r_i^2/N$, we have

$$\Psi - \Psi_0 = \frac{3NkT}{2} \left(\frac{< r^2 >}{< r_0^2 >} - 1 \right).$$

Using a very nice result of Kearsley (1989) we arrive at

$$\Psi - \Psi_0 = \frac{NkT}{2} \left(I_1 - 3 \right). \tag{33}$$

The macroscopic strain-energy density we have derived in such a way is clearly the celebrated neo-Hookean model.

The example we have just considered is a very simple one, but the steps are always the same. To summarize

- Make the assumption of entropic elasticity. A first phenomenological assumption.

- Choose a formula for the end-to-end distance of an ideal chain and compute the free energy of a single chain. A second phenomenological assumption.
- Idealize the network. This can be done in several ways but this is a basic assumption because from this choice we are able to link the macro-deformation to the micro-deformation (Cauchy–Born rule). Another phenomenological step.
- Choose the averaging procedure that enables aggregation of single chains into a full network. A last phenomenological step.

All these steps are full of *ad hoc* assumptions and at the end we obtain in any case an expression that may be contained in the full representation formula derived by using frame indifference and material symmetry in continuum mechanics. In reality, usually we obtain a subclass of materials because it is not easy to link the I_2 invariant to microscopic quantities. Thus, we obtain a strain energy in the form $W = W(I_1)$ (generalized neo-Hookean materials) a functional form that for theoretical reasons seems highly unsuitable for producing good models of experimental data.

All the models derived by studying a representative unit cell of linked chains whose ends move as if embedded in a homogeneously deforming elastic continuum (i.e. using the affine deformation assumption as the Cauchy–Born rule), as in the Arruda–Boyce (1993) and Wang–Guth (1952) models, suffer all of the above problems. Therefore, they are phenomenological, just as are the models considered in the classical continuum mechanics approach. The fact that they use a more refined model of the single chain is not enough to take into account all the regimes of deformation, and the fact that they are generalized neo-Hookean strain energies is a major obstruction to obtaining a good fit to experimental data.

4 Invariants

Some words of caution are necessary before starting to work with experimental data. Indeed, the *principal invariants* are only a possible set of invariants that may be used to express the strain energy of an isotropic material. There is an infinity of such choices but all choices are clearly equivalent. The equivalence is ensured by the fact that all sets of invariants are in one-to-one correspondence with the principal stretches $\lambda_1, \lambda_2, \lambda_3$. For a general survey about invariants in nonlinear elasticity we refer to Steigmann (2002). It is clear that in correspondence with any choice there is a different, but equivalent, representation formula for the Cauchy stress tensor (or any other stress measure).

The equivalence among the various independent sets of invariants means that for every property of the strain energy in a given set a corresponding property can be found in any other set via a suitable transformation. The last paper by Rivlin (2006) is about necessary and sufficient conditions for the strain-energy function of an isotropic elastic solid, regarded as a function of the principal deformation invariants, to be expressible in the Valanis–Landel (1967) form

$$W(I_1, I_2, I_3) = w(\lambda_1) + w(\lambda_2) + w(\lambda_3). \tag{34}$$

The contents of the paper by Rivlin are long and complex algebraic computations showing a trivial fact, but still today there are authors ignoring this equivalence and thinking that by considering a new set of invariants a completely different new theory is around the corner. The choice of invariants is only a matter of taste and convenience.

To show where the difference is in using different sets of invariants we analyze a triaxial state of deformation. To this end we consider a homogeneous deformation from a reference configuration where we have introduced the Cartesian coordinates X_i, $i = 1, 2, 3$, and vector basis $\boldsymbol{E}_i$, $i = 1, 2, 3$, into the actual configuration where we have introduced the Cartesian coordinates x_i, $i = 1, 2, 3$, and vector basis $\boldsymbol{e}_i$, $i = 1, 2, 3$. The deformation we are considering is described by

$$x_1 = \lambda_1 X_1, \quad x_2 = \lambda_2 X_2, \quad x_3 = \lambda_3 X_3. \tag{35}$$

Therefore, the $\boldsymbol{E}_i$ $(= \boldsymbol{e}_i)$, are the eigenvectors of the polar decomposition of the deformation gradient and the λ_i are the corresponding eigenvalues. If we assume the material to be incompressible $\lambda_1\lambda_2\lambda_3 = 1$. All the geometrical quantities associated with (35) are very simply computed. For example, the components of the left Cauchy–Green tensor are simply given by $(B_{ij}) = \mathrm{diag}(\lambda_1^2, \lambda_2^2, \lambda_3^2)$.

It is well known that for an isotropic material stress and strain must be coaxial, i.e. $\boldsymbol{TB} = \boldsymbol{BT}$, and therefore the eigenvectors of the stress tensor must be the $\boldsymbol{E}_i$. We denote the corresponding stress eigenvalues by t_1, t_2, t_3. This result is independent of the representation formula.

If we have a biaxial machine we can consider the stretching of a rectangular sheet of thickness H, in the undeformed state, by forces in its plane. The major faces of the sheet are free of traction and we denote by f_1 and f_2 the forces per unit length on the edges, measured in the undeformed state, imposed to stretch the sheet. This means that we have

$$t_1 = \lambda_1 f_1/H, \quad t_2 = \lambda_2 f_2/H, \quad t_3 = 0. \tag{36}$$

Now we choose a particular set of invariants and a representation formula. Let us start with the principal invariants and (23). To simplify the algebra involved in this discussion we can restrict our attention to incompressible materials. Our goal is to determine, from experimental data, the functional form of the strain energy $W = W(I_1, I_2)$. This means that if we idealize the deformation in the biaxial machine as a homogeneous deformation we must have

$$t_i = -p + 2W_1\lambda_i^2 - 2W_2\lambda_i^{-2}, \quad i = 1, 2, 3. \tag{37}$$

Because $t_3 = 0$ and $\lambda_1\lambda_2\lambda_3 = 1$ then

$$p = 2W_1\lambda_1^{-2}\lambda_2^{-2} - 2W_2\lambda_1^2\lambda_2^2.$$

Therefore, we may compute that

$$W_1 = \mathcal{F}_1(\lambda_1, \lambda_2; f_1, f_2), \quad W_2 = \mathcal{F}_2(\lambda_1, \lambda_2; f_1, f_2), \tag{38}$$

where

$$\begin{aligned} \mathcal{F}_1 &= \frac{1}{2H(\lambda_1^2 - \lambda_2^2)}\left\{\frac{\lambda_1^3 f_1}{\lambda_1^2 - \lambda_1^{-2}\lambda_2^{-2}} - \frac{\lambda_2^3 f_2}{\lambda_2^2 - \lambda_1^{-2}\lambda_2^{-2}}\right\}, \\ \mathcal{F}_2 &= \frac{1}{2H(\lambda_2^2 - \lambda_1^2)}\left\{\frac{\lambda_1 f_1}{\lambda_1^2 - \lambda_1^{-2}\lambda_2^{-2}} - \frac{\lambda_2 f_2}{\lambda_2^2 - \lambda_1^{-2}\lambda_2^{-2}}\right\}. \end{aligned} \tag{39}$$

From this simple computation we discover two important features of the biaxial (triaxial in the compressible case) experiment.

First, from any experiment it is possible to have measurements corresponding to the derivatives of the strain energy and we cannot infer directly the strain-energy density function.

Second, the experiment under scrutiny allows us to completely characterize the derivatives of the strain energy. Indeed, theoretically speaking we say that we know a function $y = f(x)$ in a certain domain $\mathcal{D}$ if we know the value y for any value $x \in \mathcal{D}$. Because we are working with experiments our knowledge is only an approximate knowledge. In the real world it is possible to sample $\mathcal{D}$ only in a finite and discrete way. In spite of this, the deformation we are considering virtually allows us to explore all the elements of $\mathcal{D}$. This is in contrast with a simple tension experiment where $t_2 = t_3 = 0$ and therefore $\lambda_1 = \lambda$, $\lambda_2 = \lambda_3 = 1/\sqrt{\lambda}$. In this case it is possible to investigate the domain of definition of the derivatives only in a subset, no matter how sophisticated are our testing machines. This is an *a priori* limitation.

If we perform a standard experiment, via our biaxial machine, we can collect certain information. For any choice $(f_{1,*}, f_{2,*})$ we measure the corresponding deformations $(\lambda_{1,*}, \lambda_{2,*})$. It is possible to perform these measurements N times and therefore to collect N sets of experimental data

$$\{(f_{1,1}, f_{2,1}, \lambda_{1,1}, \lambda_{2,2}), \ldots, (f_{1,N}, f_{2,N}, \lambda_{1,N}, \lambda_{2,N})\}.$$

In such a way it is possible to compute the corresponding values **numerically** of $\mathcal{F}_1$ and $\mathcal{F}_2$ for each measurement and then to evaluate W_1 and W_2. We point out that when evaluated $\mathcal{F}_1$ and $\mathcal{F}_2$ are just numbers.

Equations (38) are very powerful relationships. For example, if we wish to test experimentally if it is possible to represent the strain energy of a given material as $W = W(I_1)$ (i.e. our real material may be idealized as a generalized neo-Hookean material) we have to check if

$$\mathcal{F}_2(f_{1,i}, f_{2,i}, \lambda_{1,i}, \lambda_{2,i}) \approx 0, \quad \forall i = 1, \ldots, N. \tag{40}$$

This is a *universal* check on the validity of the generalized neo-Hookean assumption. This means that we have provided a way to test the assumption which does not depend on the special choice of the functional form $W(I_1)$. Clearly this check is always neglected in papers that suggest generalized neo-Hookean models.

On the other hand, we can use the same data to determine the parameters of an assumed functional form. For example, if we assume the Mooney–Rivlin model

$$W(I_1, I_2) = C_{10}(I_1 - 3) + C_{01}(I_2 - 3) \tag{41}$$

we have to minimize the expressions

$$\sum_{i=1}^{N} [C_{10} - 2\mathcal{F}_1(f_{1,i}, f_{2,i}, \lambda_{1,i}, \lambda_{2,i})]^2, \quad \sum_{i=1}^{N} [C_{02} - 2\mathcal{F}_2(f_{1,i}, f_{2,i}, \lambda_{1,i}, \lambda_{2,i})]^2.$$

Now, the question is: what happens if we change the set of invariants?

Let us imagine we are going from the principal invariants I_1 and I_2 to a new set of invariants i_1 and i_2. Then we have to move from $W(I_1, I_2)$ to $\tilde{W}(i_1, i_2)$ and this will change the representation formula (23). Using the new representation formula our problem is once again to measure the derivatives $\partial\tilde{W}/\partial i_1$ and $\partial\tilde{W}/\partial i_2$. In the framework of the experiment we are considering it is necessary only to perform the simple computation

$$\frac{\partial\tilde{W}}{\partial i_1}\frac{\partial i_1}{\partial I_1} + \frac{\partial\tilde{W}}{\partial i_2}\frac{\partial i_2}{\partial I_1} = \mathcal{F}_1, \quad \frac{\partial\tilde{W}}{\partial i_1}\frac{\partial i_1}{\partial I_2} + \frac{\partial\tilde{W}}{\partial i_2}\frac{\partial i_2}{\partial I_2} = \mathcal{F}_2. \tag{42}$$

Using Cramer's rule and denoting by $\mathcal{J}$ the Jacobian of the transformation (which we assume one-to-one) we obtain

$$\begin{bmatrix} \tilde{W}_1 \\ \tilde{W}_2 \end{bmatrix} = \mathcal{J}^{-1} \begin{bmatrix} \mathcal{F}_1 \\ \mathcal{F}_2 \end{bmatrix}.$$

Therefore, the problems we have to face are always the same. It does not matter which set of invariants we choose. The only advantage of using a new set of invariants may be a sort of preconditioning (if we need it) or, because $\mathcal{J}$ may be non uniform, a magnification of some ranges of data.

4.1 The Mooney Plot

The simple tension experiment is a simple and classical experiment of rubber mechanics. Let us consider only the incompressible case and let's apply a tensile force f to a rubber specimen. In this case, the principal Cauchy stresses are $t_1 \neq 0$ and $t_2 = t_3 \equiv 0$. It may be shown that in this case, idealizing the deformation field as homogeneous, we must have

$$\lambda_1 = \lambda, \quad \lambda_2 = \lambda_3 = 1/\sqrt{\lambda}.$$

Therefore since $f = t_1/\lambda$ from (23) we obtain

$$f = 2\left(\lambda - \frac{1}{\lambda^2}\right)\left(W_1 + \frac{1}{\lambda}W_2\right).$$

This equation may be rewritten as

$$W_1 + zW_2 = \mathcal{F}(f, z), \tag{43}$$

where $z = 1/\lambda$. The plot of the rescaled tensile forces $\mathcal{F}$ versus z is referred to as the Mooney plot.

It is clear that equation (43) is convenient when we consider the Mooney–Rivlin strain energy, because in this case (43) is given by

$$\mu_1 + z\mu_2 \equiv 2\mathcal{F}$$

and can we determine the constitutive parameters μ_i using a linear regression in the $(\mathcal{F}, z)$-plane.

The reason why the Mooney plot is convenient is clear. If you have experimental data spanning, as in the classical Treloar experiments (Treloar, 1944), the λ range $[1, 8]$ using the physical data when $\lambda \in [1, 2]$ (moderate finite deformations) then the influence of this range of variation is not as important as in Mooney plot. Indeed, the rescaling used in the Mooney plot ensures that the data in the range $\lambda \in [1, 2]$ will count 50%, but much less in the physical plane. Clearly, the same results may be obtained by using a weighted regression analysis in fitting the data.

4.2 Universal Relations

A universal relation is an equation that holds for every material in a specified class. For compressible and incompressible isotropic materials, universal relations are mainly generated by the coaxiality of strain and stress and they were first discovered by Rivlin in the late 1940s for the simple shear deformation and subsequently examined in detail for a huge class of deformations by Beatty (1987). Pucci and Saccomandi have found a general method to determine and classify universal relations, called the *universal manifold* method, which works not only with isotropic materials.

Therefore, for a general isotropic material we have three scalar universal relations given by the components of the skew-tensor $\boldsymbol{TB} - \boldsymbol{BT}$. This tensor, because of the coaxiality condition, must be a null tensor. For special subclasses of materials, such as generalized neo-Hookean materials, additional universal relations can arise. Equation (40) is an example of such an additional universal relation. For triaxial stretch the universal relations associated with coaxiality are trivial ($T_{12} = T_{13} = T_{23} \equiv 0$).

Universal relations are fundamental tools for discriminating between possible *a priori* assumptions in the modelling activity. Let us consider for example, that in the framework of an incompressible elastic material, we wish to postulate that $W = W(I_1)$, i.e. the real material is assumed to be a generalized neo-Hookean material. In this case the Cauchy stress is defined as

$$\boldsymbol{T} = -p\boldsymbol{I} + 2W_1\boldsymbol{B}.$$

Now, let us consider an arbitrary vector $\boldsymbol{u}$ and let us compute the steps

$$\boldsymbol{Tu} = -p\boldsymbol{u} + 2W_1\boldsymbol{Bu} \rightarrow \boldsymbol{Tu} \times \boldsymbol{u} = 2W_1\boldsymbol{Bu} \times \boldsymbol{u} \rightarrow (\boldsymbol{Tu} \times \boldsymbol{u}) \cdot \boldsymbol{u} = 0.$$

The last relation is indeed independent of $\boldsymbol{u}$ and it is a universal relation peculiar to generalized neo-Hookean materials. It is not dependent on the particular functional form of the strain energy density $W = W(I_1)$.

Let us consider the simple shear deformation

$$x = X + \kappa Y, \quad y = Y, \quad z = Z.$$

This is probably the simplest example of a finite deformation although it is a deformation which is difficult to produce experimentally because of the surface tractions required. In this case

$$T_{11} = -p + 2(1+\kappa^2)W_1 - 2W_2, \quad T_{12} = 2\kappa(W_1 + W_2), \quad T_{13} = T_{23} = 0,$$
$$T_{22} = -p + 2W_1 - 2(1+\kappa^2)W_2, \quad T_{33} = -p + 2W_1 - 2W_2. \tag{44}$$

Here there are three universal relations. The trivial ones $T_{13} = T_{23} = 0$ together with

$$T_{11} - T_{22} = \kappa T_{12}. \tag{45}$$

If one finds experimentally that (45) is not satisfied then one may conclude that the material under investigation is not an isotropic Cauchy elastic material.

If the specimen of material under investigation is a block and we consider that the faces of this block normal to the $Z-$direction are traction free we impose $T_{33} = 0$, i.e.

$$p = 2W_1 - 2W_2.$$

Then

$$\begin{aligned} T_{11} &= 2\kappa^2 W_1, \quad T_{12} = 2\kappa(W_1 + W_2), \\ T_{22} &= -2\kappa^2 W_2, \quad T_{13} = T_{23} = T_{33} = 0. \end{aligned}$$

Now we have three trivial universal relations $T_{13} = T_{23} = T_{33} = 0$ plus (45).

The next step is to suppose that $W = W(I_1)$, assuming that the material may be modelled as a generalized neo-Hookean material. Then we obtain

$$\begin{aligned} T_{11} &= 2\kappa^2 W_1, \quad T_{12} = 2\kappa W_1, \\ T_{22} &= T_{13} = T_{23} = T_{33} = 0. \end{aligned}$$

This means that a new trivial universal relations $T_{22} = 0$ arises and this new universal relations is peculiar to generalized neo-Hookean materials. Therefore, if one finds experimentally that (45) is satisfied but the $T_{22} \neq 0$ then one may conclude that the material under investigation is an isotropic Cauchy elastic material but not a generalized neo-Hookean material.

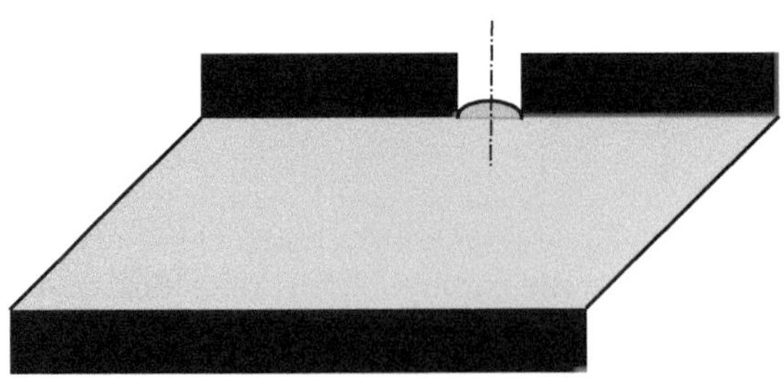

Figure 1. An example of a device for a simple shear experiment with a hole on the top surface to detect in a simple and direct way the normal stress T_{22} via a little hole in the top plate (courtesy of Michel Destrade).

This suggests a very simple experimental device for checking if a material may be represented as a generalized neo-Hookean material. It is sufficient to drill a little hole in the top plate of a simple shear device (as depicted in Figure 1) and to observe if the material bumps into the hole when sheared.

5 Some Experimental Data

The goal of a mathematical model is to describe experimental results and predict real world behaviour. To see the problem we may encounter in this direction we consider some experimental rubber data. We consider Treloar's data from his classical 1944 simple tension experiment (see Treloar, 1944, 2005, for details) and two sets of more recent data analyzed in Carrillo et al. (2010)[10]. We denote the two sets of Carrillo's data by DC9 and DC11.

The format of all the figures we use to compare data with the mechanical predictions of the models is the following one. In the first frame we consider the model in physical space (tensile force versus stretch), in the second frame the Mooney plot and the last frame is dedicated to the relative errors.

We start with a contrast of the experimental data with the neo-Hookean model. For the neo-Hookean model the Mooney plot is quite simple since $2\mathcal{F}(f, z) = \mu$ and the experimental data clearly contradict this finding in any range of deformation. To show this point we provide in Figure 2, the fitting using the neo-Hookean model with both the data of Treloar and the DC11 data.

In Rivlin's mind the bad neo-Hookean performance had to be improved by considering $W(I_1) = \mu(I_1 - 3) + f(I_2)$. A special case of this is the Mooney–Rivlin material, for which

$$2W = \mu \left[\left(\frac{1}{2} + \beta \right) (I_1 - 3) + \left(\frac{1}{2} - \beta \right) (I_2 - 3) \right],$$

where β is a dimensionless constant in $[-1/2, 1/2]$ and μ is the infinitesimal shear modulus[11].

The Mooney–Rivlin form is not the only possibility for $f(I_2)$. Other possibilities are provided by the Gent–Thomas term $f(I_2) = C_2 \log(I_2/3)$ (Gent and Thomas, 1958) or the Carroll (2009) term, for which $f(I_2) = C_2\sqrt{I_2}$. In both models C_2 is a constant.

[10] We thank Dr. Carrillo for providing us with the tabulated version of the data he uses in his paper and Professor Ogden for providing us with the tabulated version of the original Treloar data.

[11] Please note a difference of notation with respect to (41) where the constants C_{10} and C_{01} were used.

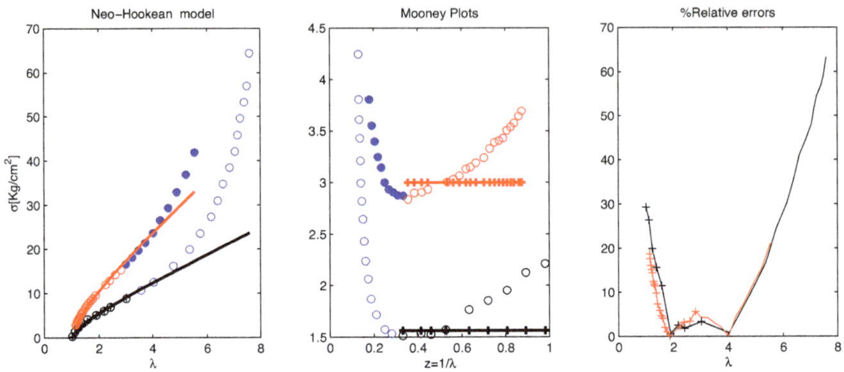

Figure 2. Result of the fitting with the neo-Hookean model of the original Treloar's data (black curve) and of DC11 (red curve).

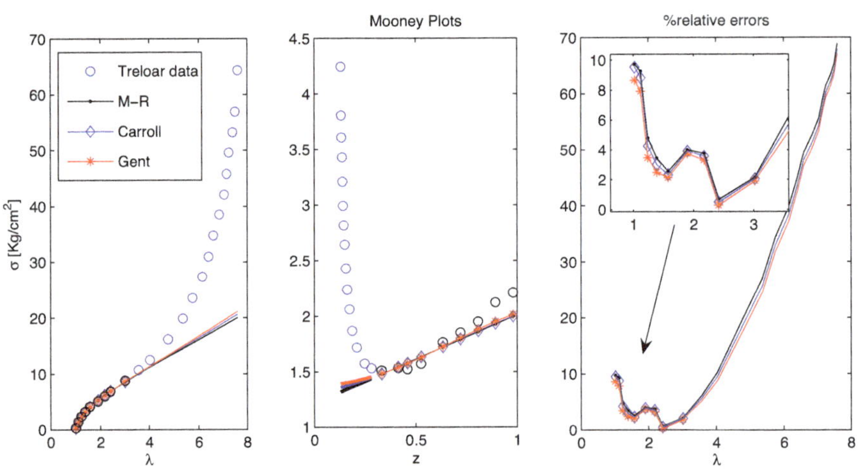

Figure 3. Treloar's data: result of the fitting with I_2 corrections by Mooney-Rivlin, Carroll and Gent-Thomas models.

The introduction of the $f(I_2)$ term improves in a substantial way the ability of the model to describe the experimental data, at least for a moderate range of deformations. On the other hand, we know that from the use of the universal relation it is simple to check if the term $f(I_2)$ is necessary to model the experimental data. From Figures 3, 4 and 5 it is possible to

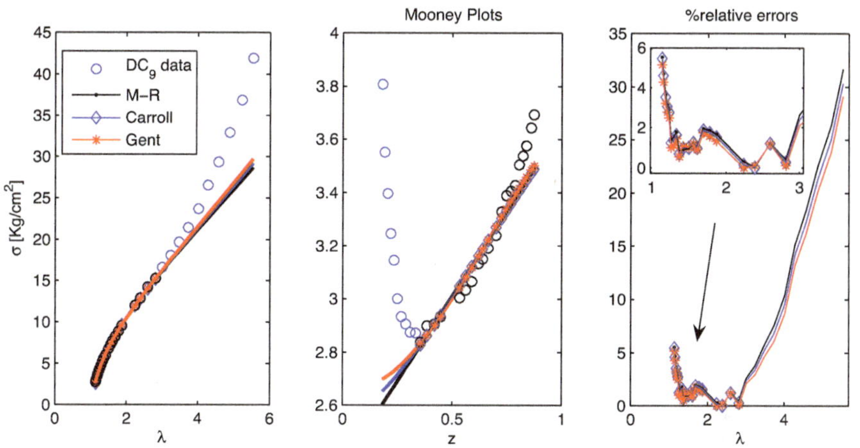

Figure 4. DC9 data: result of the fitting with I_2 corrections by Mooney–Rivlin, Carroll and Gent–Thomas models.

appreciate that all the proposed $f(I_2)$ terms are nearly equivalent in the sense that they allow a description of the data in a range that is bounded by the upturn of the Mooney plot.

Each of the models we have considered until now contains *only* two constitutive parameters. Our analysis of the weakly nonlinear theories points out that only a fourth-order weakly nonlinear theory may ensure a complete and general mechanical behaviour and fourth-order theories need three independent constitutive parameters. For this reason at least another constitutive term is missing in the previous models. An additional term may be suggested by the Pincus analysis in the form

$$\gamma(I_1 - 3)^n, \tag{46}$$

where γ is a parameter to be fixed whereas n is suggested by the Pincus scaling analysis as $n = 1.5$.

In so doing we are able to extend further the ability of the various models to describe in a careful way the data after the upturn in the Mooney plot. Once again adding (46) to any of the previous Rivlin $f(I_2)$ models the fitting results are very similar. The results of the fitting of the Mooney–Rivlin, Carroll and Gent–Thomas models with the Pincus correction are shown in Figures 6, 7 and 8 for the Treloar, DC9 and DC11 data, respectively.

The last step is to extend our model to very large deformation. First of all let us say something about the physics of what is going on at large

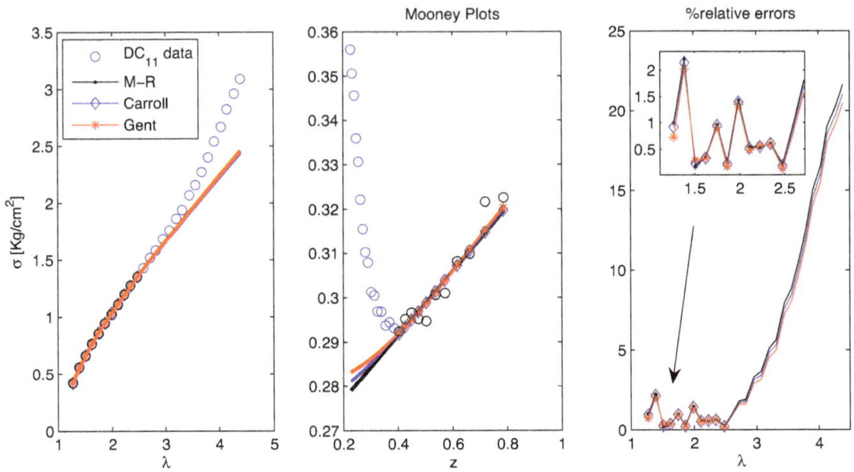

Figure 5. DC11 data: result of the fitting with I_2 corrections by Mooney–Rivlin, Carroll and Gent–Thomas models.

extensions in rubber-like materials. It is well known that polymers may be *tactic* or *atactic*, a question of stereochemistry. Strain crystallization occurs in tactic polymers (like natural rubber) and the last part of the curve representing experimental data is determined mainly by this phenomenon. To model crystallization we need to consider new *internal* variables able to represent a change in the *material* configuration. On the other hand, for atactic polymers to describe *extreme* deformations we need to incorporate the contour length of the macromolecular chains.

Several models that introduce the contour length have been proposed in statistical mechanics, the FJC and WLC models of a single chain being two examples. These models are characterized by different orders of *singularity* as the chain approaches its contour length. Here the problem is to understand if there is a way to infer from experimental data information about the order of such singularities.

To this end consider the last n_0 points of the force-extension data, that is those for high stretch values. In order to estimate how fast the data approach the vertical asymptote of the limiting chain effect, let us assume that the response to the tensile force is described by the relation

$$y = \frac{c}{(1-z)^k}, \quad z = \frac{\lambda}{\lambda_s}, \tag{47}$$

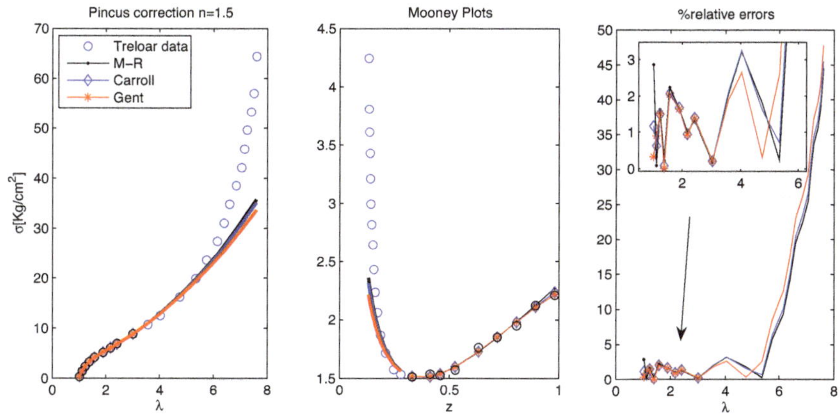

Figure 6. Treloar's data: result of the fitting with the Pincus correction with $n = 1.5$ to the Mooney–Rivlin, Carroll and Gent–Thomas models.

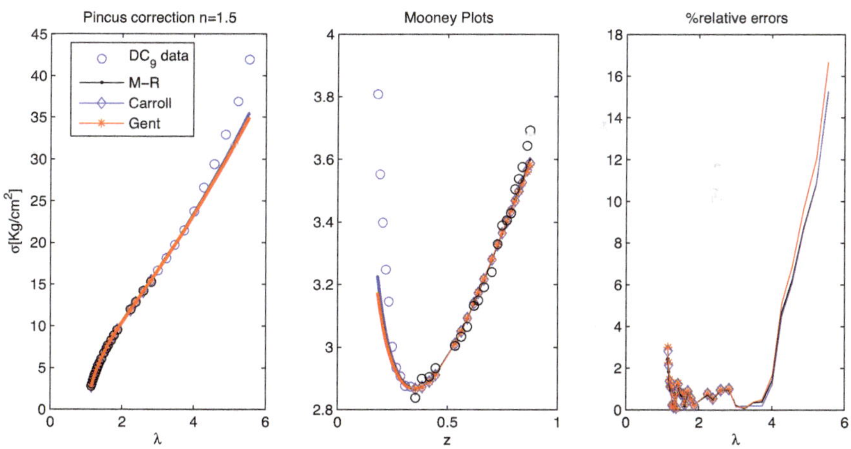

Figure 7. DC9 data: result of the fitting with the Pincus correction with $n = 1.5$ to the Mooney–Rivlin, Carroll and Gent–Thomas models.

where $\lambda = \lambda_s$ indicates the location of the unknown asymptote. In logarithmic coordinates we have

$$Y = \log_{10}(y) = \log_{10}(c) - k \log_{10}(1 - z) = q + mX, \tag{48}$$

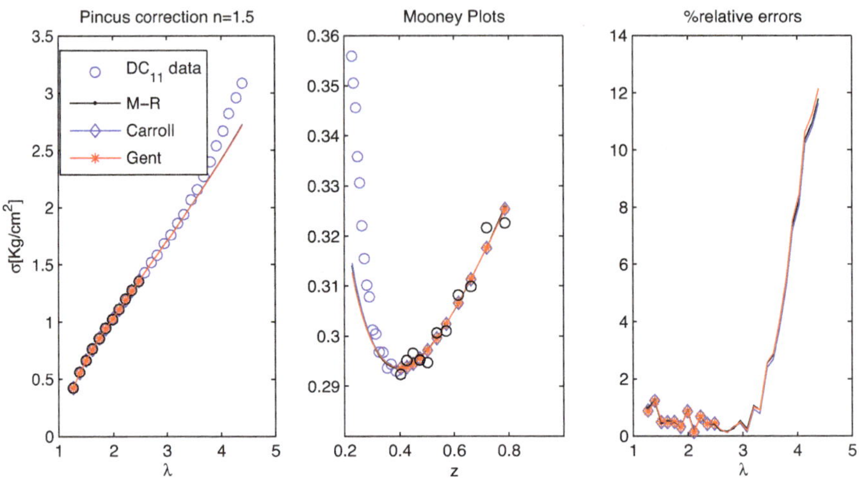

Figure 8. DC11 data: result of the fitting with the Pincus correction with $n = 1.5$ to the Mooney–Rivlin, Carroll and Gent–Thomas models.

where $X = \log_{10}(1 - z)$. This allows us to estimate q and m and then the values of $c = 10^q, k = -m$ by a standard linear fitting. Examples of what is going on is reported in Figures 9 and 10. From these empirical experiments it seems that if the experimental data are spanned over a wide range. The FJC singularity is better than the WLC singularity.

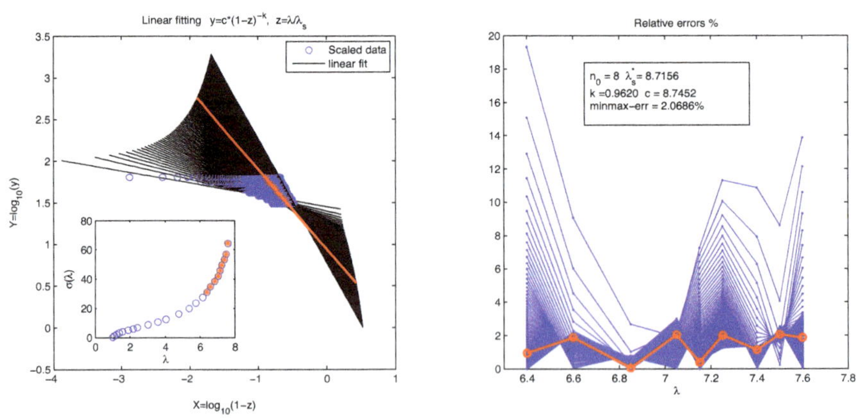

Figure 9. Treloar's data: singularity order estimate when $n_0 = 8$.

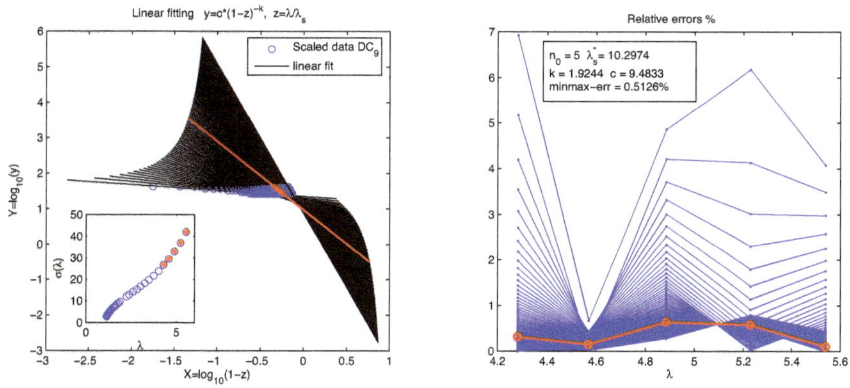

Figure 10. DC9 data: singularity order estimate when $n_0 = 5$.

In Figure 11 we show that if we consider the model

$$W(I_1, I_2) = -\mu J_m \ln\left(1 - \frac{I_1 - 3}{J_m}\right) + f(I_2) + \gamma(I_1 - 3)^n,$$

where the first term corresponds to the Gent model, J_m being a constant, with the different $f(I_2)$ terms we have introduced, the results with Treloar's data are quite good, as is shown by the relative errors.

The conclusions are that a step by step procedure which is inspired by mesoscopic considerations and some general theoretic considerations (for example, universal relations and compatibility with the full fourth-order weakly nonlinear theory) is required. Two features are particularly interesting. First the relative *stability* of the various models we obtain. The term $f(I_2)$ is necessary but, at least in simple extension, it seems that the various form proposed in the literature do not change the details in a drastic way. Second, the possibility of obtaining a reasonable relative error in the whole range of deformation.

6 Anisotropy

Fibre-reinforced materials are usually not isotropic. Therefore, more general classes of models have to be considered. The problem is that also in the simplest framework the introduction of a preferred direction increases the number of invariants. For example, for transverse isotropy, the classical approach formulated by Rivlin and co-workers (see Spencer, 1972) assumes

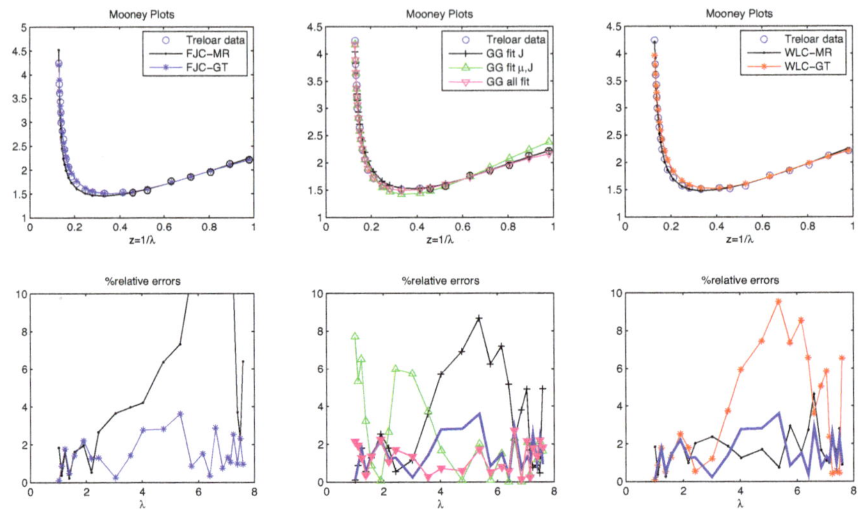

Figure 11. Treloar's data: results of the fitting using all data to identify the parameter J_m, with the other parameters fixed to those obtained by linear fitting using only $N = 9$ points and the Pincus correction with $n = 1.5$.

a strain-energy function of the form

$$W = W\left(I_1, I_2, I_4, I_5\right) \tag{49}$$

for an incompressible material, where

$$I_1 = \operatorname{tr}\boldsymbol{C}, \quad I_2 = \frac{1}{2}\left[I_1^2 - \operatorname{tr}\left(\boldsymbol{C}^2\right)\right], \quad I_4 = \boldsymbol{M}\cdot(\boldsymbol{C}\boldsymbol{M}), \quad I_5 = \boldsymbol{M}\cdot(\boldsymbol{C}^2\boldsymbol{M}), \tag{50}$$

with $\boldsymbol{C}$ denoting the right Cauchy–Green deformation tensor and $\boldsymbol{M}$ denoting the preferred direction of the material in the reference configuration. The incompressibility constraint is given by

$$I_3 = \det \boldsymbol{C} = 1. \tag{51}$$

The corresponding Cauchy stress (in this section the Cauchy stress is denoted $\boldsymbol{\sigma}$) for incompressible, homogeneous, transversely isotropic, nonlinear hyperelastic materials is given by

$$\begin{aligned}\boldsymbol{\sigma} = -p\boldsymbol{I} + 2W_1\boldsymbol{B} + 2W_2\left(I_1\boldsymbol{B} - \boldsymbol{B}^2\right) + 2W_4\boldsymbol{m}\otimes\boldsymbol{m} \\ + 2W_5\left(\boldsymbol{m}\otimes\boldsymbol{B}\boldsymbol{m} + \boldsymbol{B}\boldsymbol{m}\otimes\boldsymbol{m}\right),\end{aligned} \tag{52}$$

where $W = W(I_1, I_2, I_4, I_5)$ is the strain-energy function per unit volume, attached subscripts denoting partial differentiation with respect to the appropriate invariant, $\boldsymbol{m} = \boldsymbol{FM}$, and p is an arbitrary scalar field.

To ensure that the stress is identically zero in the undeformed configuration, it will be required that

$$2W_1^0 + 4W_2^0 = p^0, \quad W_4^0 + 2W_5^0 = 0, \tag{53}$$

where the 0 superscript indicates evaluation for $I_1 = I_2 = 3$, $I_4 = I_5 = 1$, and p^0 is the value of the arbitrary field in the reference configuration. It will also be assumed that the strain energy vanishes in the undeformed configuration, i.e. that

$$W^0 = 0. \tag{54}$$

The preferred direction is usually physically induced by the presence of fibres embedded in an elastic matrix and this will also be assumed here.

Assume now that the strains are infinitesimal. Let $\boldsymbol{u} = \boldsymbol{x} - \boldsymbol{X}$ denote the displacement of a typical particle from the reference configuration. Let $\boldsymbol{H} \equiv \partial\boldsymbol{u}/\partial\boldsymbol{X}$ and $\boldsymbol{\epsilon} \equiv (1/2)\left(\boldsymbol{H} + \boldsymbol{H}^T\right)$. Assume now that

$$\epsilon = \|\boldsymbol{\epsilon}\| << 1,$$

where $\|\cdot\|$ denotes the *sup* norm. Expanding the invariants to the second order in ϵ yields

$$\begin{aligned}
I_1 &= 3 + 2\mathrm{tr}\,\boldsymbol{\epsilon} + \mathrm{tr}\,(\boldsymbol{H}^T\boldsymbol{H}),\\
I_2 &= 3 + 4\mathrm{tr}\,\boldsymbol{\epsilon} + 2\,(\mathrm{tr}\,\boldsymbol{\epsilon})^2 - 2\mathrm{tr}\,\boldsymbol{\epsilon}^2 + 2\mathrm{tr}\,(\boldsymbol{H}^T\boldsymbol{H}),\\
I_3 &= 1 + \mathrm{tr}\,\boldsymbol{\epsilon} + H_{11}H_{22} + H_{11}H_{33} + H_{22}H_{33}\\
&\qquad\qquad - H_{12}H_{21} - H_{13}H_{31} - H_{23}H_{32},\\
I_4 &= 1 + 2\boldsymbol{M}\cdot(\boldsymbol{\epsilon M}) + \boldsymbol{M}\cdot(\boldsymbol{H}^T\boldsymbol{HM}),\\
I_5 &= 1 + 4\boldsymbol{M}\cdot(\boldsymbol{\epsilon M}) + 4\boldsymbol{M}\cdot(\boldsymbol{\epsilon}^2\boldsymbol{M}) + 2\boldsymbol{M}\cdot(\boldsymbol{H}^T\boldsymbol{HM}).
\end{aligned}$$

In the linear theory, the incompressibility constraint (51) is therefore equivalent to

$$\begin{aligned}
&\mathrm{tr}\,\boldsymbol{\epsilon} = 0,\\
&H_{11}H_{22} + H_{11}H_{33} + H_{22}H_{33} - H_{12}H_{21} - H_{13}H_{31} - H_{23}H_{32} = 0,
\end{aligned} \tag{55}$$

from which the identity

$$\mathrm{tr}\,(\boldsymbol{H}^T\boldsymbol{H}) = 2\mathrm{tr}\,(\boldsymbol{\epsilon}^2)$$

follows. Using these results and the conditions (53), (54), expanding the nonlinear strain-energy function (49) to the second order in ϵ yields

$$\begin{aligned} W &= 2\left(W_1^0 + W_2^0\right) \operatorname{tr}(\epsilon^2) + 4W_5^0 \boldsymbol{M} \cdot (\epsilon^2 \boldsymbol{M}) \\ &+ 2\left(W_{44}^0 + 4W_{55}^0 + 4W_{45}^0\right) \left[\boldsymbol{M} \cdot (\epsilon \boldsymbol{M})\right]^2 . \end{aligned} \tag{56}$$

To linearize the stress–deformation relation (52) note that, since $\boldsymbol{F} = \boldsymbol{I} + \boldsymbol{H}$, it follows that

$$\boldsymbol{B} = \boldsymbol{C} = \boldsymbol{I} + 2\epsilon$$

to the first order in ϵ. Therefore, for infinitesimal deformations, since $(55)_1$ holds,

$$I_1 = I_2 = 3.$$

It follows then that for $k = 1, 2, 4, 5$,

$$W_k = W_k^0 + 2\boldsymbol{M} \cdot (\epsilon \boldsymbol{M}) \left(W_{k4}^0 + 2W_{k5}^0\right).$$

Substitution of these linearized expressions into the nonlinear stress–deformation relation (52) yields the following linearized stress–strain relation:

$$\begin{aligned} \boldsymbol{\sigma} &= -p\boldsymbol{I} + 4\left(W_1^0 + W_2^0\right) \epsilon \\ &+ 4\left(W_{44}^0 + 4W_{45}^0 + 4W_{55}^0\right) \boldsymbol{M} \cdot (\epsilon \boldsymbol{M}) (\boldsymbol{M} \otimes \boldsymbol{M}) \\ &+ 4W_5^0 (\epsilon \boldsymbol{M} \otimes \boldsymbol{M} + \boldsymbol{M} \otimes \epsilon \boldsymbol{M}) . \end{aligned} \tag{57}$$

Thus, the theory of linear, incompressible, transversely isotropic elasticity is a three constant theory. It seems sensible to relate these constitutive constants to physical constants obtained from experiments. The usual experiments involve cuboid specimens. Assume that the fibres are aligned parallel to two faces of the cuboid. Let $\pm \boldsymbol{M}^{\perp_1}, \pm \boldsymbol{M}^{\perp_2}$ denote the normals to the other two pairs of faces. First consider simple tension tests of cuboid specimens with the direction of the applied forces in the direction of the fibres, and with E_L denoting Young's modulus. The stress along and perpendicular to the fibres is given by $(\boldsymbol{\sigma M}) \cdot \boldsymbol{M}$ and $(\boldsymbol{\sigma M}^{\perp}) \cdot \boldsymbol{M}^{\perp}$ respectively. In simple tension testing, $(\boldsymbol{\sigma M}^{\perp}) \cdot \boldsymbol{M}^{\perp} \equiv 0$. This yields

$$p = 4\left(W_1^0 + W_2^0\right) (\epsilon \boldsymbol{M}^{\perp}) \cdot \boldsymbol{M}^{\perp} = -2\left(W_1^0 + W_2^0\right) (\epsilon \boldsymbol{M}) \cdot \boldsymbol{M},$$

on using the incompressibility condition. It then follows that $(\boldsymbol{\sigma M}) \cdot \boldsymbol{M} = E_L (\epsilon \boldsymbol{M}) \cdot \boldsymbol{M}$ with

$$E_L \equiv 6W_1^0 + 6W_2^0 + 4W_{44}^0 + 16W_{45}^0 + 16W_{55}^0 + 8W_5^0. \tag{58}$$

In simple shear along the fibres, $(\boldsymbol{\sigma M}) \cdot \boldsymbol{M}^{\perp}$ is specified. If μ_L denotes the corresponding shear modulus, then $(\boldsymbol{\sigma M}) \cdot \boldsymbol{M}^{\perp} = 2\mu_L(\boldsymbol{\epsilon M}) \cdot \boldsymbol{M}^{\perp}$, where

$$\mu_L \equiv 2\left(W_1^0 + W_2^0 + W_5^0\right). \tag{59}$$

A simple inversion yields

$$\begin{aligned} 2\left(W_1^0 + W_2^0\right) &= \mu_T, \quad 2W_5^0 = \mu_L - \mu_T, \\ 4\left(W_{44}^0 + 4W_{45}^0 + 4W_{55}^0\right) &= E_L - 4\mu_L + \mu_T, \end{aligned} \tag{60}$$

and therefore the strain energy (56) and stress–strain relation (57) can be re-written in the respective forms

$$\begin{aligned} W &= \mu_T \mathrm{tr}\, \boldsymbol{\epsilon}^2 + 2\left(\mu_L - \mu_T\right) \boldsymbol{M} \cdot \left(\boldsymbol{\epsilon}^2 \boldsymbol{M}\right) \\ &\quad + \frac{1}{2}\left(E_L - 4\mu_L + \mu_T\right)\left[\boldsymbol{M} \cdot (\boldsymbol{\epsilon M})\right]^2, \\ \boldsymbol{\sigma} &= -p\boldsymbol{I} + 2\mu_T \boldsymbol{\epsilon} + \left(E_L - 4\mu_L + \mu_T\right) \boldsymbol{M} \cdot (\boldsymbol{\epsilon M}) \boldsymbol{M} \otimes \boldsymbol{M} \\ &\quad + 2\left(\mu_L - \mu_T\right)\left(\boldsymbol{\epsilon M} \otimes \boldsymbol{M} + \boldsymbol{M} \otimes \boldsymbol{\epsilon M}\right), \end{aligned} \tag{61}$$

which are those usually introduced where, in standard reference books, the notation $\beta \equiv E_L - 4\mu_L + \mu_T$ is used.

It is interesting to notice that in modelling transversely isotropic materials, it is usual to ignore the I_2, I_5 invariants and to adopt the assumption that $W = W(I_1, I_4)$; see, for example, among many others, Humphrey and Yin (1987) and Humphrey et al. (1990) or the anisotropic character of the material has been considered via a strain energy depending on a single combination of I_4 and I_5 (see, for example, Ciarletta et al, 2011). In a recent series of papers these constitutive assumptions has been deeply criticized (Destrade et al., 2013) and the *unphysical* features of the constitutive assumption $W = W(I_1, I_4)$ have been stressed in detail. From the above considerations, it is clear that if $W_5^0 = 0$ we have that in the linear limit $\mu_L = \mu_T$ which is clearly an unphysical prediction.

In a recent paper investigating the stress and strain coaxiality properties for transverse isotropic materials,[12] using the tool of universal relations, we have laid the theoretical foundations for a simple experimental program to establish in a definitive way that all the invariants are necessary to obtain meaningful models for fibre-reinforced materials and this applies in the full nonlinear setting.

Using such findings it is very hard to believe that constitutive equations based on only one anisotropic invariant may be able to model successfully

[12] Two tensor $\boldsymbol{A}$ and $\boldsymbol{B}$ are coaxial if $\boldsymbol{AB} = \boldsymbol{BA}$. See Saccomandi and Vianello (1997) for coaxiality in the framework of transversely isotropic materials

real world fibre-reinforced materials. This means that an experimental program and a step-by-step modelling procedure for anisotropic materials such as the one we have outlined in the previous section for isotropic models is very complex due to the number of invariants involved. There has been an intensive research activity in this direction but still we are far away from understanding in a proper way how it is possible to derive feasible and reliable models for anisotropic elastic materials.

7 DNA

In the previous sections we have tried to design a *rational* pathway to a phenomenological theory of rubber-like elasticity. Another possible approach is to try a *bottom-up* approach. We start from some micro-mechanical point of view and we try to derive from this approach a macroscopic theory. Sometimes this method of investigation is called a multi-scale approach and usually the result is always a phenomenological theory, but now the model contains more information about the mesoscopic structure of the material. Often this is the best we can do in the framework of amorphous solids.

Using this approach in the last years we were able to provide several models that have tried to explain the mechanical behaviour of different soft materials (spider silk, arterial walls, polymers reinforced with filler) and complex phenomena like damage (see, for example, De Tommasi et al. 2008, 2010). Here I focus on DNA mechanics.

DNA is a double stranded molecule composed of two poly-nucleotide strands that are bound together by hydrogen bonds between complementary nucleotide bases. Under normal conditions the strands wind around the DNA axis as two identical right-handed helices with diameter of 2 nm and pitch of 3.57 nm. During various important intracellular biological processes, such as transcription or replication, mechanical forces, exerted by various proteins, act on DNA and cause its bending, twisting, stretching, or shearing.

Macroscopic properties of DNA can be studied using a range of techniques, including atomic force microscopy (AFM) observations, electron microscopy visualization, FRET measurements of distances between parts of a molecule, or single molecule manipulation. The latter technique can subject a single DNA molecule to a mechanical loading, consisting of stretching and twisting, by micro-manipulation of objects to which DNA is attached, such as glass needles, glass beads, or magnetic beads.

Because molecular simulations of DNA are still a problem from a computational point of view a plethora of *mesoscopic* models have been proposed. For example, it is possible to use an ideal elastic rod model in which the hydrogen bonding remains unperturbed, and DNA is assumed to be homogeneous, isotropic, intrinsically straight, unshearable and inextensible. The mechanical response of such a model can be characterized by two parameters: a bending modulus and a twisting modulus. The statistical mechanics equivalent of this ideal elastic rod is the worm-like chain (WLC) model. This model fits stretching experiments well up to a tensile force of 10 pN. More sophisticated rod theories that take into account of, for example, sequence-dependent geometrical and physical properties of DNA may be considered.

Another possibility is to use discrete models and here we can range from minimal models (Ising-like models) to much more complicated models taking account of sequence-dependent degrees of freedom. The topic is so complicated that any approach can be useful.

An important simple mesoscopic model of DNA was proposed by Peyrard and Bishop (1989) (the PB model). They assumed that each strand can be represented as a set of point masses, each corresponding to a nucleotide. The main characteristics of the model are:

- The longitudinal displacements of bases are neglected.
- The DNA is homogeneous.
- Two neighbouring nucleotides on the same strand are connected by a harmonic potential.
- The hydrogen bonding between two base pairs on opposite strands is represented by a Morse potential. It represents not only the attraction between base pairs but also the repulsion between the two phosphate groups on the opposite strands.

To write down the equations of the model it is sufficient to consider the Hamiltonian

$$H = \sum_n \left\{ \frac{1}{2}\left(\dot{u}_n^2 + \dot{v}_n^2\right) + \frac{1}{2}K\left[(u_n - u_{n+1})^2 + (v_n - v_{n+1})^2\right] \right.$$
$$\left. + D(e^{[-a(u_n - v_n)/\sqrt{2}]} - 1)^2 \right\}, \qquad (62)$$

where u_n and v_n are the displacements of the nucleotides from equilibrium for the first chain and the second chain, respectively, m is the reduced mass, K the intra-strand stiffness and D and a the Morse parameters. A simplification of this model is possible if we consider the reduced coordinates $x_n = (u_n + v_n)/\sqrt{2}$ and $y_n = (u_n - v_n)/\sqrt{2}$ and we ignore the x_n mode.

The Peyrard and Bishop model has been successfully applied to study the melting of DNA. Modelling DNA at a base pair level as a regular lattice has been sufficient to probe the theory with experiments and to understand complex and important phenomena of DNA denaturation. This has been a great success, but it is clear that many improvements of the PB model are possible.

A *vexata* question in the role of the PB model has been the harmonic intra-strand *stacking* potential (see Saccomandi and Sgura, 2006). Indeed, it is well known from the experimental data that if you wish to pull apart two nearest bases in the same strand you need a lot of energy. The sugar-phosphate backbones are fairly rigid. Then the stiffnesses of the *springs* connecting the particles on the same strand are usually several orders of magnitude greater than the *springs* introduced to model the hydrogen bonds. In this picture it is clear that longitudinal modes are of *second order* with respect to, for example, torsional or other modes. The problem is that in a mesoscopic model the forces we introduce are not direct representations of the real world interactions but a peculiar *synthesis* of those real interactions projected in the lower dimensional space of the model.

The recent single molecule experiments on DNA show clearly that the elasticity of DNA as a macromolecule is mainly *entropic*. This means that as usual in polymer mechanics the sugar-phosphate backbone of the single strand of the DNA must be modelled as a sequence of jointed rigid rods or beams. If we use the beam picture we have already noticed that we obtain the worm-like chain model to describe the overall mechanics of the macromolecule. Following this model DNA is quite flexible, because a huge number of its configurations may realize a given end-to-end distance and it does not matter how stiff are the sugar-phosphate bonds. At the mesoscopic level the longitudinal motions of DNA are not connected to the sugar-phosphate elasticity but to the entropic elasticity of the chains. The rigidity of the sugar-phoshate bond is a source of *enthalpic* energy.

To understand the central importance of a nonlinear stacking interaction we consider the continuum limit of the PB model. This limit is describe by the semi-linear Klein–Gordon equation

$$Y_{\tau\tau} - S_L Y_{XX} + \{[e^{-Y} - 1]^2\}_Y = 0, \tag{63}$$

where $Y = ay$ is the dimensionless stretching of the base pair, $\tau = \sqrt{Ds/mt}$ is a dimensionless time and $S_L = K/Da^2$. In the static case we obtain in a simple and direct way the exact solution

$$Y(X) = \ln\left(e^{(X-X_0)\sqrt{2/S_L}} + 1\right).$$

This is a domain wall solution. Here X_0 is an integration constant that determines the position of the wall and the solution describes a DNA configuration where for $X < X_0$ the molecule is closed, whereas for $X >> X_0$ the base-pair separation grows linearly with space and the molecule is fully denaturated.

Now we just propose a simple modification of equation (63), namely

$$Y_{\tau\tau} - S_L Y_{XX} - S_{NL} Y^2 Y_{XX} + \{[e^{-Y} - 1]^2\}_Y = 0, \tag{64}$$

where S_{NL} is the nonlinear stacking term. It is interesting to consider the *extreme* nonlinear case, i.e. $S_L = 0$. Now the static domain wall solution is given by the solution of the ordinary differential equation

$$S_{NL} Y^2 Y_{XX} - \{[e^{-Y} - 1]^2\}_Y = 0,$$

which may be integrated easily once to give

$$S_{NL} Y_X^4 = C + \{[e^{-Y} - 1]^2\},$$

where C is an integration constant. By considering the asymptotic behaviour at $X \to -\infty$, which requires $Y = 0$ and $Y_x = 0$, we obtain $C = 0$. Introducing

$$z = \left(\frac{4}{S_{NL}}\right)^{1/4} (X - X_0),$$

where once again X_0 is an integration constant, we obtain the exact solutions

$$Y_+(z) = \ln\left(\frac{1}{8} + \frac{1}{2}e^{-z} + \frac{1}{2}e^{-2z}\right) + z, \quad \text{if } z > \ln 2, \qquad 0 \text{ if } z \leq \ln 2$$

and

$$Y_-(z) = \ln\left(\frac{1}{8} + \frac{1}{2}e^{z} + \frac{1}{2}e^{2z}\right) - z, \quad \text{if } z > -\ln 2, \qquad 0 \text{ if } z \leq -\ln 2.$$

The peculiarity of such solutions is that they have compact support. Mathematically speaking this is due the nonlinear character of the intra-strand elastic potential. This fact shows the importance of the modelling issues in a macromolecule such as DNA. The mechanics of DNA in tension needs to be discussed in more detail than has been the case in the past. About this point a deep discussion can be opened. Indeed, for a large number of scientists the best we can do is to continue on the pathway of the mesoscopic modelling activity we have just sketched or to resort to complete and complex molecular simulations. Other scientist may, on the other hand, find it necessary to use a continuum approach, but this approach is only dictated in some sense by the geometrical problem.

In this framework a theory of rods (or birods, a particular choice of braid systems; see, for example, Mohaker and Maddocks, 2005) seems to be sufficient. In my opinion single-molecule experiments on DNA suggest

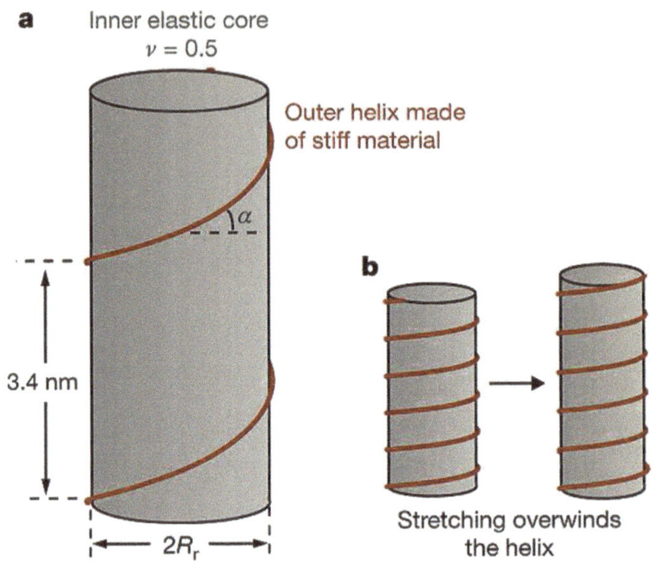

Figure 12. a. We model DNA as an elastic rod (grey) wrapped helically by a stiff wire (red). The inner core of radius R_r is assumed to have a Poisson's ratio $\nu = 0.5$. The outer wire is fixed to the inner rod helically with a pitch of 3.4 nm, and contributes to the overall mechanical properties because it resists stretching and compression. The outer helix increases the torsional rigidity and yields a twist–stretch coupling that depends upon the helix angle, α. **b**. Stretching generates an overwinding of the helix because the inner rod decreases in diameter as it is stretched. The outer helix is then able to wrap a larger number of times over the length of the molecule. In this figure, a shallow helix angle was used in order to exaggerate the overwinding effect seen with DNA (from DNA overwinds when stretched, Jeff Gore, Zev Bryant, Marcelo Nllmann, Mai U. Le, Nicholas R. Cozzarelli and Carlos Bustamante, Nature 442, 836–839).

that a complete three-dimensional mechanical description of DNA may be necessary. Indeed, simple physical intuition predicts that DNA should unwind under tension, as it is pulled towards a denatured structure. This means that there is a coupling between twisting and stretching degrees of freedom. A coupling that in nonlinear elasticity is natural and well known: the Poynting effect.

It is interesting to read the description of the so called *toy model* of DNA proposed in (Gore et al., 2006). The model is summarized in the Figure 12 and in my opinion, after reading this cartoon, it is clear that there is plenty of room for nonlinear elasticity to help with understanding the mechanics of DNA.

A typical force–extension curve for double-stranded DNA (dsDNA) exhibits three portions. During the first portion there is an entropic stretching regime (usually modelled by the worm-like chain), followed by a force plateau in the region of 65 picoNewtons, while in the last portion there is a sharp transition from the usual B-form to a new overstretched form, usually designated S-DNA. The structure of S-DNA remains the subject of debate but it should not be confused with single-stranded DNA (ssDNA).

Here I try to develop a model of the DNA filament using a nonstandard version of the phenomenological theory of nonlinear elasticity where the stress is determined as a function of the deformation gradient calculated with respect to a varying reference configuration in such a way that it is possible to introduce some micro-mechanical considerations. The micro-mechanical interpretation of this theory was first proposed by Eckart and then developed by Rajagopal and Wineman (1992); see, also, De Tommasi et al. (2008).

Let us consider a motion $\boldsymbol{x} = \boldsymbol{x}(\boldsymbol{X}, t)$, where $\boldsymbol{x}$ is the current position of a particle located at $\boldsymbol{X}$ in the undeformed configuration at time $t = 0$. The deformation gradient is given by $\boldsymbol{F}(\boldsymbol{X}, t) = \partial \boldsymbol{x}/\partial \boldsymbol{X}$ and the left Cauchy–Green tensor by $\boldsymbol{B} = \boldsymbol{F}\boldsymbol{F}^T$. We assume that there is a range of deformations for which the material behaves like an incompressible, isotropic elastic material, i.e. the Cauchy stress $\boldsymbol{T} = -p\boldsymbol{I} + \boldsymbol{T}^{(\mathrm{E},1)}$, where $-p\boldsymbol{I}$ is the indeterminate part of the stress due to the constraint of incompressibility ($\det \boldsymbol{F} = 1$) and the extra stress takes the form

$$\boldsymbol{T}^{(\mathrm{E},1)} = 2W_1^{(1)}\boldsymbol{B} - 2W_2^{(1)}\boldsymbol{B}^{-1}. \tag{65}$$

The strain-energy function $W^{(1)} = W^{(1)}(I_1, I_2)$ is a function of the principal invariants $I_1 = \operatorname{tr} \boldsymbol{B}$ and $I_2 = \operatorname{tr}(\boldsymbol{B}^{-1})$, where $W_i^{(1)} = \partial W^{(1)}/\partial I_i$, $i = 1, 2$. An activation criterion is needed to determine when the microstructural change begins. This is provided by introducing a scalar deformation state parameter s. Here, we suppose that $s = s(I_1, I_2)$ depends on the deformation through I_1 and I_2 for consistency with the requirement of isotropy, although more general forms of s may easily be adopted. For $s < s_a$, the threshold value of s, no conversion has yet occurred, i.e. all the material is in its original form and the extra stress is given by (65). On the other hand, for a value of the state parameter $\widehat{s}$ beyond s_a microstructural changes have occurred and the reference configuration has changed. This implies that the stress is now a function of the relative deformation gradient for the material formed at state $\widehat{s}$ given by $\widehat{\boldsymbol{F}} = \partial \boldsymbol{x}/\partial \widehat{\boldsymbol{x}}$, where $\widehat{\boldsymbol{x}}$ is the position of the particle in the configuration corresponding to deformation state $\widehat{s}$. In Figure 13

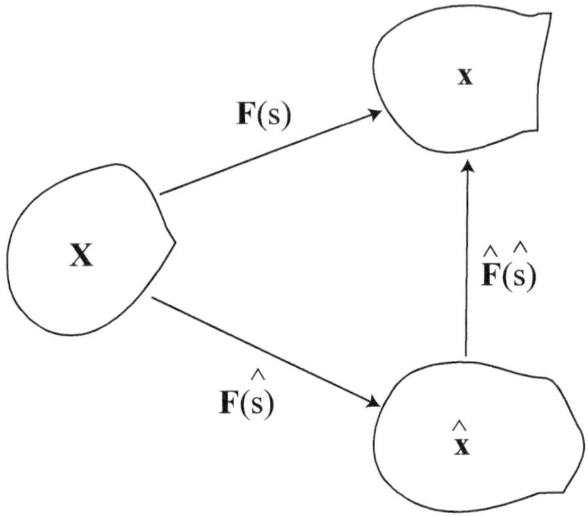

Figure 13. Schematic illustration of a material with the original reference configuration (left), the evolving reference configuration at the deformation state $\widehat{s}$ and the current configuration.

the original reference configuration, the configuration at $\widehat{s}$ and the current configuration are depicted. The associated Cauchy–Green tensor is given by $\widehat{\boldsymbol{B}} = \widehat{\boldsymbol{F}}\widehat{\boldsymbol{F}}^T$.

We shall assume that the new material formed at the state $\widehat{s}$ is still elastic, isotropic and incompressible such that the extra Cauchy stress at state s in this new configuration formed at the deformation state $\widehat{s}$ is given by

$$\boldsymbol{T}^{(\mathrm{E},2)} = 2W_1^{(2)}\widehat{\boldsymbol{B}} - 2W_2^{(2)}\widehat{\boldsymbol{B}}^{-1}. \tag{66}$$

Here $W^{(2)} = W^{(2)}(\widehat{I}_1, \widehat{I}_2)$ is the strain-energy function of the newly formed material, relative to the reference configuration at $\widehat{s}$. Another important simplifying assumption is that a single function $W^{(2)}$ governs the strain energy during the continuous microstructural change. The total current stress is taken as the superposition of the contributions from the material remaining in its original configuration and from all the new material formed at deformation states $\widehat{s} \in [s_a, s]$, i.e.

$$\boldsymbol{T}(s) = -p\boldsymbol{I} + b(s)\boldsymbol{T}^{(\mathrm{E},1)} + \int_{s_a}^{s} a(\widehat{s})\boldsymbol{T}^{(\mathrm{E},2)} d\widehat{s}, \tag{67}$$

where $a(\widehat{s})$ is a *conversion function*, which measures the 'rate' of conversion with respect to the parameter $\widehat{s}$, and

$$b(s) = 1 - \int_{s_a}^{s} a(\widehat{s}) d\widehat{s}$$

is the *volume fraction* of the material in the original configuration remaining at state s.

For $s < s_a$, as already indicated, no conversion has yet occurred, so all the material is therefore in its original state and the total extra stress is given by (65). As the deformation proceeds and the state parameter s increases beyond s_a the conversion function $a(s)$ determines the amount of material transformed into the new phase.

We require the conversion function to satisfy $a(s) = 0$ when $s \leq s_a$ and $a(s) > 0$ for $s > s_a$, while $b(s) = 1$ when $s \leq s_a$ and $0 \leq b(s) < 1$ for $s > s_a$. Thus, to complete the model, constitutive equations for $W^{(1)}$ and $W^{(2)}$, the activation criterion s_a and the conversion function $a(s)$ have to be prescribed. Our model is three dimensional and fully consistent with the theory of continuum mechanics. To illustrate the ideas quantitatively we begin with a prototype that is empirical and one-dimensional, and we then show how to recast the theory in three-dimensional form.

A one-dimensional model. Let x denote the one-dimensional extension. On the same basis as illustrated in Section 2, for the one-dimensional force f we have

$$f(x) = b(x) f^{(1)}(x) + \int_{x_a}^{x} a(\widehat{x}) f^{(2)}(\widehat{x})\, d\widehat{x}. \tag{68}$$

In (68) the various quantities $a(x), b(x), x_a$ have the same meaning as before, with x replacing s. Since the process of conversion is continuous we have

$$b(x) = 1 - \int_{x_a}^{x} a(\widehat{x}) d\widehat{x}, \quad x \geq x_a. \tag{69}$$

The constitutive assumptions we introduce are: for $f^{(1)}$, a logistic modification of the original one-dimensional Fung model widely used in biomechanics, i.e.

$$f^{(1)}(x) = \frac{\mu_1}{2} \frac{\exp[\beta(x - x_0)]}{\exp[\beta(x - x_0)] + \gamma}, \tag{70}$$

where the material constants $\mu_1 > 0$, $x_0 > 0$ and $\beta > 0$ have dimensions of force, length and 1/length, respectively, and $\gamma > 0$ is a dimensionless constant; and, for $f^{(2)}$, the WLC interpolation formula

$$f^{(2)}(x) = \mu_2 \left[\frac{1}{4} (1-z)^{-2} - \frac{1}{4} + z \right]. \tag{71}$$

Here $\mu_2 = kT/l_p$, where l_p is the persistence length, k is Boltzmann's constant and T the absolute temperature (degrees Kelvin), $z = x/l_c$ and l_c is the contour length of the molecule. We have chosen a logistic Fung model to capture the first portion of the force–extension curve (i.e. to capture the strain-hardening phenomenon) but without introducing a singularity such as that in the WLC formula. The WLC model is used to capture the sharp increase in force at the end of the curve just after the plateau. It is clear that the modelling of the plateau zone depends on how the reference configuration evolves, and this may be controlled by the choice of the conversion function $a(x)$. Usually, in the context of rubber mechanics, very simple models for the conversion function are adopted (for example, quadratic or piecewise linear functional forms). Here we use a functional form suggested by statistical mechanics, namely a probability distribution function computed by considering two possible states for a chain composed of a fixed number of base pairs with a given fixed difference in the energy between the two states. For this purpose, let c_1, c_2 be real constants and

$$g(x) = \frac{c_1\, e^{-c_1\,(x-c_2)}}{[1+e^{-c_1\,(x-c_2)}]^2}.$$

Let δ be a positive constant and define

$$a(x) = \delta(g(x) - g(x_a)), \qquad x \in [x_a, x_c], \tag{72}$$

with $a(x) = 0$ otherwise. Here, we are assuming that the conversion has been completed when x reaches the value x_c, and this imposes the continuity requirement $a(x_c) = 0$, which leads to $c_2 = (x_a + x_c)/2$.

We denote by C the total fraction of the material that can undergo conversion. Then,

$$C = \int_{x_a}^{x_c} a(\widehat{x})d\widehat{x}. \tag{73}$$

From the definition of C in (73), we calculate the value of δ by

$$\delta = \frac{C}{\frac{e^{-c_1\,(x_a+x_c)/2} - e^{-c_1\,(x_c-x_a)/2}}{[1+e^{-c_1\,(x_c-x_a)/2}][1+e^{-c_1\,(x_a+x_c)/2}]} - g(x_a)(x_c - x_a)},$$

and then the constitutive parameters to be identified in this empirical model are $\mu_1, \beta, \gamma, \mu_2$ and l_c. Moreover, we have to fix the activation criterion and therefore we also need values for x_a, x_c, c_1 and C. At this stage the only *a priori* information about these parameters is that $C \in [0, 1]$.

Three dimensions. In three dimensions the single molecule force–extension experiment is idealized as a simple tension test, for which the deformation is given by

$$x = \frac{1}{\sqrt{\lambda}} X, \quad y = \frac{1}{\sqrt{\lambda}} Y, \quad z = \lambda Z, \tag{74}$$

where λ is the stretch in the axial (i.e. z) direction. The current deformation gradient is given in matrix form by $\boldsymbol{F}(\lambda) = \mathrm{diag}(1/\sqrt{\lambda}, 1/\sqrt{\lambda}, \lambda)$, and the corresponding Cauchy–Green matrix is $\boldsymbol{B}(\lambda) = \mathrm{diag}(1/\lambda, 1/\lambda, \lambda^2)$. Hence,

$$I_1 = \lambda^2 + 2\lambda^{-1}, \quad I_2 = \lambda^{-2} + 2\lambda. \tag{75}$$

Since this is a one-parameter deformation, it is straightforward to establish that there is a one-to-one correspondence between the activation parameter s and the stretch λ, and we write $s = s(\lambda) = I_1(\lambda) - 3$. In fact, it is easy to see that for $\lambda \geq 1$ this is a monotone increasing function and can be inverted. Hence, it is possible to identify an *activation stretch*, which we denote by λ_a, instead of a generic activation parameter s_a; then, $s_a = s(\lambda_a) = I_1(\lambda_a) - 3$. The deformation gradient at state $\widehat{\lambda}$ is denoted by $\widehat{\boldsymbol{F}}(\widehat{\lambda}) = \boldsymbol{F}(\lambda)\boldsymbol{F}^{-1}(\widehat{\lambda})$ and we compute

$$\begin{aligned} \widehat{\boldsymbol{F}}(\lambda) &= \mathrm{diag}\left(\sqrt{\widehat{\lambda}/\lambda}, \sqrt{\widehat{\lambda}/\lambda}, \lambda/\widehat{\lambda}\right), \\ \widehat{\boldsymbol{B}}(\lambda) &= \mathrm{diag}\left(\widehat{\lambda}/\lambda, \widehat{\lambda}/\lambda, \lambda^2/\widehat{\lambda}^2\right). \end{aligned} \tag{76}$$

It follows that $\widehat{I}_1 = \lambda^2/\widehat{\lambda}^2 + 2\widehat{\lambda}/\lambda$, and similarly for $\widehat{I}_2$.

If we consider the class of elastic materials referred to as generalized neo-Hookean materials, with $W = W(I_1)$, then from (65) we obtain the principal components of the Cauchy stress tensor in the form

$$t_i = 2\lambda_i^2 W_1 - p, \quad i = 1, 2, 3. \tag{77}$$

The requirement that the lateral surfaces of the specimen undergoing the simple extension are traction free, $t_1 = t_2 = 0$, yields

$$p = 2\lambda^{-1} W_1. \tag{78}$$

Generalizing these results to the case (67), we see that the tensile force per unit deformed cross-sectional area necessary to achieve the stretch is given by the Cauchy stress component

$$\begin{aligned} t_3(\lambda) &= 2b(\lambda)\left(\lambda^2 - \lambda^{-1}\right) W_1^{(1)} \\ &\quad + 2\int_{s_a}^{s} a(\widehat{s}) \left(\frac{\lambda^2}{\lambda(\widehat{s})^2} - \frac{\lambda(\widehat{s})}{\lambda}\right) W_1^{(2)}\left(\lambda(\widehat{s})\right) d\widehat{s}, \end{aligned} \tag{79}$$

where

$$b(\lambda) = 1 - \int_{s_a}^{s} a(\widehat{s})d\widehat{s}. \tag{80}$$

The corresponding stress per unit undeformed area of cross section is $F(\lambda) = \lambda^{-1}t_3(\lambda)$.

At this point it is necessary to complement (79) with the constitutive equations. We need a constitutive equation for the strain-energy function of the material before the conversion starts, i.e. $W^{(1)}$, to model the first portion of the force–extension curve. Then, we also need a constitutive equation for the function $W^{(2)}$ that governs the mechanical behaviour of the newly formed material. This choice is important for modelling the "last" portion of the force–extension curve. The overstretching plateau, as already pointed out, is modelled by the choice of the conversion function $a(s)$. For the strain energy $W^{(1)}$ in the first regime we consider a modification of the strain-energy function, here denoted W^F, based on the original exponential model of Fung for biological tissues given by

$$W_1^F = \frac{\mu_1}{2}\exp[\beta(I_1 - 3)], \tag{81}$$

where $\mu_1 > 0$ is a constant, here with dimensions of stress, and $\beta > 0$ is a dimensionless constant. As for the one-dimensional case, we need to modify this function because a saturation phenomenon has to be taken into account. The mechanical behaviour characterizing the strain stiffening of the DNA molecule in the first portion of the force–extension curve cannot influence what happens in the plateau zone. For this reason we consider a logistic modification of the (three-dimensional) Fung model (81) analogous to that used for one dimension. This is given by

$$W_1^{(1)} = \frac{\mu_1}{2}\frac{\exp[\beta(I_1 - 3)]}{\exp[\beta(I_1 - 3)] + \gamma}, \tag{82}$$

so that

$$W^{(1)} = \frac{\mu_1}{2\beta}\ln\left(\exp(\beta(I_1 - 3)) + \gamma\right), \tag{83}$$

where $\gamma > 0$ is again a dimensionless constant. The function (82) reduces to the neo-Hookean material $W = \mu_1(I_1 - 3)/2$ when $\gamma = 0$.

For the strain-energy function $W^{(2)}$ in the second portion of the deformation range we consider the phenomenological model first proposed by Gent (1996) and given by

$$W^{(2)}(\widehat{I}_1) = -\frac{\mu_2}{2}J_m \ln\left(1 - \frac{\widehat{I}_1 - 3}{J_m}\right), \quad \widehat{I}_1 < J_m + 3, \tag{84}$$

where μ_2 is the shear modulus for infinitesimal deformations and $J_m\ (>0)$ is the limiting value of $\widehat{I}_1 - 3$ associated with limiting chain extensibility. In the limit as the chain extensibility parameter tends to infinity ($J_m \to \infty$), (84) also reduces to the classical neo-Hookean model. The model (84) can be easily connected with the FJC model. For (84) the response function is given by

$$W_1^{(2)} = \frac{\mu_2}{2} \frac{J_m}{J_m - (\widehat{I}_1 - 3)}, \tag{85}$$

so that the stress has a singularity as $\widehat{I}_1 \to J_m + 3$.

The three-dimensional model we have proposed contains several constitutive parameters that have to be found by using a nonlinear least squares fitting procedure. The parameters needed to fix the strain-energy functions are μ_1, β, γ and μ_2, J_m. Moreover, we use the same activation criterion as was used for the one-dimensional model in (72). Hence, to fix the activation criterion we need to identify the interval with ends $x_a \to \lambda_a$ and $x_c \to \lambda_c$ and the parameters c_1, C. Note that this criterion may easily be reformulated in a way compatible with three-dimensional elasticity in terms of the invariant $I_1 - 3$. Equation (79) provides a formula for the Cauchy stress, but it is the nominal stress $F(\lambda) = \lambda^{-1} t_3(\lambda)$ (force per unit reference cross-sectional area) that is needed for the data fitting. We therefore transform the data set (x_i, f_i) into the data set (λ_i, f_i), where $\lambda_i = 1 + x_i/l_c$, with the contour length l_c identified in the one-dimensional case. To match the dimensions of the force f in the data the stress $F(\lambda) = \lambda^{-1} t_3(\lambda)$ has to be multiplied by the reference cross-sectional area, which is unknown. However, this is just a multiplicative factor that is accounted for by incorporating it into the constants μ_1 and μ_2, which then have dimensions of force as in the one-dimensional situation.

Let us fix C and let us try to identify the other parameters such that in the space $(\lambda, F(\lambda))$ the three-dimensional model "sees" the overstretching transition (plateau) in the original data as its Maxwell line. Curves obtained from the fitting procedure, for different (fixed) values of C, are reported in Figure 14. In this framework, the idea of best fit depends on the kind of device used for the experiments. In a hard device, that is following the data on the plateau, the best fit is obtained for $C = 0.2$. In a soft device, it is reasonable to consider as absolute minimizers those having the plateau as their Maxwell line, that is such that the areas below and under the plateau are equal. As shown in Figure 14, this is attained for larger values of C.

It is very hard to understand in the mechanics of biomolecules what it is truly important in explaining the biological function of the protein. Therefore it is easy to argue that our approach is not necessary and not significant.

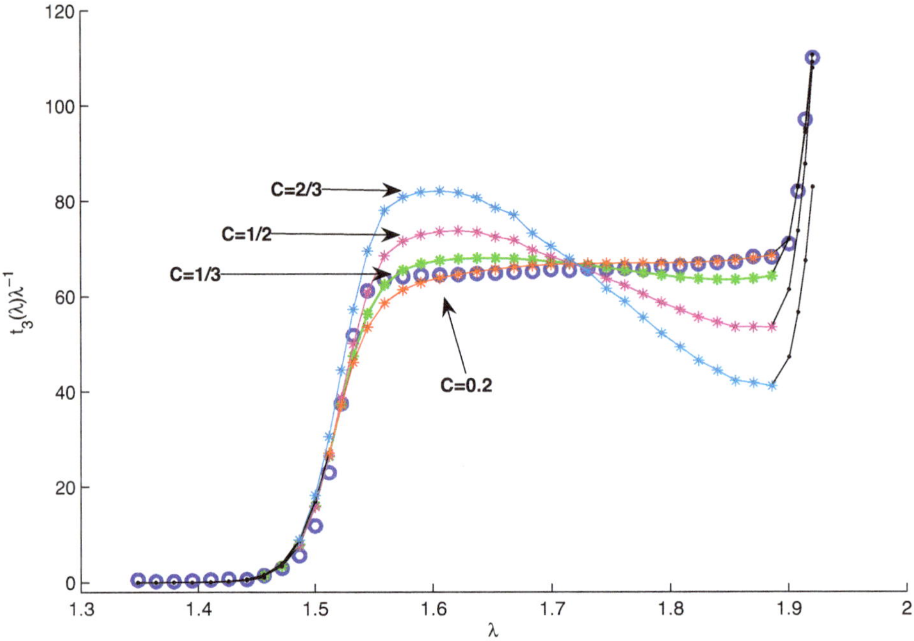

Figure 14. Plot of $\lambda^{-1}t_3(\lambda)$ vs λ. Circles are experimental data. The curves are obtained from the fitting procedure, for different (fixed) values of C, reported in the plot. The stars indicate the range of values for which the conversion is active. Note that the dimensions of the stress have been converted into 'force' by multiplication by the unknown reference cross-sectional area, which is incorporated into the parameters μ_1 and μ_2. Details may be found in Ogden et al. (2011).

In any case we think that these kinds of models allow us to understand in more detail some aspects that one-dimensional empirical theory cannot focus on in the right way. For example, what is meant by the term *mixture* or how, from a mechanical point of view, should the term overstretching be considered.

Bibliography

S. S. Antman. *Nonlinear Problems of Elasticity*, 2nd edition. Springer, New York, 2005.

E. M. Arruda and M. C. Boyce. A three-dimensional constitutive model for the large stretch behavior of rubber elastic materials. *J. Mech. Phys. Solids.*, 41:389–412, 1993.

M. F. Beatty, M.F. Topics in finite elasticity: Hyperelasticity of rubber, elastomers, and biological tissues–with examples. *Appl. Mech. Rev.*, 40:1699–1734, 1987.

J.-M. Y. Carrillo and A. V. Dobrynin. Effect of the electrostatic interactions on stretching of semiflexible and biological polyelectrolytes. *Macromolecules*, 43:2589–2604, 2010.

M. M. Carroll. Must elastic materials be hyperelastic. *Math. Mech. Solids*, 14:369–376, 2009.

C. Chui, E. Kobayashi, X. Chen, T. Hisada, and I. Sakuma. Combined compression and elongation experiments and non-linear modelling of liver tissue for surgical simulation. *Med. Biol. Eng. Comp.*, 42:787–798, 2004.

P. Ciarletta, I. Izzo, S. Micera, and F. Tendick T. Stiffening by fiber reinforcement in soft materials: a hyperelastic theory at large strains and its application. *J. Biomech. Behav. Biomed. Mat.*, 4:1359–1368, 2011.

M. Destrade, B. Mac Donald, J. G. Murphy, and G. Saccomandi. At least three invariants are necessary to model the mechanical response of incompressible, transversely isotropic materials. *Comp. Mech.*, 52:959–969, 2013.

D. De Tommasi, G. Puglisi, and G. Saccomandi. Localized versus diffuse damage in amorphous materials. *Phys. Rev. Lett.*, 100:085502, 2008.

D. De Tommasi, G. Puglisi, and G. Saccomandi. Damage, self-healing and hysteresis in spider silk. *Biophys. J.*, 98:1941–1948, 2010

Y. C. Fung. Elasticity of soft tissues in simple elongation. *Am. J. Physiol.*, 213:1532–1544, 1967.

A. N. Gent. A new constitutive relation for rubber. *Rubber Chem. Technol.*, 69:59–61, 1996.

A. N. Gent and A. G. Thomas. Forms for the stored (strain) energy function for vulcanized rubber. *J. Polym. Sci.*, 28:625–628, 1958.

J. Gore, Z. Bryant, M. Nöllmann, M. U. Le, N. R. Cozzarelli, and C. Bustamante. DNA overwinds when stretched. *Nature*, 442:836–839, 2006.

J. D. Humphrey, F. C. P. Yin. A new constitutive formulation for characterizing the mechanical behavior of soft tissues. *Biophys. J.*, 52:563–570, 1987.

J. D. Humphrey, R. K. Strumpf, and F. C. P. Yin. Determination of a constitutive relation for passive myocardium: I. A new functional form. *J. Biomech. Eng.*, 112:333–339, 1990.

E. A. Kearsley. Strain invariants expressed as average stretches. *J. Rheol.*, 33:757–760, 1989.

M. Moakher and J. H. Maddocks. A double-strand elastic rod theory. *Arch. Ration. Mech. Anal.*, 177:53–91, 2005.

F. D. Murnaghan. Finite deformations of an elastic solid. *Am. J. Math.*, 59:235–260, 1937.

W. Noll. A new mathematical theory of simple materials. *Arch. Ration. Mech. Anal.*, 48:1–50, 1972.

R. W. Ogden. *Non-Linear Elastic Deformations*. Ellis Horwood, Chichester, 1984, reprinted by Dover, New York, 1997.

R. W. Ogden, G. Saccomandi, and I. Sgura. Phenomenological modeling of DNA overstretching. *J. Nonlin. Math. Phys.*, 18 (Supplementary Issue 2):411–427, 2011.

P. Pincus. Excluded volume effects and stretched polymer chains. *Macromolecules*, 9:386–388, 1976.

M. Peyrard and A. R. Bishop. Statistical mechanics of a nonlinear model for DNA denaturation. *Phys. Rev. Lett.*, 62:2755–2758, 1989.

E. Pucci and G. Saccomandi. On universal relations in continuum mechanics. *Cont. Mech. Thermodyn.*, 9:61–72, 1997.

K. R. Rajagopal. The elasticity and elasticity. *Zeits. Angew. Math. Phys. (ZAMP)*, 58:309–317, 2007.

K. R. Rajagopal and A. S. Wineman. A constitutive equation for nonlinear solids which undergo deformation induced microstructural changes. *Int. J. Plasticity*, 8:385–395, 1992.

R. S. Rivlin. The relation between the Valanis–Landel and classical strain-energy functions. *Int. J. Non-Lin. Mech.*, 41:141–145, 2006.

M. B. Rubin. On evolution equations for elastic deformation and the notion of hyperelasticity. *Int. J. Eng. Sci.*, 47:76–82, 2009.

G. Saccomandi and I. Sgura. The relevance of nonlinear stacking interactions in simple models of double-stranded DNA. *J. R. Soc. Interface*, 3:655–667, 2006.

G. Saccomandi and M. Vianello. A universal relation characterizing transversely hemitropic hyperelastic materials. *Math. Mech. Solids*, 2:181–188, 1997.

A. J. M. Spencer. *Deformations of Fibre-Reinforced Materials*. Clarendon Press, Oxford, 1972.

D. J. Steigmann. Invariants of the stretch tensors and their application to finite elasticity theory. *Math. Mech. Solids*, 7:393–404, 2002.

L. R. G. Treloar. Stress-strain data for vulcanised rubber under various types of deformation. *Trans. Faraday Soc.*, 40:59–70, 1944.

L. R. G. Treloar. *The Physics of Rubber Elasticity*, 3rd edition. Clarendon Press, Oxford, 2005.

C. Truesdell and W. Noll. *The Non-linear Field Theories of Mechanics*, 3rd edition. Springer, Berlin, Heidelberg, New York, 2004.

K. C. Valanis and R. F. Landel. The strain-energy function of a hyperelastic material in terms of the extension ratios. *J. Appl. Phys.*, 38:2997–3002, 1967.

M. C. Wang and E. Guth. Statistical theory of networks of non-Gaussian flexible chains. *J. Chem. Phys.*, 20:1144–1157, 1952.

Modeling Bioactive Materials

Luis Dorfmann*, Chris Paetsch‡

*Department of Civil and Environmental Engineering
Tufts University, Medford, MA 02155, USA
E-mail: luis.dorfmann@tufts.edu

‡Bose Corporation
Framingham, MA 01701, USA
E-mail: christopher_paetsch@bose.com

Abstract. In this chapter we examine the mechanical properties of the ventral interior lateral muscle of the tobacco hornworm caterpillar, Manduca sexta. Uniaxial loading–unloading shows that the tissue is capable of large nonlinear pseudo-elastic deformations, has a hysteretic behavior and displays stress softening during the first few cycles of repeated loading. The data are used to develop a constitutive model that accounts for the observed material response, including a continuous transition from passive to active states. We consider the material initially incompressible, pseudo-elastic and transversely isotropic with preferred orientation parallel to the longitudinal direction of the muscle. The constitutive formulation is then generalized to account for slightly compressible materials and adapted for finite element implementation. The corresponding incremental equations and the associated isochoric and volumetric parts of the elasticity tensor are derived. To account for pseudo-elasticty, the isochoric part is modified accordingly. The simulation of a bio-actuated micro-pump is used to validate and demonstrate the capability of the model.

1 Introduction

The use of tissue engineered muscle constructs as bio-actuators in demonstration projects has become increasingly common in recent years. Both the ability to scale in size (Feinberg et al., 2007; Hollerbach et al., 1991) and to use *in vivo* energy supplies makes bio-actuation an attractive option over conventional means. One promising application includes small

L. Dorfmann, R. W. Ogden (Eds.), *Nonlinear Mechanics of Soft Fibrous Materials*,
CISM International Centre for Mechanical Sciences
DOI 10.1007/978-3-7091-1838-2_4 © CISM Udine 2015

scale pumps, or micro-pumps, for their potential application in lab-on-a-chip systems (Park et al., 2007; Tanaka et al., 2007a,b) or as a multiplexed heart-like organ (Birla et al., 2008, 2009; Kubo et al., 2007; Sekine et al., 2012). Of particular note, micro-pumps may also be used as drug delivery implants (Tanaka et al., 2007a), that operate *in vivo* without externally supplied energy or actuation signals. One form of micro-pump design employs a cardiomyocyte sheet with a dome-shaped substrate, allowing the contracting membrane to pump the fluid contained within, similar to a reciprocating displacement-type pump (Park et al., 2007; Tanaka et al., 2007b, 2011). As an improvement over their preliminary design, Tanaka et al. (2007a) developed a spherical drug delivery device to overcome the inefficiencies associated with more conventional displacement-type pumps. Specifically, a hollow spherical PDMS substrate is formed with an inlet and outlet provided through a capillary attached at opposing ends. A sheet of rat cardiomyocytes overlaying the PDMS generates pulsating, spontaneous contractions.

While mammalian-derived cells are desirable for their characteristic synchronized and autonomous contractions, the time scale for device functionality is only on the order of a few days (Tanaka et al., 2007a). Alternatively, engineered tissue is derived from insects, such as those used in a prototype bio-actuator (Akiyama et al., 2010) or in micro-robots (Akiyama et al., 2012a,b), which survive a wider range of environmental conditions for a longer period of time compared to mammalian cells. Cultured cells derived from the myoblasts of Manduca sexta are developed by Baryshyan et al. (2012), and resemble the sheets of rat cardiomyocytes utilized in the aforementioned studies. These tissues are comprised of fewer fibers, making alignment *in vitro* simpler, and do not require vasculature to survive.

In anticipation of functionalized bio-actuators derived from the muscle tissue of Manduca sexta, we propose a model to assist in the design of a micro-pump for drug delivery similar to that of Tanaka et al. (2007a). However, to simplify the attachment of the muscular film, we propose replacing the spherical structure with a cylindrical shape. In order to determine an appropriate geometric configuration, we propose using numerical simulation to investigate the major design parameters, namely the orientation of the muscle fibers and the configuration of the PDMS substrate, subsequently providing a design with improved performance.

One of the few examples of a bio-actuated micro-pump simulation is the numerical analysis in Park et al. (2005, 2007), where the muscle tissue and polymeric substrate are both considered linear elastic and the contraction is accounted for by applying a force gradient at the interface of the muscle and substrate. This approach, while it is advantageous for its simplicity and ability to be implemented in most commercially available finite element codes, it does not account for the nonlinear behavior of the material, and applying a contraction force at each muscle cell may require significant effort. Other work involving numerical simulation of active muscle tissue includes Böl et al. (2009), in which a finite element model of a thin muscle film attached to a PDMS substrate is created. The active muscle fibers are represented by discrete truss elements requiring each fiber be defined, which can be a time consuming process, especially for non-rectangular geometries. Shim et al. (2012) model a similar biohybrid film, utilizing a continuum material model of rat cardiomyocytes in an explicit finite element code. The explicit solver simplifies the model implementation, but at the cost of accuracy and, possibly, computational efficiency. To overcome these disadvantages, we propose utilizing the commercially available, implicit, nonlinear finite element code, Abaqus, to implement the constitutive model presented in this chapter. This approach allows for high accuracy simulations and changes to the geometry or fiber orientation can be made efficiently, thus a large number of bio-actuated micro-pump designs may be considered.

This chapter begins with a review of experimental data on the mechanical response of the ventral interior lateral (VIL) muscle of Manduca sexta under both passive and stimulated conditions. The data show that the muscle in the passive and stimulated states demonstrates nonlinear deformations during loading and unloading and stress softening during the first few cycles of repeated loading. After four loading–unloading cycles, the stress–stretch response becomes reproducible and the material is said to be preconditioned. Next, we show the rate-dependent response of an unstimulated Manduca muscle during uniaxial loading–unloading for engineering strain rates of $\dot{\varepsilon} = 0.0144, 0.072, 0.36$ and $1.8\ \mathrm{s}^{-1}$. For completeness we also include a brief discussion on specimen preparation, measurement of the cross-sectional area and muscle stimulation.

The Manduca muscle can be considered a nonlinear pseudo-elastic composite with interacting actin and myosin fibers embedded in an isotropic base material. These fibers introduce a preferred direction and the material is said to be transversely isotropic. During tetanic stimulation the fibers increase the cross bridge formations between actin and myosin, which then induce an increase in the macroscopic material stiffness.

The underlying micromechanical behavior of muscle contraction can be incorporated in a continuum based model by adopting a multiplicative decomposition of the deformation gradient, which is known as the active strain approach. Specifically, macroscopic deformation is considered the result of both elastic deformation and shortening of the sarcomere, and can be formally expressed as $\mathbf{F} = \hat{\mathbf{F}}\mathbf{F}_0$, where $\mathbf{F}$ is the deformation gradient, $\hat{\mathbf{F}}$ is the elastic deformation of the cross-bridges and $\mathbf{F}_0$ describes the change in configuration from passive to active states in the absence of applied mechanical loads. A similar approach has been taken to account for softening in polymeric materials (Rajagopal and Wineman, 1992) and growth in soft biological tissues (Rodriguez et al., 1994).

An early user of the active strain approach for muscle tissue, Nardinocchi and Teresi (2007) modeled the isotropic contraction of cardiac myocytes, while their later work also included electromechanical coupling and anisotropic behavior (Cherubini et al., 2008). Deformation in the fiber direction is defined through strain invariants in their subsequent study (Nardinocchi and Teresi, 2013). Similar models of cardiac tissue include energy functions based on neo-Hookean (Nobile et al., 2012) and Mooney-Rivlin (Ambrosi et al., 2011) materials, where the anisotropic nature of the tissue is accounted for through the definition of the active strain tensor. Based on the passive model developed by Holzapfel and Ogden (2009b), the active strain approach is used by Rossi et al. (2012, 2013) and Ruiz-Baier et al. (2013) to develop an anisotropic, exponential energy function. For many of these studies, models are implemented in finite element codes, which have been used to (i) validate the approach against experimental data (Rossi et al., 2013), (ii) compare two competing models (Rossi et al., 2012) or (iii) investigate the influence of particular parameters on the overall cardiac behavior (Nardinocchi and Teresi, 2013; Ruiz-Baier et al., 2013).

The multiplicative decomposition of the deformation gradient has also been used to model skeletal and smooth muscles. Stålhand et al. (2008) proposed a 1D mechanochemical model of smooth muscle cells, which was extended to 3D by Stålhand et al. (2011). The active contribution to the free energy is scaled by an actomyosin overlap function and is dependent on the active stretch. Kroon (2010) suggested a similar, but simpler, model of smooth muscle behavior and the resulting constitutive relation is compared to data found in the literature. These studies all base the biochemical activation on a four-state model proposed by Hai and Murphy (1988), which accounts for the actin and myosin interactions in smooth muscle cells. Ehret et al. (2011) developed models of skeletal tissue using the active strain approach with an energy function that depends, in part, on a generalized strain invariant and incorporate weighting factors of the passive and active portions. The active contribution was later coupled to include dependence on the electric potential (Böl et al., 2012). Weickenmeier et al. (2014) investigated slightly compressible behavior of skeletal muscle modifying an incompressible model proposed by Ehret et al. (2011). Sharifimajd and Stålhand (2013) proposed a skeletal muscle model in 1D with the active stretch additively decomposed (Stålhand et al., 2008, 2011). Numerical simulations using various bioactive models have been conducted to simulate the response of a rat tibialis anterior muscle (Hernández-Gascón et al., 2013), the human biceps (Böl et al., 2011) and the human masseter muscle (Weickenmeier et al., 2014).

In this chapter we adopt the active strain approach to model the continuous transition of Manduca muscle between passive and active states. We start by assuming that a muscle can be treated as a transversely isotropic nonlinear hyperelastic material with preferred orientation parallel to the longitudinal direction of the muscle. Data show that loading and subsequent unloading stress–stretch curves, both in the passive and active states, do not coincide indicating energy dissipation. The objective is to develop a constitutive model that accounts for the observed material response, including a continuous transition from passive to active states and energy dissipation during cyclic loading. Initially we assume that the material is incompressible, which is a convenient idealization and is used frequently in practical applications of biological material models. As a matter of fact, many biological tissues do not change their volume for deformations within the physiological range. We define the necessary notation to describe the kinematics, summarize appropriate stress and deformation tensors along with their associated invariants. We also review the main aspects of the theory of pseudo-elasticity introduced originally by Ogden and Roxburgh (1999).

The constitutive formulation is then generalized to account for slightly compressible materials and adapted for finite element implementation. The compact form of the Kirchhoff stress tensor is used as the starting point from where the corresponding increments and the associated isochoric and volumetric parts of the elasticity tensor are derived. To account for pseudo-elasticity, the isochoric part is modified accordingly. To validate and demonstrate the capability of numerical simulation, a model of a bio-actuated micro-pump is analyzed.

2 Experimental Methods

Mathematical and numerical modeling of bioactive materials requires the use of constitutive equations, which in their simplest forms must account for the passive, active and transitional states. The challenge is to select or develop an appropriate formulation and to experimentally determine the values of associated model parameters. To this end, experimental data of bioactive materials are essential representing, for example, the uniaxial loading–unloading response in the passive and active states. In this chapter we use the tobacco hornworm Manduca sexta as the biological model, which is a relatively simple system to study, allows access to live tissue and ease of testing and provides biologists with a greater understanding of the neuro-mechanical behavior of caterpillars. Specifically, we examine the mechanical properties of the ventral interior lateral (VIL) muscle of the third abdominal segment (A3) of Manduca sexta under both passive and stimulated conditions. We focus on A3 VIL muscle since it is one of the largest larval muscles comprising 14 muscle fibers (Dorfmann et al., 2007).

Unlike either amorphous or crystalline materials, muscles are complex composites. Each muscle fiber contains aligned actin and myosin filaments within an amorphous matrix material composed of proteins, lipids and polysaccharides. Cross-bridge cycling between actin and myosin proteins is the source of muscle contraction and the fundamental building block of force generation at the molecular level. Cross-bridging begins with the head of myosin molecules attached to the actin filament. While still connected, the myosin head flexes and rotates, thereby sliding the actin molecule along the myosin filament. Afterwards, the myosin detaches from the actin molecule, freeing the myosin head to return to the inflexed position and to reconnect again with the actin molecule further up the filament. During activation, the entire process is repeated continuously, inducing contractile properties to the muscle fiber. Most studies focus on the active state; however, there is a growing appreciation of the importance of the passive properties of these materials for their roles as brakes and dampers.

In contrast to most vertebrate muscles, Manduca muscles do not act upon large attachment structures of jointed skeletons. They are attached to infoldings of the body wall (apodemes) such that muscle activity has a direct action on overt movements. Furthermore, because Manduca significantly changes shape, the muscles undergo finite deformations and their passive properties are known to be particularly important (Dorfmann et al., 2008). We show that these tissues are capable of large nonlinear pseudo-elastic deformations, have a hysteretic behavior and display stress softening during the first few cycles of repeated loading. One important aspect of muscle function is that the mechanical properties also depend on multiple time-dependent processes. For active muscle, part of the stress softening is due to the time course of the contractile force developed by the muscle during stimulation. Owing to the diminishing of contractile force over time, data for stimulated muscle at low strain rates, which would require lengthy stimulation, are not biologically representative or experimentally feasible.

The Manduca muscles are organized in repeated segments along the body length and in several overlapping layers. As with all striated muscles, their activity is controlled by depolarization initiated release of transmitters from motor nerve terminals. The amount and time course of transmitter release is a function of the pattern and frequency of action potentials in the motoneuron. Trains of action potentials (tetany) stimulate muscles to develop force, or to shorten, by cycling cross bridge formation between actin and myosin fibers, (Huxley, 2000). During each step of a crawl in Manduca, the ventral interior lateral muscle in each body segment is activated by a brief burst (< 1 s) of action potentials with a mean frequency of about 20 Hz. During other movements the muscle can be activated by bursts of action potentials at 30 to 40 Hz for several seconds.

2.1 Specimen Preparation

Following the method described by Bell and Joachim (1976), larvae were raised on an artificial diet at 26°C under long-day conditions (17 hours light/7 hours dark). Animals on the second day of the fifth instar were used for testing using the following procedure:

1. Preliminary dissections were made to map the location of the attachment points of the muscle, VIL, in the third abdominal segment to exterior markers; see Figure 1.
2. Animals were anaesthetized by chilling on ice, and the length of VIL was measured using the markers determined in Step 1. A full-length dorsal incision was made and the gut, head and posterior half of the terminal segment removed.
3. The cuticle, intact muscles and nerves were exposed in physiological saline, with the interior of the cuticle facing upward. During dissection the saline was perfused with air approximately every five minutes.
4. VIL was dissected out leaving a small portion of attached cuticle at both ends.
5. The two motor neurons innervating VIL, project through the dorsal nerve (Levine and Truman, 1985). Therefore, the closed nerve and segmental ganglion were kept intact, and all other nerves were severed.
6. Experiments were completed within 2 hours of the time of the initial incision.

2.2 Measurement of Cross-sectional Area

Microscopic analyses of VIL revealed a total of fourteen individual 4–5 mm long fibers. The direct measurement of the cross-sectional area of the muscle is either invasive or at least risks injury to the individual fibers. Therefore, we determined the undeformed cross-sectional area of the muscles using a relationship derived from invasive measurements of separate muscle preparations.

Since individual fibers are difficult to distinguish clearly in intact muscles, we removed fibers from resting-length single-muscle preparations until the remaining fibers could be clearly imaged from overhead. Since the fibers are not circular but rather elliptical in cross-section, an overhead image provides only the major axis dimension. We therefore imaged the cut ends of stained fibers after rotating into a vertical position. Dimensions were calculated in NIH ImageJ software using 169 μm diameter polymer beads for calibration (Duke Scientific Corp., Palo Alto, CA, USA) included in the images as a reference. The ratio of major to minor axis lengths was used to calculate the cross-sectional area of fibers from overhead image dimensions. Mean cross-sectional area was multiplied by the number of fibers and the resulting value of 0.265 mm^2 used to calculate the nominal stress in uniaxial unloading tests.

2.3 Strain Cycling and Force Measurement

Muscles were pinned by the attached cuticle at each end in a horizontal bath of saline. One end was pinned to the edge of an elastomer platform in the bath, while the other was secured by a hook to an Aurora 300B-LR lever-arm ergometer (Aurora Scientific Inc., Aurora, Ontario). The shaft of the hook and the movement of the lever arm were horizontal, and the hook was attached to the lever arm by a small tunnel of epoxy that allowed free movement of the hook only in the horizontal plane.

Lever arm motion was controlled and force recorded by DMC software (Aurora Scientific Inc.) via a National Instruments NiDaq 6024E interface. Lever arm position was updated at 1800 Hz and force and position were

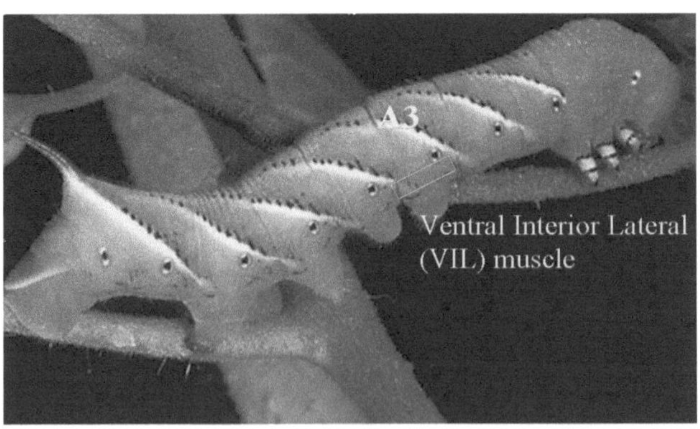

Figure 1. Image of a caterpillar Manduca sexta and the muscle location.

recorded at 2000 Hz. Data were filtered and exported in DMA software (Aurora Scientific Inc.).

Bath temperature was maintained at 25°C by a thermostatically controlled Peltier device built into the stage supporting the bath. Saline was continuously exchanged and kept aerated.

During tetanic stimulation, VIL develops force comparatively slowly, reaching peak force up to several seconds after initiation of stimulus; see Figure 2. We found considerable variation between preparations, with peak force rise time ranging from 1 to 6 seconds. Once reaching a peak value, force under sustained stimulation declines, again with much variation in time course between muscle preparations. Generally, preparations that develop force more slowly also sustained force more uniformly throughout the final 8 seconds of a 10 seconds tetanus. Initiating strain cycling 2 seconds after initiating stimulus allowed even more slowly contracting preparations to reach more than 90% of peak force before strain cycling began.

For measurements of VIL under tetanic stimulus, a suction electrode was

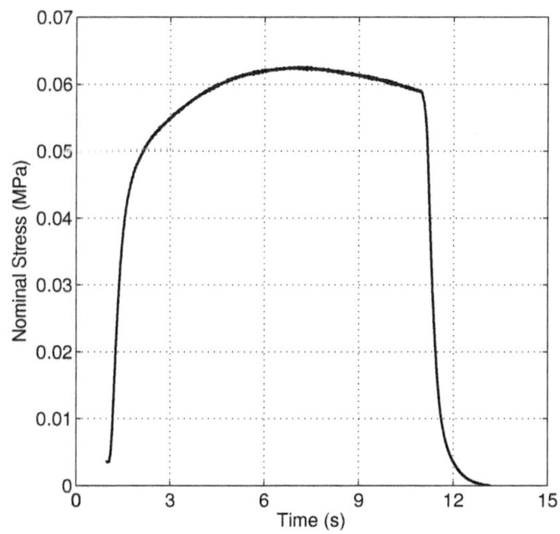

Figure 2. Time course of force developed by VIL in vitro held at the length measured in the resting animal and given a 10 seconds electrical stimulus via the muscle's motor neuron. In this preparation, from which the data in Figure 3 were obtained, the stimulated muscle sustained > 90% of peak force for the final 8 seconds of the stimulus train, reaching peak force 5.9 seconds after initiation of stimulus. Reproduced from Dorfmann et al. (2007).

applied to the dorsal nerve. Because peak force response time for VIL was 1.5 - 6 seconds, a supramaximal stimulus (40 volt, 40 Hz) was initiated 2 seconds before strain cycling began and sustained throughout strain cycling. The biological data shown in the next section are a representative example of 5 tested specimens.

3 Experimental Results

3.1 Active and Passive Responses

The contractile properties of striated muscles are similar across species. In general, muscle force is proportional to its cross-sectional area, and speed is proportional to its length. Force generation in most striated muscles is limited to a narrow range of movements (with strains of 10–20% being typical). This arises because thick and thin myofilaments overlap decreases as fibers stretch, while the Z-disk prevents actin and myosin from penetrating into adjacent sarcomeres during contraction; see, for example, Holmes and Geeves (2000). This non-continuous nature of the axial fibers results in a narrow peak of maximum force development centered on the resting length of the muscle; on each side of this peak the maximum stimulated isometric force is significantly less.

Figure 3 reports experimental data on the passive and stimulated responses of the Manduca muscle for a constant strain rate $\dot{\varepsilon} = 0.0144$ s^{-1}. The specimen is first stretched to reestablish the resting length of the muscle in the animal, which generates residual stresses in the material. Next, from this configuration, the muscle is subjected to 4 loading–unloading cycles in the passive state up to a maximum extension $\lambda = 1.15$. The corresponding data are shown in the upper image in Figure 3. Afterwards, at the end of the fourth unloading, a tetanic stimulation is administrated producing a change in material properties. During the first 2 seconds of stimulation, with the muscle length held constant, a buildup of force is noted. Under sustained stimulation the muscle is then subjected to additional loading–unloading cycles; see the graph in the lower image. The data clearly show large nonlinear pseudo-elastic deformations, a dependence on the amount of stimulation, a hysteric response during loading–unloading and stress softening during the first few cycles of repeated loading.

The lower graph in Figure 3 shows the gradual decrease in stiffness during loading of stimulated VIL, which likely represents a progressive decrease in available overlap for cross-link formation. This interpretation is supported by the relatively linear loading stiffness of a passive muscle shown in the upper graph of Figure 3. In a passive muscle bridge formation between actin and myosin would be expected to be minimal.

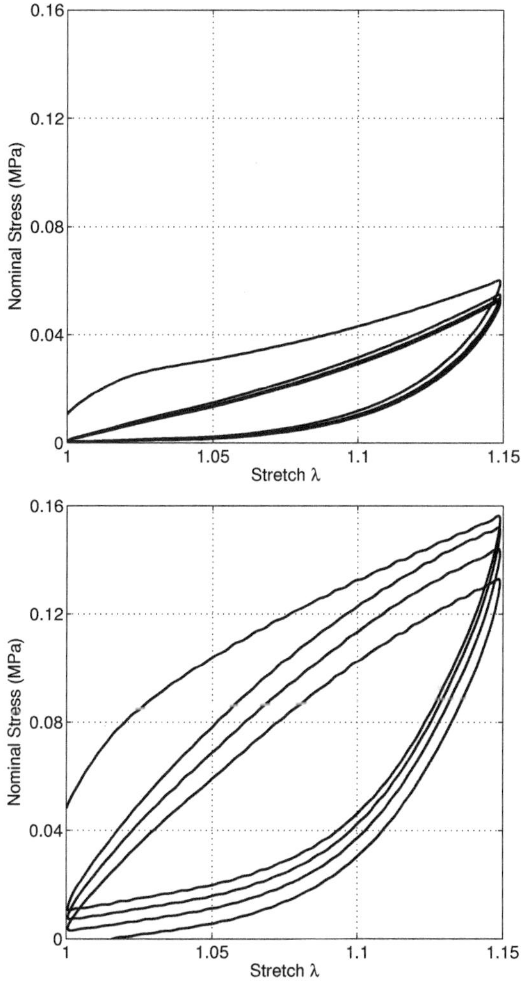

Figure 3. Pre-conditioning of a caterpillar muscle with maximum stretch of $\lambda = 1.15$. Results of a muscle in the passive state are reported in the top image, while those for stimulated condition are shown in the lower image. Data for the stimulated condition were obtained during the final 8 seconds of a 10 seconds stimulus, corresponding to the period during which the same muscle preparation sustained force within 90% of peak values under tetanic isometric stimulation. Reproduced from Dorfmann et al. (2007).

Interestingly, the loading hysteresis for VIL is not particularly large despite the application of a 15% strain. At least part of the explanation may lie in the extended range over which Manduca larval muscles can develop force. This ability is presumably an adaptation for long movements without the advantage of a rigid lever system. Larval muscles of holometabolous insects such as Manduca have several structural features that suggest they are supercontracting (Duch et al., 2000; Goldstein and Burdette, 1971; Hardie, 1976; Rheuben and Kammer, 1980; Royuela et al., 2000; Schwartz and Ruff, 2002). For example, muscles in Manduca caterpillars are striated but have long sarcomeres, very high tetanus-to-twitch force ratios and much slower force development (Rheuben and Kammer, 1980). They also develop force at lengths from 0.5 to 1.5 times resting length (Garmirian and Trimmer, unpublished), a range associated with supercontracting muscles in tsetse flies, which have both supercontracting and non-supercontracting visceral muscles (Rice, 1970). Although it is not known whether Manduca larval muscles have the perforated or incomplete Z-disks that characterize supercontracting muscle (Herrel et al., 2002), the Z-disks are at least described as 'poorly defined and irregularly arranged' (Rheuben and Kammer, 1980). Supercontracting muscles are found in both vertebrates and invertebrates and are often associated with hydrostatic movements (e.g., Osborne (1967); Rice (1970); Candia Carnevali (2005)).

One prediction of this interpretation of the deformation-dependent loading stiffness is that conventional striated (non-supercontracting) muscle will exhibit a more marked loading hysteresis when subjected to such large strain. This is because in striated muscle the sarcomeres are aligned and the degree of actin and myosin overlap throughout the muscle will be determined directly by the degree of stretch. In contrast, in Manduca muscles the sarcomeres are less distinct, which could lead to a greater heterogeneity in fiber overlap.

3.2 Loading Rate Dependency

Manduca muscles in the passive and active states were shown to dissipate energy with each loading–unloading cycle and stress soften on repeated loading. An outstanding issue in muscle properties is the mechanism by which passive force changes with the deformation rate and how it recovers after unloading (Proske and Morgan, 1999; Mutungi and Ranatunga, 2000). Both actin/myosin cross-bridge breakage and reformation, and the unfolding of gap-filament proteins (e.g., titin) have been proposed as likely mechanisms (Bagni et al., 2004; Campbell and Lakie, 1998; Granzier and Labeit, 2004; Granzier and Wang, 1993a,b). Intramuscular collagenous structural elements (Gosline et al., 2002) and muscle junctions (Lieber et al., 2000) may also contribute to properties observed in experimental muscle preparations. Complicating the matter still further is the possible variation in the relative contribution of different structural elements in different locomotory muscles, even in the same species (Prado et al., 2005). These different molecular mechanisms are important because they influence the assumptions and validity of most mechanical models used to describe muscle behavior.

In the previous section, the stress–deformation characteristics of Manduca muscles were determined under constant rates of loading and unloading, both in the passive state and during tetanic stimulation. In this section we summarize experimental data on the rate-dependent response of an unstimulated muscle of Manduca sexta for strain rates of $\dot{\varepsilon} = 0.0144, 0.072, 0.36$ and $1.8\ \mathrm{s}^{-1}$. For each value of $\dot{\varepsilon}$ the muscle is subjected to a total of five loading–unloading cycles of preconditioning up to a preselected maximum extension $\lambda_{\mathrm{max}} = 1.18$. The last loading–unloading cycle will be used to evaluate and compare the mechanical behavior of the muscle. These strain rates were chosen to represent the range of muscle stretching speeds experienced during normal Manduca behavior. When the caterpillar is exploring the substrate, VIL strain rates are less than $0.1\ \mathrm{s}^{-1}$ but during the rapid defensive strike reflex the rate can be as high as $2\ \mathrm{s}^{-1}$ for a brief period (Walters et al., 2001).

The experimental methods have been described in Section 2.2. The departures in the current experimental procedure from methods described in Section 2.2 are that (a) values of engineering strain rates $\dot{\varepsilon} = 0.0144$, 0.072, 0.36 and $1.8\ \mathrm{s}^{-1}$ are imposed, and (b) all measurements are of an unstimulated muscle. We note that the stress deformation response is, in general, dependent on time and intensity of stimulation (Dorfmann et al., 2007). However, in these experiments we are not concerned with phenomena such as stimulation, recovery time and viscous effects.

To assess the effect of loading rates on the mechanical response of an unstimulated A3 VIL muscle, several series of periodic loading–unloading uniaxial extension tests were carried out at a constant temperature of 25°C. The resting (reference) length, given by the initial distance of the pinned connections at each end of the muscle, was found to be 5.5 mm and was used to determine the longitudinal strain. Changes in the distance between these connections were measured with an accuracy of 1 μm. The tensile force was measured by using an Aurora 300B-LR lever-arm ergometer with an accuracy of less than 0.3 mN. Finally, following the methods summarized in Section 2.2, we found the reference cross-sectional area to be 0.4 mm^2. This information was used to determine the nominal stress as the ratio of the axial force to the cross-sectional area measured in the reference configuration.

During each of the tests, the muscle was subjected to five cycles of pre-conditioning with constant strain rate $\dot{\varepsilon}$ up to a pre-selected extension with stretch $\lambda = 1.18$. The pre-conditioning was performed in order to monitor the progression of stress softening and to determine the ultimate stress–deformation response for stretches up to $\lambda = 1.18$. Figure 4 shows the nominal stress versus stretch λ for the muscle in unstimulated state with engineering strain rates of $\dot{\varepsilon} = 0.0144$ and $\dot{\varepsilon} = 0.072$, Figure 5 shows the nominal stress as a function of stretch for strain rates of 0.36 s^{-1} and 1.8 s^{-1}. Figure 6 compares the response of the muscle during the fifth loading–unloading cycle for each of the four loading rates and the rate-dependence is noted. Specifically, the following observations are made:

- The initial configuration, corresponding to the resting length of the muscle in the animal, is not stress free. We take the corresponding geometric configuration to be the reference configuration from which to measure any subsequent extension generated by the application of mechanical loads.
- The stresses corresponding to the same strain level increase with the loading rate during loading and unloading.
- There are large differences in the stresses corresponding to the same strain level under loading and unloading during the first cycle in periodic tests with a fixed strain amplitude. The differences increase with the loading rate.
- There is a reduction in the stress at a given strain on each successive loading. The reduction is largest on the first and second loading–unloading cycles and becomes rather small after about 5 cycles. Again, the reduction in the stress increases with the loading rate.
- After five pre-conditioning loading–unloading cycles the stress–stretch responses are essentially repeatable and additional stress softening is negligible.
- Energy lost during cyclic loading is often presented as an *efficiency value* or *work loop efficiency*. Using data from the fifth loading–unloading cycle to evaluate the work loop efficiency, no dependence on the loading rate is observed. The efficiency values for cyclic loading with engineering strain rates of $\dot{\varepsilon} = 0.0144, 0.072, 0.36$ and 1.8 s^{-1} are 0.0025, 0.0029, 0.0027 and 0.0027 Nmm, respectively.

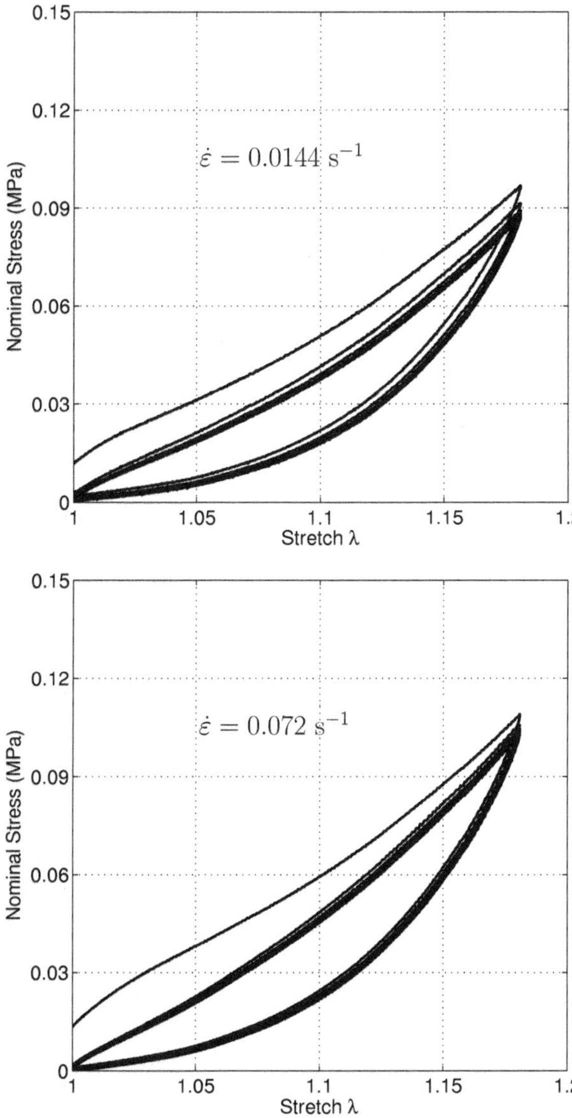

Figure 4. Cyclic tension tests of a passive caterpillar muscle with maximum stretch of $\lambda_{\max} = 1.18$ and engineering strain rates of $\dot{\varepsilon} = 0.0144\ \mathrm{s}^{-1}$ and $\dot{\varepsilon} = 0.072\ \mathrm{s}^{-1}$. Reproduced from Dorfmann et al. (2008).

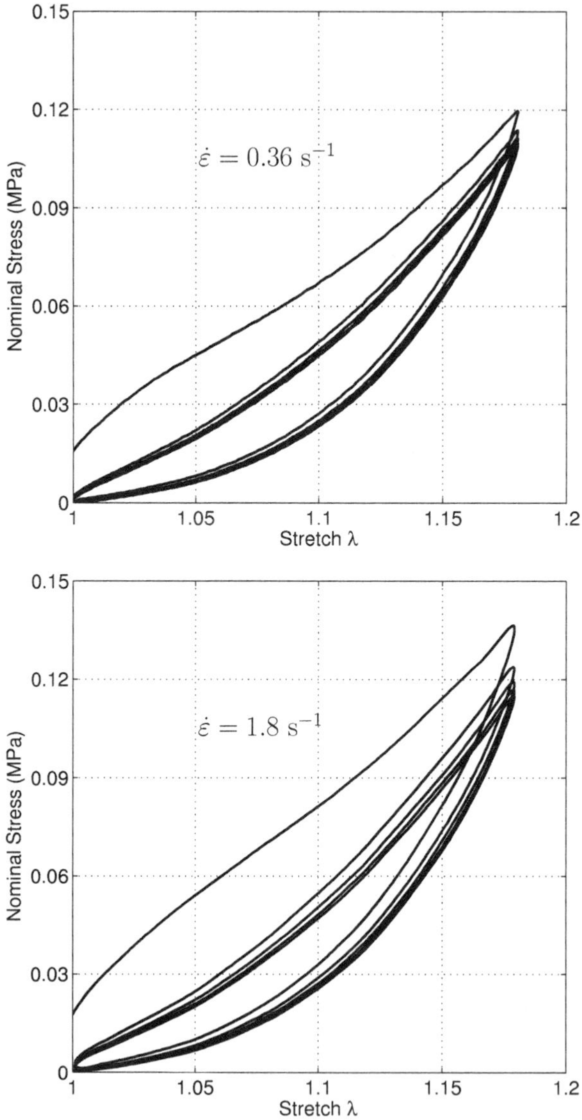

Figure 5. Cyclic tension tests of a passive caterpillar muscle with maximum stretch of $\lambda_{\max} = 1.18$ and engineering strain rates of $\dot{\varepsilon} = 0.36\ \mathrm{s}^{-1}$ and $\dot{\varepsilon} = 1.8\ \mathrm{s}^{-1}$. Reproduced from Dorfmann et al. (2008).

3.3 Transitioning Between Activation States

In Figure 7, we include experimental data of the ventral interior lateral muscle of the Manduca sexta in both the passive and the stimulated state. The muscle is first subjected to repeated loading–unloading cycles in the passive state up to a maximum extension $\lambda = 1.11$. During the third unloading, a tetanic stimulation is administrated, producing an increase in the macroscopic material stiffness. Woods et al. (2008) show that the stiffness in the muscle fiber depends on the amount of tetanic stimulation and deformation. Deformation-dependent stiffness can be interpreted as a decrease in the overlap of actin-myosin filaments as the muscle fibers stretch. This is clearly shown by the active response in Figure 7, where the increase in stiffness reduces with increasing deformation. The data show passive to active transition during unloading resulting in an increase in stress with reducing stretch. The stimulation phase can be adapted such that transition from passive to active occurs during loading with stress increasing with elongation. We note, for completeness, that a reduction in stiffness with increasing

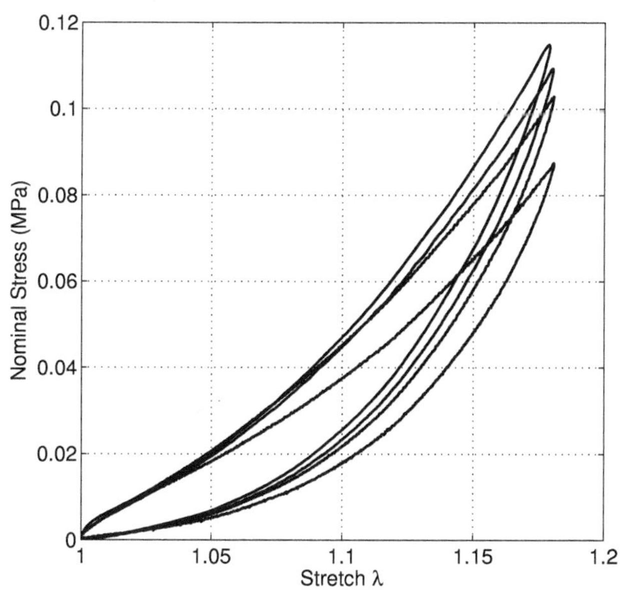

Figure 6. Comparison of the rate-dependent responses of a preconditioned Manduca muscle. Engineering strain rates are $\dot{\varepsilon} = 0.0144$ s^{-1}, $\dot{\varepsilon} = 0.072$ s^{-1}, $\dot{\varepsilon} = 0.36$ s^{-1} and $\dot{\varepsilon} = 1.8$ s^{-1}. Reproduced from Dorfmann et al. (2008).

deformation for a stimulated muscle is shown by the stress–deformation response in Figure 3.

4 Basic Equations

The mechanical behavior of hard tissues, such as bones and teeth, can be described using infinitesimal deformation theories. Under this condition, no distinction is made between the current deformed configuration and the undeformed natural reference state. In contrast, soft tissue can sustain finite deformations and therefore nonlinear theories must be used to describe the mechanical response. Existing finite deformation models may be distinguished based on a micromechanical or phenomenological approach. Micromechanical models, Cowin (2004) used the term mechanistic models, are less developed, rely on biological mechanisms, provide insight into chemical processes and interaction and may also describe the rearrangements of the microstructure. Phenomenological models, on the other hand, provide

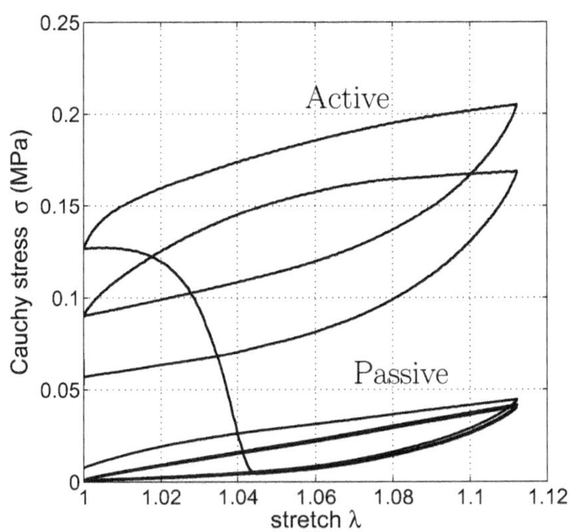

Figure 7. Periodic uniaxial loading–unloading data of a caterpillar muscle showing the Cauchy stress versus the stretch λ. Tetanic stimulation is administrated during the third unloading producing an increasing in stress with decreasing stretch. Reproduced with permission from Paetsch et al. (2012).

a descriptive and predictive capability for the mechanical response of soft tissue. They are based on a continuum mechanics approach and, in the simplest situation, are capable of simulating the nonlinear elastic behavior of isotropic materials.

Microscopic analysis and histology show that the microstructure of muscles is highly complex, containing aligned actin and myosin fibers embedded within an amorphous material (Rheuben and Kammer, 1980). In addition, muscles are now known to contain molecularly massive elastic proteins such as titin (Tskhovrebova and Trinick, 2003; Tskhovrebova et al., 1997) with folded domains that result in asymmetric, time-dependent stretching and shortening during load cycling. We do not attempt to incorporate the mechanical response and the interactions of the microstructural constituents, but focus on a purely phenomenological theory with no dependence on time.

We start by assuming that a muscle can be treated as a transversely isotropic nonlinear hyperelastic material with preferred orientation parallel to the longitudinal direction of the muscle. Experimental data and a detailed discussion on the mechano-chemical interaction of a Manduca muscle are given in Dorfmann et al. (2007). The objective is to develop a constitutive model that accounts for the observed material response, including a continuous transition from passive to active state. We assume a quasi-static context and do not allow material properties and deformation to depend on time.

In this section, we introduce the necessary notations to describe the kinematics, summarizing appropriate stress and deformation tensors along with their associated invariants. We also briefly summarize the main aspects of the theory of pseudo-elasticity originally introduced by Ogden and Roxburgh (1999). The interested reader, for all noteworthy details of the theory of nonlinear mechanics of continua, is referred to the classic book by Ogden (1997).

4.1 Kinematics

To describe a deformation, we denote the stress-free reference configuration of the subject by $\mathcal{B}_0$ and identify a generic material point by its position vector $\mathbf{X}$ relative to an arbitrarily chosen origin. Application of mechanical forces deforms the body, so that the point $\mathbf{X}$ occupies the new position $\mathbf{x} = \chi(\mathbf{X})$ in the deformed configuration $\mathcal{B}$. The vector field χ describes the deformation of the body and assigns to each point $\mathbf{X}$ a unique position $\mathbf{x}$ in $\mathcal{B}$ and vice versa. In other words, the deformation function χ is a one-to-one mapping with suitable regularity properties.

The deformation gradient tensor $\mathbf{F}$ relative to $\mathcal{B}_0$, is defined by

$$\mathbf{F} = \mathrm{Grad}\,\mathbf{x}, \tag{1}$$

Grad being the gradient operator with respect to $\mathbf{X}$. We also adopt the standard notation $J = \det \mathbf{F}$, with the convention $J > 0$.

The deformation gradient can be decomposed according to the unique polar decomposition

$$\mathbf{F} = \mathbf{RU} = \mathbf{VR}, \tag{2}$$

where $\mathbf{R}$ is a proper orthogonal tensor and $\mathbf{U}$ and $\mathbf{V}$ are positive definite and symmetric, respectively the right and left stretch tensors. These can be expressed in spectral form. For $\mathbf{U}$, for example, we have the spectral decomposition

$$\mathbf{U} = \sum_{i=1}^{3} \lambda_i \mathbf{u}^{(i)} \otimes \mathbf{u}^{(i)}, \tag{3}$$

where the principal stretches $\lambda_i > 0, i = 1, 2, 3$, are the eigenvalues of $\mathbf{U}$, $\mathbf{u}^{(i)}$ are the (unit) eigenvectors of $\mathbf{U}$, and $\otimes$ denotes the tensor product. The volume change of many biological tissues is very small and we therefore assume that during all admissible deformations no change in volume occurs, i.e. the deformations are isochoric. Using equations (2) and (3), we have

$$J = \det \mathbf{F} = \det \mathbf{U} = \det \mathbf{V} = \lambda_1 \lambda_2 \lambda_3 \equiv 1. \tag{4}$$

Using the polar decomposition (2), we define

$$\mathbf{C} = \mathbf{F}^{\mathrm{T}}\mathbf{F} = \mathbf{U}^2, \qquad \mathbf{B} = \mathbf{F}\mathbf{F}^{\mathrm{T}} = \mathbf{V}^2, \tag{5}$$

which denote the right and left Cauchy–Green deformation tensors respectively. According to the theory of invariants (Spencer, 1971), there exist three principal invariants for $\mathbf{C}$, equivalently $\mathbf{B}$, defined by

$$I_1 = \mathrm{tr}\,\mathbf{C}, \quad I_2 = \frac{1}{2}\left[(\mathrm{tr}\,\mathbf{C})^2 - \mathrm{tr}\,(\mathbf{C}^2)\right], \quad I_3 = \det \mathbf{C} = J^2, \tag{6}$$

where tr is the trace of a second-order tensor. Alternatively, in terms of principal stretches, the invariants I_1, I_2, I_3 are expressed as

$$I_1 = \lambda_1^2 + \lambda_2^2 + \lambda_3^2, \quad I_2 = \lambda_1^2\lambda_2^2 + \lambda_2^2\lambda_3^2 + \lambda_3^2\lambda_1^2, \quad I_3 = \lambda_1^2\lambda_2^2\lambda_3^2. \tag{7}$$

Histological analyses of a muscle suggest anisotropy in the microstructural composition associated with the basic contractile units that consist principally of highly organized thin and thick filaments. In a macroscopic sense, the muscle has a preferred orientation and can therefore be regarded as transversely isotropic. Let the unit vector $\mathbf{M}$ define the preferred direction in the reference configuration $\mathcal{B}_0$. Additional invariants, denoted I_4 and I_5 and associated with the direction $\mathbf{M}$, are given by

$$I_4 = \mathbf{M}\cdot\mathbf{C}\,\mathbf{M} = \mathbf{FM}\cdot\mathbf{FM} = \mathbf{m}\cdot\mathbf{m}, \tag{8}$$

$$I_5 = \mathbf{M}\cdot\mathbf{C}^2\mathbf{M} = \mathbf{m}\cdot\mathbf{B}\,\mathbf{m}, \tag{9}$$

where we have introduced the notation $\mathbf{m} = \mathbf{FM}$ to define the mapping of $\mathbf{M}$ under the deformation $\mathbf{F}$.

There is a simple geometric interpretation for the invariant I_4. The square root of I_4 provides the stretch of the material in the direction $\mathbf{M}$. There is no similar kinematical interpretation for I_5.

4.2 Hyperelasticity

The theory of hyperelasticity characterizes the elastic response of a body by a strain energy function W defined per unit volume in the reference configuration $\mathcal{B}_0$. For a homogeneous material, W depends only on the deformation gradient $\mathbf{F}$ and we write $W = W(\mathbf{F})$. In this subsection, we restrict attention to incompressible materials, subject to the constraint (4). Therefore, the nominal stress tensor $\mathbf{S}$ and the symmetric Cauchy stress tensor $\boldsymbol{\sigma}$ are given, respectively, by

$$\mathbf{S} = \frac{\partial W}{\partial \mathbf{F}} - p\mathbf{F}^{-1}, \qquad \boldsymbol{\sigma} = \mathbf{F}\frac{\partial W}{\partial \mathbf{F}} - p\mathbf{I}, \tag{10}$$

where p is an arbitrary hydrostatic pressure. Equations (10) show that for an incompressible material, the Cauchy stress $\boldsymbol{\sigma}$ and the nominal stress $\mathbf{S}$ are related by $\boldsymbol{\sigma} = \mathbf{FS}$.

The unit vector $\mathbf{M}$ describes the preferred direction of a transversely isotropic material in the reference configuration $\mathcal{B}_0$. The material response is indifferent to an arbitrary rotation about the direction $\mathbf{M}$ and is not altered by a change of direction from $\mathbf{M}$ to $-\mathbf{M}$. Following the analysis of such materials given by Spencer (1971) and Ogden (2001), we define a transversely isotropic material as one for which the strain energy W is an isotropic function of the two tensors, $\mathbf{C} = \mathbf{F}^{\mathrm{T}}\mathbf{F}$ and $\mathbf{M}\otimes\mathbf{M}$. For an incompressible material, $I_3 \equiv 1$ and the form of W is reduced to dependence on the four independent invariants I_1, I_2, I_4, I_5. We write $W = W(I_1, I_2, I_4, I_5)$.

In order to obtain the explicit expressions of the nominal stress tensor $\mathbf{S}$ and the associated Cauchy stress tensor $\boldsymbol{\sigma}$, the derivatives of the strain invariants with respect to $\mathbf{F}$ are needed. Following standard derivation rules, these are given by

$$\frac{\partial I_1}{\partial \mathbf{F}} = 2\mathbf{F}^{\mathrm{T}}, \qquad \frac{\partial I_2}{\partial \mathbf{F}} = 2(I_1\mathbf{F}^{\mathrm{T}} - \mathbf{F}^{\mathrm{T}}\mathbf{F}\mathbf{F}^{\mathrm{T}}), \tag{11}$$

$$\frac{\partial I_4}{\partial \mathbf{F}} = 2\mathbf{M} \otimes \mathbf{FM}, \qquad \frac{\partial I_5}{\partial \mathbf{F}} = 2(\mathbf{M} \otimes \mathbf{FCM} + \mathbf{CM} \otimes \mathbf{FM}). \tag{12}$$

A direct calculation of (10), using equations (11) and (12) leads to

$$\begin{aligned} \mathbf{S} &= 2(W_1 + I_1 W_2)\mathbf{F}^{\mathrm{T}} - 2W_2\mathbf{C}\,\mathbf{F}^{\mathrm{T}} + 2\,W_4\mathbf{M} \otimes \mathbf{FM} \\ &\quad + 2\,W_5(\mathbf{M} \otimes \mathbf{FCM} + \mathbf{CM} \otimes \mathbf{FM}) - p\,\mathbf{F}^{-1}, \end{aligned} \tag{13}$$

$$\begin{aligned} \boldsymbol{\sigma} &= 2(W_1 + I_1 W_2)\mathbf{B} - 2W_2\mathbf{B}^2 + 2\,W_4\mathbf{m} \otimes \mathbf{m} \\ &\quad + 2\,W_5(\mathbf{m} \otimes \mathbf{Bm} + \mathbf{Bm} \otimes \mathbf{m}) - p\,\mathbf{I}, \end{aligned} \tag{14}$$

where the abbreviations $W_i = \partial W/\partial I_i, i = 1, 2, 4, 5$, have been introduced. When the dependence on I_4 and I_5 in equations (13) and (14) is omitted, the associated expressions for an isotropic material are obtained.

5 Model Development

5.1 Kinematics of Activation

In Section 3.3 we presented uniaxial stress–stretch data of a Manduca muscle transitioning from a passive to an active state. The mechanical response of the muscle tissue exhibits two primary characteristics: the active muscle is no longer stress free when returned to its reference configuration and the loading and unloading paths do not coincide. Paetsch et al. (2012) proposed a phenomenological constitutive model which accounts for these two primary characteristics of muscle tissue. The model was modified and adapted for finite element implementation and used to simulate the nonlinear behavior of bioactive devices (Paetsch and Dorfmann, 2013). In the following, we briefly present some key attributes of the constitutive model while details are provided in the aforementioned references.

To understand the modeling approach, it is helpful to review the basic muscle physiology. Tubular sarcomeres provide the basic contractile units of muscle fibers and are comprised of overlapping actin and myosin proteins. During activation, an increased concentration of calcium ions (Ca^+) causes an increased interaction between the actin and myosin proteins, resulting in a ratchet-like motion and developing tension in the fiber. When the stimulus is removed and Ca^+ concentrations drop, the interactions are reduced along with the tension, and the muscle returns to its inactive state.

The change in muscle conformation due to both activation and application of mechanical loads is therefore considered to occur in two stages.

First, take the unstimulated, passive tissue to occupy the natural (stress-free) configuration denoted $\mathcal{B}_0$. After the introduction of a stimulus, with no applied tractions or body forces, the muscle body occupies a new natural (stress-free) configuration, denoted $\hat{\mathcal{B}}$, and the change in conformation of the body is described by the active deformation tensor, $\mathbf{F}_0$. It is important to note that $\mathbf{F}_0$ is dependent on the physiology of the muscle tissue and not the result of a gradient operation. Secondly, after the application of mechanical loads, the body moves to the current configuration $\mathcal{B}$. To account for this behavior, we introduce a multiplicative decomposition of the deformation gradient $\mathbf{F}$ such

$$\mathbf{F} = \hat{\mathbf{F}}\mathbf{F}_0, \tag{15}$$

where $\hat{\mathbf{F}}$ is related to the deformation of the body resulting from the applied load, is measured relative to the active natural configuration $\hat{\mathcal{B}}$ and can be interpreted as the elastic deformation of the cross-bridges formed during stimulation. The decomposition of the deformation tensor is similar to the approach taken by Rajagopal and Wineman (1992) and Demirkoparan et al. (2009) for the formation and breaking of cross-links in polymeric materials and has been used for muscle tissue by Cherubini et al. (2008), and Nardinocchi and Teresi (2007).

The introduction of $\hat{\mathbf{F}}$ leads to the definition of the related right and left Cauchy–Green tensors, respectively

$$\hat{\mathbf{C}} = \mathbf{F}_0^{-\mathrm{T}}\mathbf{C}\mathbf{F}_0^{-1}, \qquad \hat{\mathbf{B}} = \mathbf{F}\mathbf{C}_0^{-1}\mathbf{F}^{\mathrm{T}}, \tag{16}$$

where we have made use of the connection $\mathbf{C}_0 = \mathbf{F}_0^{\mathrm{T}}\mathbf{F}_0$. Invariants analogous to those given in equations (8) and (9) may also be defined in terms of $\mathbf{F}_0$ and $\hat{\mathbf{F}}$, and we write

$$I_{4_0} = \mathbf{M} \cdot \mathbf{C}_0\mathbf{M}, \qquad \hat{I}_4 = \hat{\mathbf{M}} \cdot \hat{\mathbf{C}}\hat{\mathbf{M}}, \tag{17}$$

where $\hat{\mathbf{M}}$ is a unit vector in the active natural configuration $\hat{\mathcal{B}}$ and is given by

$$\hat{\mathbf{M}} = I_{4_0}^{-1/2}\mathbf{F}_0\mathbf{M}. \tag{18}$$

Equation $(17)_1$ provides the square of the stretch of muscle fibers in $\hat{\mathcal{B}}$ relative to the passive natural configuration $\mathcal{B}_0$ and, similarly, $\hat{I}_4$ defines the square of the stretch of the fibers in $\mathcal{B}$ relative to the active natural configuration $\hat{\mathcal{B}}$. By substituting (15) into the expression for I_4 given in (8), it can be shown that the multiplicative decomposition of the deformation gradient carries over to I_4 as

$$I_4 = \hat{I}_4 I_{4_0}. \tag{19}$$

The remaining invariants are provided here for reference and we write

$$I_{5_0} = \mathbf{M} \cdot \mathbf{C}_0^2 \mathbf{M} = I_{4_0} \hat{\mathbf{M}} \cdot \mathbf{B}_0 \hat{\mathbf{M}}, \tag{20}$$

where $\mathbf{B}_0 = \mathbf{F}_0 \mathbf{F}_0^{\mathrm{T}}$ and, similarly

$$\hat{I}_5 = \hat{\mathbf{M}} \cdot \hat{\mathbf{C}}^2 \hat{\mathbf{M}} = I_{4_0}^{-1} \mathbf{m} \cdot \hat{\mathbf{B}} \mathbf{m}, \tag{21}$$

where, again, we have used $\mathbf{m} = \mathbf{FM}$, the fiber direction in the current configuration (which is not in general a unit vector).

In the subsequent developments, the derivatives of the additional invariants related to the active and passive fibers relative to the deformation gradient are required, including the related stress tensors. The non-zero derivatives are given by

$$\frac{\partial \hat{I}_4}{\partial \mathbf{F}} = \frac{\partial \left(I_4 I_{4_0}^{-1} \right)}{\partial \mathbf{F}} = 2 I_{4_0}^{-1} \mathbf{M} \otimes \mathbf{FM}, \tag{22}$$

$$\frac{\partial \hat{I}_5}{\partial \mathbf{F}} = 2 I_{4_0}^{-1} \left(\mathbf{M} \otimes \mathbf{F} \mathbf{C}_0^{-1} \mathbf{C} \mathbf{M} + \mathbf{C}_0^{-1} \mathbf{C} \mathbf{M} \otimes \mathbf{FM} \right), \tag{23}$$

where the intermediate configuration $\hat{\mathcal{B}}$ is independent of the deformation $\mathbf{F}$.

The nominal stress tensor for transversely isotropic materials with fibers having an additional stress-free reference configuration $\hat{\mathcal{B}}$ can now be expressed as

$$\begin{aligned} \mathbf{S} = {} & 2(W_1 + I_1 W_2)\mathbf{F}^{\mathrm{T}} - 2W_2 \mathbf{C}\,\mathbf{F}^{\mathrm{T}} + 2\, I_{4_0}^{-1} W_{\hat{4}} \mathbf{M} \otimes \mathbf{FM} \\ & + 2\, I_{4_0}^{-1} W_{\hat{5}} (\mathbf{M} \otimes \mathbf{F} \mathbf{C}_0^{-1} \mathbf{C} \mathbf{M} + \mathbf{C}_0^{-1} \mathbf{C} \mathbf{M} \otimes \mathbf{FM}) - \hat{p}\, \mathbf{F}^{-1}, \end{aligned} \tag{24}$$

where $W_{\hat{i}}, \hat{i} = 4, 5$, are the partial derivative of the strain energy function with respect to the invariants $\hat{I}_4$ and $\hat{I}_5$ and the value of $\hat{p}$ is in general different from the value of p used in equation (13). The corresponding expression for the Cauchy stress tensor is

$$\begin{aligned} \boldsymbol{\sigma} = {} & 2(W_1 + I_1 W_2)\mathbf{B} - 2W_2 \mathbf{B}^2 + 2\, I_{4_0}^{-1} W_{\hat{4}} \mathbf{m} \otimes \mathbf{m} \\ & + 2\, I_{4_0}^{-1} W_{\hat{5}} (\mathbf{m} \otimes \hat{\mathbf{B}} \mathbf{m} + \hat{\mathbf{B}} \mathbf{m} \otimes \mathbf{m}) - \hat{p}\, \mathbf{I}, \end{aligned} \tag{25}$$

which replaces expression (14).

5.2 Energy Dissipation

Like many other biological tissues, muscles exhibit softening during the strain cycle, where the stress is less during unloading than during the corresponding loading path, a behavior that is also found in polymer-like materials where the applied stretch results in changes to the underlying molecular network (Dorfmann and Ogden, 2004; Dorfmann et al., 2007). The hysteretic behavior observed during loading and unloading is accounted for in Paetsch et al. (2012) and Paetsch and Dorfmann (2013) using the theory of pseudo-elasticity and is summarized here for reference. The theory was originally developed for modeling the behavior of filled rubber (Ogden and Roxburgh, 1999; Dorfmann and Ogden, 2003, 2004), applied to a biological tissue by Franceschini et al. (2006) and eventually adapted for muscle tissue by Dorfmann et al. (2007). We begin by introducing a pseudo-energy function, per unit volume, of the form

$$W = W(\mathbf{F}, \eta), \tag{26}$$

where the scalar parameter η modifies the response of the material. The behavior is no longer elastic and we therefore refer to $W(\mathbf{F}, \eta)$ as a *pseudo-energy* function. Importantly, as shown below, a connection is made between the scalar parameter η and the current deformation gradient $\mathbf{F}$. This allows a continuous change of the strain energy function during the deformation and captures the observed stress-strain behavior.

The additional variable η may be inactive or active; active η modifies the strain energy function and therefore introduces a change in the material response. Let the variable η be inactive during loading and assign it a unit value. We therefore have

$$W_0(\mathbf{F}) = W(\mathbf{F}, 1) \tag{27}$$

for the resulting strain energy function. In (27) and in what follows, the zero subscript, i.e. W_0, is associated with the situation in which η is inactive. For an incompressible material the nominal stress associated with inactive η is then denoted by $\mathbf{S}_0$ and given by

$$\mathbf{S}_0 = \frac{\partial W_0}{\partial \mathbf{F}}(\mathbf{F}) - p_0\mathbf{F}^{-1}, \qquad \det\mathbf{F} = 1. \tag{28}$$

Let the beginning of unloading be the signal to activate η. Then, the nominal stress, again for incompressible materials, is given by

$$\mathbf{S} = \frac{\partial W}{\partial \mathbf{F}}(\mathbf{F}, \eta) + \frac{\partial W}{\partial \eta}(\mathbf{F}, \eta)\frac{\partial \eta}{\partial \mathbf{F}}(\mathbf{F}) - p\mathbf{F}^{-1}, \quad \det\mathbf{F} = 1. \tag{29}$$

This expression simplifies considerably if we select a pseudo-energy function $W(\mathbf{F}, \eta)$ such that

$$\frac{\partial W}{\partial \eta}(\mathbf{F}, \eta) = 0, \tag{30}$$

which defines η in terms of $\mathbf{F}$; see Ogden and Roxburgh (1999). We may write the solution to equation (30) formally as

$$\eta = \hat{\eta}(\mathbf{F}). \tag{31}$$

Then, the expression for the nominal stress (29) becomes

$$\mathbf{S} = \frac{\partial W}{\partial \mathbf{F}}(\mathbf{F}, \eta) - p\mathbf{F}^{-1}, \quad \det \mathbf{F} = 1, \tag{32}$$

whether or not η is active. When η is active the right-hand side is evaluated for η given by (31).

Thus far we have not specified the explicit form of $W(\mathbf{F}, \eta)$, nor the form of the function $\hat{\eta}(\mathbf{F})$. Following the approach taken in Ogden and Roxburgh (1999), Ogden (2001), Dorfmann and Ogden (2003) and Dorfmann and Ogden (2004) we use the expression

$$W(\mathbf{F}, \eta) = \eta W_0(\mathbf{F}) + \phi(\eta), \tag{33}$$

where the function ϕ accounts for the energy dissipated during cyclic loading. For consistency with (27), the function ϕ, for inactive η, must satisfy the condition $\phi(1) = 1$ and from equation (30) we have the additional requirement

$$\phi'(\eta) = -W_0(\mathbf{F}). \tag{34}$$

For a more general discussion of the theory of pseudo-elasticity the reader is referred to Lazopoulos and Ogden (1998) and Ogden and Roxburgh (1999), and for the use of pseudo-elasticity for muscle tissue is given in Dorfmann et al. (2007) and Dorfmann et al. (2008).

5.3 Compressible Materials

The internal constraint of incompressibility is adopted, in general, when constitutive equations of biological materials are derived. The use of the internal constraint is justified because many biological tissues are almost incompressible. However, the formulation based on compressible materials is more convenient when a finite element implementation is to be performed, which is the case in the present study. Therefore, in the following we relax the constraint of incompressibility (4) and assume the general case of compressible materials with $J \neq 1$. Following the developments given by Ogden (1976, 1978) we split the deformation gradient into a dilational (volume-changing) and isochoric (volume-preserving) contribution in the form

$$\mathbf{F} = (J^{1/3}\mathbf{I})\bar{\mathbf{F}} = J^{1/3}\bar{\mathbf{F}}, \tag{35}$$

where $J^{1/3}\mathbf{I}$ describes the volume-changing part and $\bar{\mathbf{F}}$ is the volume-preserving part of the deformation. It follows that

$$\det \bar{\mathbf{F}} = 1. \tag{36}$$

Recall the polar decomposition of the deformation gradient $\mathbf{F} = \mathbf{RU}$. Inserting into equation (35) defines the modified stretch tensor $\bar{\mathbf{U}}$

$$\bar{\mathbf{F}} = J^{-1/3}\mathbf{RU} = \mathbf{R}(J^{-1/3}\mathbf{U}) = \mathbf{R}\bar{\mathbf{U}}. \tag{37}$$

The eigenvalues of the modified stretch tensor $\bar{\mathbf{U}}$ can be obtained by writing $\mathbf{U}$ in spectral form as shown by equation (3). We have

$$\bar{\mathbf{U}} = J^{-1/3}\mathbf{U} = J^{-1/3}\sum_{i=1}^{3}\lambda_i \mathbf{N}^{(i)} \otimes \mathbf{N}^{(i)} = \sum_{i=1}^{3}\bar{\lambda}_i \mathbf{N}^{(i)} \otimes \mathbf{N}^{(i)}, \tag{38}$$

where the modified principal stretches are defined as

$$\bar{\lambda}_i = J^{-1/3}\lambda_i, \tag{39}$$

which satisfy the incompressibility condition

$$\bar{\lambda}_1\bar{\lambda}_2\bar{\lambda}_3 = 1. \tag{40}$$

The right Cauchy–Green deformation tensor $\mathbf{C}$ can now be split into an isochoric and a dilatational component as

$$\mathbf{C} = \mathbf{F}^{\mathrm{T}}\mathbf{F} = J^{2/3}\,\bar{\mathbf{F}}^{\mathrm{T}}\bar{\mathbf{F}} = (J^{2/3}\mathbf{I})\,\bar{\mathbf{C}}, \tag{41}$$

where the volume preserving deformation is given by $\bar{\mathbf{C}}$ with $\det \bar{\mathbf{C}} = 1$ and the volume-changing contribution is specified by $J^{2/3}\mathbf{I}$. The principal invariants of $\bar{\mathbf{C}}$ are given by

$$\bar{I}_1 = \mathrm{tr}\,\bar{\mathbf{C}}, \quad \bar{I}_2 = \frac{1}{2}\left[(\mathrm{tr}\,\bar{\mathbf{C}})^2 - \mathrm{tr}\left(\bar{\mathbf{C}}^2\right)\right], \quad \bar{I}_3 \equiv 1, \tag{42}$$

where we recall that tr denotes the trace of the second-order tensor.

As discussed earlier, muscle is considered a fiber-reinforced material with a preferred direction given by its internal structure, namely fascicles comprised of myofibril bundles. The direction of the fiber distribution is locally specified by the unit vector $\mathbf{M}$ in the reference configuration $\mathcal{B}_0$. To define a decoupled energy function that accounts for the dilatational and isochoric contributions of the total deformation, it is convenient to modify the invariants given in equations (8) and (9) as

$$\bar{I}_4 = \mathbf{M}\cdot\bar{\mathbf{C}}\mathbf{M}, \quad \bar{I}_5 = \mathbf{M}\cdot\bar{\mathbf{C}}^2\mathbf{M}. \tag{43}$$

For details of the kinematics of isotropic and transversely isotropic materials, we refer to, for example, Ogden (2001).

In Section 5.1 we introduced the kinematic variables necessary to describe the activation process. We now assume that the transition from the natural configuration $\mathcal{B}_0$ to $\hat{\mathcal{B}}$, generated by the introduction of an electrophysiological stimulus and with no applied mechanical loads, occurs at constant volume. The change in conformation is then given by the active deformation tensor $\bar{\mathbf{F}}_0$ and the corresponding right and left Cauchy–Green tensors are given by

$$\bar{\mathbf{C}}_0 = \bar{\mathbf{F}}_0^{\mathrm{T}} \bar{\mathbf{F}}_0, \quad \bar{\mathbf{B}}_0 = \bar{\mathbf{F}}_0 \bar{\mathbf{F}}_0^{\mathrm{T}}. \tag{44}$$

The corresponding invariants related to the fiber direction are

$$\bar{I}_{4_0} = \mathbf{M} \cdot \bar{\mathbf{C}}_0 \mathbf{M}, \quad \bar{I}_{5_0} = \mathbf{M} \cdot \bar{\mathbf{C}}_0^2 \mathbf{M}, \tag{45}$$

where, in the absence of applied mechanical loads, the square root of $\bar{I}_{4_0}$ still represents the stretch in the fiber direction resulting from the transition between passive and active states. When muscle fibers contract we have $\bar{I}_{4_0} < 1$. The decomposition of the total deformation gradient $\mathbf{F}$ into dilatational and isochoric contributions allows us to rewrite equation (15) in the form

$$\bar{\mathbf{F}} = \hat{\bar{\mathbf{F}}} \bar{\mathbf{F}}_0. \tag{46}$$

We use the decomposition (46) to define the elastic deformation of active muscle fibers in the current configuration $\mathcal{B}$ relative to the intermediate configuration $\hat{\mathcal{B}}$, which is given by $\hat{\bar{\mathbf{F}}} = \bar{\mathbf{F}} \bar{\mathbf{F}}_0^{-1}$. The corresponding right and left Cauchy–Green tensors, denoted respectively by $\hat{\bar{\mathbf{C}}}$ and $\hat{\bar{\mathbf{B}}}$, are defined as

$$\hat{\bar{\mathbf{C}}} = \bar{\mathbf{F}}_0^{-\mathrm{T}} \bar{\mathbf{C}} \bar{\mathbf{F}}_0^{-1}, \quad \hat{\bar{\mathbf{B}}} = \bar{\mathbf{F}} \bar{\mathbf{C}}_0^{-1} \bar{\mathbf{F}}^{\mathrm{T}}, \tag{47}$$

with the corresponding invariants given by

$$\hat{\bar{I}}_4 = \hat{\mathbf{M}} \cdot \hat{\bar{\mathbf{C}}} \hat{\mathbf{M}}, \quad \hat{\bar{I}}_5 = \hat{\mathbf{M}} \cdot \hat{\bar{\mathbf{C}}}^2 \hat{\mathbf{M}}, \tag{48}$$

where we recall from (18) that the unit vector $\hat{\mathbf{M}}$ specifies the fiber direction in the intermediate configuration. Using the decomposition (46) allows us to write $\bar{I}_4$, which is defined in $(43)_1$, in the alternative form

$$\bar{I}_4 = \hat{\bar{I}}_4 \bar{I}_{4_0}, \tag{49}$$

with no similar decomposition possible for $\hat{\bar{I}}_5$.

In general, $\bar{\mathbf{F}}_0$ may depend on $\mathbf{F}$. However, for simplicity, we assume that $\bar{\mathbf{F}}_0$ depends only on the internal structure of the muscle's sarcomere. For reference we include the derivatives of $\hat{\bar{I}}_4$ and $\hat{\bar{I}}_5$ with respect to $\mathbf{F}$ as

$$\frac{\partial \hat{\bar{I}}_4}{\partial \mathbf{F}} = \frac{\partial \left(\bar{I}_4 \bar{I}_{4_0}^{-1}\right)}{\partial \mathbf{F}} = 2J^{-1/3} \bar{I}_{4_0}^{-1} \left[\mathbf{M} \otimes \bar{\mathbf{F}}\mathbf{M} - \frac{1}{3} \bar{I}_4 \bar{\mathbf{F}}^{-1}\right], \tag{50}$$

$$\begin{aligned} \frac{\partial \hat{\bar{I}}_5}{\partial \mathbf{F}} = & \, 2J^{-1/3} \bar{I}_{4_0}^{-1} \left[\mathbf{M} \otimes \bar{\mathbf{F}} \bar{\mathbf{C}}_0^{-1} \bar{\mathbf{C}}\mathbf{M} + \bar{\mathbf{C}}_0^{-1} \bar{\mathbf{C}}\mathbf{M} \otimes \bar{\mathbf{F}}\mathbf{M} \right. \\ & \left. - \frac{2}{3} \left(\mathbf{M} \cdot \bar{\mathbf{C}} \bar{\mathbf{C}}_0^{-1} \bar{\mathbf{C}}\mathbf{M}\right) \bar{\mathbf{F}}^{-1}\right]. \end{aligned} \tag{51}$$

The pseudo-energy function, which for incompressible materials has the form (33), modifies to

$$W(\mathbf{F}, \eta) = \eta \bar{W}_0(\bar{\mathbf{F}}) + \phi(\eta) + U(J), \tag{52}$$

where we note that stress softening is now attributed to the isochoric component $\bar{W}_0(\bar{\mathbf{F}})$. The volumetric energy contribution during loading–unloading is given by $U(J)$ and the function $\phi(\eta)$ still accounts for the energy dissipated during cyclic loading. During primary loading, η is inactive and equal to unity with $W(\mathbf{F}, 1) = \bar{W}_0(\bar{\mathbf{F}}) + \phi(1) + U(J)$, where $\phi(1) \equiv 0$. During unloading, η is active and scales the isochoric energy contribution as $\eta \bar{W}_0$, where $\eta \leqslant 1$.

6 Numerical Implementation

Consider the deformation of a body from the reference configuration $\mathcal{B}_0$ to the current configuration $\mathcal{B}$, which is given by $\mathbf{x} = \boldsymbol{\chi}(\mathbf{X})$. Constitutive equations in Abaqus must be given in incremental form and we consider a linearized increment in $\mathbf{x}$, denoted $\dot{\mathbf{x}}$, which for convenience we denote by $\mathbf{u}(\mathbf{x})$

$$\mathbf{u}(\mathbf{x}) = \mathbf{u}\left(\boldsymbol{\chi}(\mathbf{X})\right) = \dot{\mathbf{x}}(\mathbf{X}). \tag{53}$$

The corresponding incremental deformation gradient has the form

$$\dot{\mathbf{F}} = \frac{\partial \mathbf{u}}{\partial \mathbf{x}} \frac{\partial \mathbf{x}}{\partial \mathbf{X}} = \mathbf{l}\mathbf{F}, \tag{54}$$

where $\mathbf{l} = \operatorname{grad} \mathbf{u}$ is the incremental displacement gradient. Here and henceforth a superimposed dot represents an increment in the quantity concerned. Equation (54) has Cartesian components

$$\dot{F}_{i\alpha} = \frac{\partial u_i}{\partial x_j} \frac{\partial x_j}{\partial X_\alpha} = l_{ij} F_{j\alpha}. \tag{55}$$

The incremental displacement gradient $\mathbf{l}$ can be decomposed into the sum of a symmetric and an antisymmetric part as $\mathbf{l} = \mathbf{d} + \mathbf{w}$, where

$$\mathbf{d} = \frac{1}{2}\left(\mathbf{l} + \mathbf{l}^{\mathrm{T}}\right), \quad \mathbf{w} = \frac{1}{2}\left(\mathbf{l} - \mathbf{l}^{\mathrm{T}}\right), \tag{56}$$

which are, respectively, symmetry and antisymmetric.

Constitutive equations in Abaqus must be of the form connecting the increment of the Kirchhoff stress, denoted $\dot{\overline{J\boldsymbol{\sigma}}}$, and the rate of deformation $\mathbf{d}$. In general, the stress increment resulting from the evaluation of $\dot{\overline{J\boldsymbol{\sigma}}}$ is not objective and, for continuum elements, Abaqus uses the Jaumann rate of the Kirchhoff stress (Simulia, 2009). Using the notation defined by Prot and Skallerud (2009) we write the Jaumann rate of $J\boldsymbol{\sigma}$ as

$$(J\boldsymbol{\sigma})^{\nabla J} = \dot{\overline{J\boldsymbol{\sigma}}} - J(\mathbf{w}\boldsymbol{\sigma} - \boldsymbol{\sigma}\mathbf{w}) = \mathbb{C}^{\nabla J} : \mathbf{d}, \tag{57}$$

where the superscript $^{\nabla J}$ is used to denote the increment determined from the Jaumann rate and $\mathbb{C}^{\nabla J}$ is the corresponding symmetric fourth-order elasticity tensor. In equation (57) we have introduced the operator : defined, for a pair of second-order tensors $\mathbf{A}$ and $\mathbf{B}$, as the double contraction $A_{ij}B_{ij}$. By equating the second and third terms with the fourth term in (57) and solving for $\dot{\overline{J\boldsymbol{\sigma}}}$, we obtain

$$\dot{\overline{J\boldsymbol{\sigma}}} = J\left(\boxed{J^{-1}\mathbb{C}^{\nabla J}} : \mathbf{d} + \mathbf{w}\boldsymbol{\sigma} - \boldsymbol{\sigma}\mathbf{w}\right). \tag{58}$$

Equations (57) and (58) can be written in component forms as

$$(J\sigma_{ij})^{\nabla J} = \dot{\overline{J\sigma_{ij}}} - J(w_{im}\sigma_{mj} - \sigma_{im}w_{mj}) = \mathbb{C}^{\nabla J}_{ijkl} d_{kl}$$

and

$$\dot{\overline{J\sigma_{ij}}} = J(\boxed{J^{-1}\mathbb{C}^{\nabla J}_{ijkl}}\, d_{kl} + w_{im}\sigma_{mj} - \sigma_{im}w_{mj}).$$

Abaqus (Simulia, 2009) requires that the fourth-order elasticity tensor be implemented as

$$\mathbb{C} = J^{-1}\mathbb{C}^{\nabla J}, \tag{59}$$

and we refer to the Abaqus Theory Manual for a detailed presentation. Using a decoupled energy function gives expressions for the isochoric and volumetric components of the Cauchy stress tensor. We denote the increments of the corresponding Kirchhoff stress rates as $\dot{\overline{J\bar{\boldsymbol{\sigma}}}}$ and $\dot{\overline{J\boldsymbol{\sigma}_{\mathrm{vol}}}}$. Equation (58) is then additively decomposed as

$$\dot{\overline{J\bar{\boldsymbol{\sigma}}}} = J(J^{-1}\bar{\mathbb{C}}^{\nabla \mathrm{J}} : \mathbf{d} + \mathbf{w}\bar{\boldsymbol{\sigma}} - \bar{\boldsymbol{\sigma}}\mathbf{w}), \tag{60}$$

and

$$\overline{J\boldsymbol{\sigma}_{\text{vol}}}^{\cdot} = J(J^{-1}\mathbb{C}_{\text{vol}}^{\nabla \text{J}} : \mathbf{d}), \tag{61}$$

where $\bar{\mathbb{C}}^{\nabla \text{J}}$ is the isochoric part of the elasticity tensor and $\mathbb{C}_{\text{vol}}^{\nabla \text{J}}$ is the volumetric part. The final form of the elasticity tensor is then given by

$$\mathbb{C} = J^{-1}(\bar{\mathbb{C}}^{\nabla \text{J}} + \mathbb{C}_{\text{vol}}^{\nabla \text{J}}). \tag{62}$$

6.1 Illustrative Example

It is instructive to provide a step-by-step presentation on how the incremental form of the Kirchhoff stress and the fourth-order elasticity tensor of a compressible transversely isotropic material are derived. In this section, for purpose of illustration, we consider the reduced energy function

$$W = \bar{W}(\bar{I}_1, \bar{I}_4) + U(J), \tag{63}$$

where $\bar{W}(\bar{I}_1, \bar{I}_4)$ and $U(J)$ represent the isochoric and volumetric contributions, respectively. For reference, we recall that the invariants $\bar{I}_1$ and $\bar{I}_4$ are defined in (42) and (43) and J is the determinant of $\mathbf{F}$. Furthermore, for this illustration, we provide all results in Cartesian components.

The expression for the Cauchy stress $\boldsymbol{\sigma}$, which for constrained materials is given in equation $(10)_2$, is replaced by

$$\boldsymbol{\sigma} = J^{-1}\mathbf{F}\frac{\partial W}{\partial \mathbf{F}}, \tag{64}$$

and to clarify notation, the component form is given as

$$\sigma_{ij} = J^{-1}F_{i\alpha}\frac{\partial W}{\partial F_{j\alpha}}, \tag{65}$$

where the switched order of the indices for the derivative with respect to the non-symmetric tensor $\mathbf{F}$ is noted, consistent with Ogden (1997). The Cauchy stress, using the energy form (63), is calculated as

$$\sigma_{ij} = J^{-1}F_{i\alpha}\frac{\partial W}{\partial F_{j\alpha}} = J^{-1}F_{i\alpha}\left[\bar{W}_1\frac{\partial \bar{I}_1}{\partial F_{j\alpha}} + \bar{W}_4\frac{\partial \bar{I}_4}{\partial F_{j\alpha}} + \frac{\mathrm{d}U}{\mathrm{d}J}\frac{\mathrm{d}J}{\mathrm{d}F_{j\alpha}}\right], \tag{66}$$

where $\bar{W}_k = \partial \bar{W}/\partial \bar{I}_k, k = 1, 4$ and

$$\begin{aligned}
\frac{\partial \bar{I}_1}{\partial F_{j\alpha}} &= 2J^{-1/3}\left[\bar{F}_{j\alpha} - \frac{1}{3}\bar{I}_1\bar{F}_{\alpha j}^{-1}\right], \\
\frac{\partial \bar{I}_4}{\partial F_{j\alpha}} &= 2J^{-1/3}\left[\bar{F}_{j\beta}M_\beta M_\alpha - \frac{1}{3}\bar{I}_4\bar{F}_{\alpha j}^{-1}\right], \\
\frac{\mathrm{d}J}{\mathrm{d}F_{j\alpha}} &= JF_{\alpha j}^{-1}.
\end{aligned} \tag{67}$$

Substituting expressions (67) into (66) gives the components of the Kirchhoff stress tensor

$$J\sigma_{ij} = 2\bar{W}_1(\bar{B}_{ij} - \frac{1}{3}\bar{B}_{rr}\delta_{ij}) + \boxed{2\bar{W}_4(m_i m_j - \frac{1}{3}m_r m_r \delta_{ij})} + J\frac{\mathrm{d}U}{\mathrm{d}J}\delta_{ij}, \quad (68)$$

where the boxed quantity represents the fiber contribution and $\mathbf{m} = \bar{\mathbf{F}}\mathbf{M}$, in index notation $m_j = \bar{F}_{j\beta}M_\beta$. Note that $\mathbf{m}$ is now defined differently form its earlier definition $\mathbf{m} = \mathbf{F}\mathbf{M}$.

This equation can be written in a more compact form by introducing the deviatoric operator defined as $\mathrm{dev}(\mathbf{A}) = \mathbb{P} : \mathbf{A}$, where the fourth-order projection tensor $\mathbb{P}$ has components $\mathbb{P}_{ijkl} = \delta_{ik}\delta_{jl} - (1/3)\delta_{ij}\delta_{kl}$ and $\mathbf{A}$ is a second-order symmetric tensor. The component form of $\mathbb{P} : \mathbf{A}$ is $\mathbb{P}_{ijkl}A_{kl}$. Equation (68) can then be written in the more compact form

$$J\boldsymbol{\sigma} = 2\bar{W}_1 \ \mathrm{dev}\bar{\mathbf{B}} + 2\bar{W}_4 \ \mathrm{dev}\,(\mathbf{m}\otimes\mathbf{m}) + \frac{\mathrm{d}U}{\mathrm{d}J}\mathbf{I}. \quad (69)$$

In the remainder of this section, for illustrative purposes, we focus on the fiber contribution outlined in (68) and use the notation

$$J\bar{\sigma}_{ij} = 2\bar{W}_4(m_i m_j - \frac{1}{3}m_r m_r \delta_{ij}) = 2\bar{W}_4\mathbb{P}_{ijkl}m_k m_l. \quad (70)$$

The corresponding increment is then simply

$$\dot{\overline{J\bar{\sigma}_{ij}}} = 2\mathbb{P}_{ijkl}\left(\dot{\bar{W}}_4 m_k m_l + \bar{W}_4\dot{\overline{m_k m_l}}\right). \quad (71)$$

We separately derive the expressions of the incremental quantities $\dot{\bar{W}}_4$ and $\dot{\overline{m_k m_l}}$. We begin with the increment of $\bar{W}_4$

$$\begin{aligned}
\dot{\bar{W}}_4 &= \bar{W}_{44}\frac{\partial \bar{I}_4}{\partial F_{j\alpha}}\dot{F}_{j\alpha} = \bar{W}_{44}\frac{\partial \bar{I}_4}{\partial F_{j\alpha}}l_{jk}F_{k\alpha}, \\
&= 2\bar{W}_{44}\left(m_j m_k - \frac{1}{3}m_m m_m \delta_{jk}\right)l_{jk}, \\
&= 2\bar{W}_{44}m_m m_n\left(\delta_{mj}\delta_{nk} - \frac{1}{3}\delta_{nm}\delta_{jk}\right)l_{jk}, \\
&= 2\bar{W}_{44}m_m m_n\left(\delta_{mj}\delta_{nk} - \frac{1}{3}\delta_{nm}\delta_{jk}\right)d_{jk}, \quad (72)
\end{aligned}$$

where $\bar{W}_{44} = \partial^2 \bar{W} / \partial \bar{I}_4^2$. To derive the increment $\dot{\overline{m_k m_l}} = \dot{m}_k m_l + m_k \dot{m}_l$ we proceeds in steps. First, we calculate

$$\begin{aligned}\dot{m}_l &= \frac{\partial m_l}{\partial F_{j\alpha}} \dot{F}_{j\alpha}, \\ &= \frac{\partial \left(J^{-1/3} F_{l\beta} M_\beta\right)}{\partial F_{j\alpha}} \dot{F}_{j\alpha} = \frac{\partial \left(J^{-1/3} F_{l\beta} M_\beta\right)}{\partial F_{j\alpha}} l_{jr} F_{r\alpha}, \\ &= l_{lr} m_r - \frac{1}{3} l_{rr} m_l = \left(\delta_{sl} m_r - \frac{1}{3} \delta_{sr} m_l\right) l_{sr},\end{aligned}$$

and then we determine

$$\dot{m}_k = l_{kn} m_n - \frac{1}{3} l_{nn} m_k = \left(\delta_{mk} m_n - \frac{1}{3} \delta_{mn} m_k\right) l_{mn}.$$

By combining these results we get the increment $\dot{\overline{m_k m_l}}$ as

$$\begin{aligned}\dot{\overline{m_k m_l}} &= (m_r m_l \,\delta_{sk} + m_r m_k \,\delta_{sl})\, l_{sr} - \frac{2}{3} m_k m_l \,\delta_{sr}\, l_{sr}, && (73) \\ &= (m_r m_l \,\delta_{sk} + m_r m_k \,\delta_{sl})\, d_{sr} - \frac{2}{3} m_k m_l \,\delta_{sr}\, d_{sr} && (74) \\ &+ (m_r m_l \,\delta_{sk} + m_r m_k \,\delta_{sl})\, w_{sr} - \frac{2}{3} m_k m_l \,\delta_{sr}\, w_{sr},\end{aligned}$$

where we have used the decomposition of the incremental displacement gradient $\mathbf{l} = \operatorname{grad} \mathbf{u}$ into symmetric and skew-symmetric parts (56), i.e. $l_{sr} = d_{sr} + w_{sr}$.

Note that the first term in equation (74) contracts over the indices s and r and d_{sr} is symmetric, and therefore only the symmetric portion of $(m_r m_l \,\delta_{sk} + m_r m_k \,\delta_{sl})$ is retained. The fourth term of the same equation has a skew-symmetric tensor w_{sr} contracted with the symmetric tensor δ_{sr} and therefore $w_{rr} = 0$. Equation (74) can then be written in the form

$$\dot{\overline{m_k m_l}} = (\mathbb{A}_{klrs} - \frac{2}{3} m_k m_l \,\delta_{rs}) d_{sr} + (m_r m_l \,\delta_{sk} + m_r m_k \,\delta_{sl})\, w_{sr}, \quad (75)$$

where

$$\mathbb{A}_{klrs} = \frac{1}{2} \left(m_r m_l \,\delta_{sk} + m_r m_k \,\delta_{sl} + m_s m_l \,\delta_{rk} + m_s m_k \,\delta_{rl}\right). \quad (76)$$

Inserting equations (75) and (72) into (71), we obtain the increment of the Kirchhoff stress from the fiber contribution. It has the form

$$\begin{aligned}\overline{J\bar{\sigma}_{ij}}^{\cdot} &= 4\bar{W}_{44}\left(m_i m_j m_s m_r - \frac{1}{3}\bar{I}_4\delta_{ij}m_s m_r - \frac{1}{3}\bar{I}_4 m_i m_j\delta_{sr} + \frac{1}{9}\bar{I}_4^2\delta_{ij}\delta rs\right)d_{sr}\\ &+ 2\bar{W}_4\left(\mathbb{A}_{ijrs} - \frac{2}{3}\,\delta_{ij}m_s m_r - \frac{2}{3}m_i m_j\delta_{sr} + \frac{2}{9}\bar{I}_4\delta_{ij}\delta rs\right)d_{sr}\\ &+ 2\bar{W}_4\mathbb{P}_{ijkl}m_r m_l\,w_{kr} + 2\bar{W}_4\mathbb{P}_{ijkl}m_r m_k\,w_{lr}.\end{aligned} \tag{77}$$

We now compare the explicit expression (77) with the formulation (60) and, making use of equation (70), gives the fourth-order elasticity tensor of the fiber contribution. For convenience of reference we include equation (60) as

$$\overline{J\bar{\boldsymbol{\sigma}}}^{\cdot} = J(J^{-1}\bar{\mathbb{C}}^{\nabla J} : \mathbf{d} + \mathbf{w}\bar{\boldsymbol{\sigma}} - \bar{\boldsymbol{\sigma}}\mathbf{w}), \tag{78}$$

and the elasticity tensor of the fiber contribution has the component form

$$\begin{aligned}\bar{\mathbb{C}}^{\mathrm{fib}\,\nabla J}_{ijsr} &= 4\bar{W}_{44}\left(m_i m_j m_s m_r - \frac{1}{3}\bar{I}_4\delta_{ij}m_s m_r - \frac{1}{3}\bar{I}_4 m_i m_j\delta_{sr} + \frac{1}{9}\bar{I}_4^2\delta_{ij}\delta rs\right)\\ &+ 2\bar{W}_4\left(\mathbb{A}_{ijrs} - \frac{2}{3}\,\delta_{ij}m_s m_r - \frac{2}{3}m_i m_j\delta_{sr} + \frac{2}{9}\bar{I}_4\delta_{ij}\delta rs\right).\end{aligned} \tag{79}$$

7 Incremental Formulation

7.1 Fourth-Order Elasticity Tensor

The compact form of the Kirchhoff stress tensor $J\boldsymbol{\sigma}$ is given in Section 6.1. We now use this equation as the starting point from which the corresponding increment and the associated elasticity tensor will be derived. For convenience of reference we show the expression of the Kirchhoff stress tensor (69) as

$$J\boldsymbol{\sigma} = 2\left[\bar{W}_1\,\mathrm{dev}\bar{\mathbf{B}} + \bar{W}_4\,\mathrm{dev}\,(\mathbf{m}\otimes\mathbf{m})\right] + \frac{\mathrm{d}U}{\mathrm{d}J}\mathbf{I}. \tag{80}$$

Consider again a compressible and transversely isotropic material with the energy given in (63), but we now report all relevant results in coordinate-free notation. The increment of the isochoric part of $J\boldsymbol{\sigma}$, denoted $J\bar{\boldsymbol{\sigma}}$ (not to be confused with (70), which only accounts for the fiber contribution), has the form

$$\overline{J\,\bar{\boldsymbol{\sigma}}}^{\cdot} = 2\left[\boxed{\dot{\bar{W}}_1}\,\mathrm{dev}\bar{\mathbf{B}} + \bar{W}_1\boxed{\overline{\mathrm{dev}\bar{\mathbf{B}}}^{\cdot}} + \boxed{\dot{\bar{W}}_4}\,\mathrm{dev}\,(\mathbf{m}\otimes\mathbf{m}) + \bar{W}_4\boxed{\overline{\mathrm{dev}\,(\mathbf{m}\otimes\mathbf{m})}^{\cdot}}\right], \tag{81}$$

where each of the boxed quantities will be derived separately. We begin with the increment of $\bar{W}_1(\bar{I}_1, \bar{I}_4)$ written in the form

$$\dot{\bar{W}}_1 = \bar{W}_{11}\dot{\bar{I}}_1 + \bar{W}_{14}\dot{\bar{I}}_4. \tag{82}$$

The increments of the isochoric invariants $\bar{I}_1$ and $\bar{I}_4$ are

$$\dot{\bar{I}}_1 = 2\bar{\mathbf{F}} : \dot{\bar{\mathbf{F}}}, \quad \dot{\bar{I}}_4 = 2\dot{\bar{\mathbf{F}}}\mathbf{M} \cdot \mathbf{m}, \tag{83}$$

where we recall that $\mathbf{m} = \bar{\mathbf{F}}\mathbf{M}$ and where the increment of the isochoric deformation gradient $\bar{\mathbf{F}}$ is given by

$$\dot{\bar{\mathbf{F}}} = \mathbf{l}\bar{\mathbf{F}} - \frac{1}{3}(\mathbf{I} : \mathbf{d})\bar{\mathbf{F}} = \mathbf{l}\bar{\mathbf{F}} - \frac{1}{3}\mathrm{div}\,\mathbf{u}\,\bar{\mathbf{F}}. \tag{84}$$

Substituting expressions (83) and (84) into (82) gives

$$\dot{\bar{W}}_1 = 2\left[\bar{W}_{11}\mathrm{dev}\,\bar{\mathbf{B}} + \bar{W}_{14}\mathrm{dev}\,(\mathbf{m}\otimes\mathbf{m})\right] : \mathbf{d}, \tag{85}$$

where we have used the property $\bar{\mathbf{B}} : \mathbf{l} = \bar{\mathbf{B}} : \mathbf{d}$. Following similar steps, we find

$$\dot{\bar{W}}_4 = \bar{W}_{14}\dot{\bar{I}}_1 + \bar{W}_{44}\dot{\bar{I}}_4 = 2\left[\bar{W}_{14}\mathrm{dev}\,\bar{\mathbf{B}} + \bar{W}_{44}\mathrm{dev}\,(\mathbf{m}\otimes\mathbf{m})\right] : \mathbf{d}, \tag{86}$$

where we have used the shorthand notations

$$\bar{W}_{11} = \partial^2\bar{W}/\partial\bar{I}_1^2, \quad \bar{W}_{14} = \partial^2\bar{W}/\partial\bar{I}_1\partial\bar{I}_4, \quad \bar{W}_{44} = \partial^2\bar{W}/\partial\bar{I}_4^2. \tag{87}$$

We now derive the increment of $\mathrm{dev}\,\bar{\mathbf{B}}$, which is given by

$$\overline{\mathrm{dev}\,\dot{\bar{\mathbf{B}}}} = \overline{\mathbb{P} : \dot{\bar{\mathbf{B}}}} = \mathbb{P} : \dot{\bar{\mathbf{B}}} = \mathrm{dev}\,(\dot{\bar{\mathbf{B}}}), \tag{88}$$

where we recall that $\mathbb{P}$ is the fourth-order projection tensor

$$\mathbb{P}_{ijkl} = \delta_{ik}\delta_{jl} - (1/3)\delta_{ij}\delta_{kl}. \tag{89}$$

The increment of $\dot{\bar{\mathbf{B}}}$ is

$$\dot{\bar{\mathbf{B}}} = \mathbb{H} : \mathbf{e} + \mathbf{w}\bar{\mathbf{B}} - \bar{\mathbf{B}}\mathbf{w}, \tag{90}$$

where $\mathbf{e} = \mathrm{dev}\,\mathbf{d}$ and

$$\mathbb{H}_{ijkl} = \frac{1}{2}(\delta_{ik}\bar{B}_{jl} + \delta_{il}\bar{B}_{jk} + \bar{B}_{ik}\delta_{jl} + \bar{B}_{il}\delta_{jk}). \tag{91}$$

Inserting into (88) gives

$$\overline{\operatorname{dev}\bar{\mathbf{B}}}^{\boldsymbol{\cdot}} = \operatorname{dev}(\mathbb{H} : \mathbf{e}) + \mathbf{w}\bar{\mathbf{B}} - \bar{\mathbf{B}}\mathbf{w}. \tag{92}$$

Following similar steps we obtain the relation

$$\overline{\operatorname{dev}(\mathbf{m} \otimes \mathbf{m})}^{\boldsymbol{\cdot}} = \operatorname{dev}(\mathbb{A} : \mathbf{e}) + \mathbf{w}(\mathbf{m} \otimes \mathbf{m}) - (\mathbf{m} \otimes \mathbf{m})\mathbf{w}, \tag{93}$$

where

$$\mathbb{A}_{ijkl} = \frac{1}{2}(\bar{m}_k\bar{m}_j\delta_{il} + \bar{m}_l\bar{m}_j\delta_{ik} + \bar{m}_i\bar{m}_k\delta_{jl} + \bar{m}_i\bar{m}_l\delta_{jk}). \tag{94}$$

Substituting equations (85), (86), (92) and (93) into (81), comparing the obtained expression with (60), gives the isochoric tangent stiffness related to the Jaumann rate as

$$\begin{aligned} \bar{\mathbb{C}}^{\nabla \mathrm{J}} &= 2\big[2\big(\bar{W}_{11}\operatorname{dev}\bar{\mathbf{B}} \otimes \operatorname{dev}\bar{\mathbf{B}} + \bar{W}_{14}\operatorname{dev}\bar{\mathbf{B}} \otimes \operatorname{dev}(\mathbf{m} \otimes \mathbf{m}) \\ &+ \bar{W}_{14}\operatorname{dev}(\mathbf{m} \otimes \mathbf{m}) \otimes \operatorname{dev}\bar{\mathbf{B}} + \bar{W}_{44}\operatorname{dev}(\mathbf{m} \otimes \mathbf{m}) \otimes \operatorname{dev}(\mathbf{m} \otimes \mathbf{m})\big) \\ &+ \bar{W}_1\left(\mathbb{H} - \frac{2}{3}(\mathbf{I} \otimes \bar{\mathbf{B}} + \bar{\mathbf{B}} \otimes \mathbf{I}) + \frac{2}{9}\bar{I}_1\mathbf{I} \otimes \mathbf{I}\right) \\ &+ \bar{W}_4\left(\mathbb{A} - \frac{2}{3}(\mathbf{I} \otimes \mathbf{m} \otimes \mathbf{m} + \mathbf{m} \otimes \mathbf{m} \otimes \mathbf{I}) + \frac{2}{9}\bar{I}_4\mathbf{I} \otimes \mathbf{I}\right)\Big]. \end{aligned} \tag{95}$$

The remaining increment involves the volumetric part of the Cauchy stress $J\boldsymbol{\sigma}_{\mathrm{vol}}$, which has the form

$$\overline{J\boldsymbol{\sigma}_{\mathrm{vol}}}^{\boldsymbol{\cdot}} = \dot{J}\left(\frac{\mathrm{d}U}{\mathrm{d}J} + J\frac{\mathrm{d}^2U}{\mathrm{d}J^2}\right)\mathbf{I}, \tag{96}$$

where the increment of J is given by $\dot{J} = J\operatorname{div}\mathbf{u}$. It follows that the volumetric part of the tangent stiffness is given by

$$\mathbb{C}_{\mathrm{vol}}^{\nabla \mathrm{J}} = J\left(\frac{\mathrm{d}U}{\mathrm{d}J} + J\frac{\mathrm{d}^2U}{\mathrm{d}J^2}\right)\mathbf{I} \otimes \mathbf{I}. \tag{97}$$

The final form of the elasticity tensor (62), which is required to implement the material model into Abaqus is obtained by combining (95) and (97) as

$$\mathbb{C} = J^{-1}\left(\bar{\mathbb{C}}^{\nabla \mathrm{J}} + \mathbb{C}_{\mathrm{vol}}^{\nabla \mathrm{J}}\right). \tag{98}$$

We note that energy dissipation using the theory of pseudo-elasticity, described briefly in Section 5.2, is not yet accounted for in the numerical implementation. The tangent stiffness matrix given in (98) needs to be amended as shown next.

7.2 Fourth-Order Pseudo-Elasticity Tensor

The pseudo-energy function for compressible materials is given in (52) and repeated here for ease of reference

$$W(\mathbf{F}, \eta) = \eta \bar{W}_0(\bar{\mathbf{F}}) + \phi(\eta) + U(J). \tag{99}$$

Applying (64) gives the corresponding Cauchy stress for loading and unloading in the form

$$\boldsymbol{\sigma} = \eta \bar{\boldsymbol{\sigma}}_0 + \boldsymbol{\sigma}_{\mathrm{vol}}, \tag{100}$$

where the scalar parameter η is acting on the isochoric stress component only. We recall from Section 5.2 that during loading the variable η is inactive with a constant value of unity. During unloading η is turned on and its value is given by (30).

To derive an expression for the fourth-order pseudo-elastic tensor, we focus on the isochoric part of the stress tensor (100) written as

$$J\bar{\boldsymbol{\sigma}} = \eta J \bar{\boldsymbol{\sigma}}_0, \tag{101}$$

which is consistent with the formulation given in (60). During unloading the increment of (101) is

$$\overline{\eta J \bar{\boldsymbol{\sigma}}_0}^{\,\cdot} = \dot{\eta}(J\bar{\boldsymbol{\sigma}}_0) + \eta \overline{J \bar{\boldsymbol{\sigma}}_0}^{\,\cdot} \tag{102}$$

where η is dependent on the deformation through $\bar{W}_0$; see equation (34). Therefore

$$\dot{\eta} = \frac{\mathrm{d}\eta}{\mathrm{d}\bar{W}_0} \dot{\bar{W}}_0, \tag{103}$$

with the increment of the energy $\bar{W}_0$ given by

$$\dot{\bar{W}}_0 = \mathrm{tr}\left(\frac{\partial \bar{W}_0}{\partial \bar{\mathbf{F}}} \dot{\mathbf{F}} \right) = J \mathrm{tr}\left(\bar{\boldsymbol{\sigma}}_0 \mathbf{d} \right), \tag{104}$$

where we recall that $\dot{\mathbf{F}} = \mathbf{l}\mathbf{F}$; see equation (54). Therefore, the isochoric part of the elasticity tensor $\bar{\mathbb{C}}^{\nabla \mathrm{J}}$, defined in (98), is amended as

$$\bar{\mathbb{C}}^{\nabla \mathrm{J}} = \frac{\mathrm{d}\eta}{\mathrm{d}\bar{W}_0} (J\bar{\boldsymbol{\sigma}}_0) \otimes (J\bar{\boldsymbol{\sigma}}_0) + \eta \bar{\mathbb{C}}_0^{\nabla \mathrm{J}}. \tag{105}$$

The first term accounts for the rate the material is softening, the second gives the scaled primary response. The specific form of $\mathrm{d}\eta / \mathrm{d}\bar{W}_0$ depends on the definition of η.

7.3 An Energy Function of Bioactive Materials

Following the suggestions in Holzapfel et al. (2000), Holzapfel and Ogden (2009a) and Lin et al. (2009), we neglect the dependence of W on $\hat{\bar{I}}_5$. Therefore, let the decoupled energy function be given by

$$W = \bar{W}_{\text{iso}}(\bar{I}_1) + \bar{W}_{\text{fib}}(\bar{I}_4, \hat{\bar{I}}_4) + U(J), \tag{106}$$

where the isochoric contribution is provided by the isotropic part $\bar{W}_{\text{iso}}$ and by the passive and active fibers $\bar{W}_{\text{fib}}$, and U represents the volumetric change. The arguments of $\bar{W}_{\text{fib}}$, $\bar{I}_4$ and $\hat{\bar{I}}_4$, are the invariants related to $\bar{\mathbf{F}}\bar{\mathbf{F}}^{\text{T}}$ and $\hat{\bar{\mathbf{F}}}\hat{\bar{\mathbf{F}}}^{\text{T}}$, respectively. Recall that $\bar{I}_4$ provides the square of the fiber stretch relative to the passive natural configuration $\mathcal{B}_0$ and $\hat{\bar{I}}_4$ gives the square of the fiber stretch relative to the active natural configuration $\hat{\mathcal{B}}$ defined in $(43)_1$ and $(48)_1$, respectively.

The isotropic connective tissue is taken to be a neo-Hookean material

$$\bar{W}_{\text{iso}} = \frac{\mu}{2}(\bar{I}_1 - 3), \tag{107}$$

where μ is the shear modulus. The volumetric energy function, U, is taken as the penalty function (Prot and Skallerud, 2009)

$$U(J) = \frac{\kappa}{2}(J-1)^2, \tag{108}$$

where κ is the bulk modulus. Since muscle tissue is generally regarded as incompressible, κ is selected to be sufficiently large to minimize changes in volume.

The isochoric contribution of the active and passive fibers is additively decomposed into respective terms and weighted by the fraction of active fibers. It is expressed here as

$$\bar{W}_{\text{fib}}(\bar{I}_4, \hat{\bar{I}}_4) = (1-\alpha)\bar{W}_{\text{pass}}(\bar{I}_4) + \alpha\bar{W}_{\text{act}}(\hat{\bar{I}}_4), \tag{109}$$

where the first term provides the passive fiber energy, which is dependent on the stretch of the fiber relative to the passive natural configuration. The second term provides the energy of the active fiber with dependence on the stretch of the fiber relative to the active natural configuration through $\hat{\bar{I}}_4$, and α is a scalar representing the volume fraction of active fibers. Following Paetsch et al. (2012), the value of α is assumed independent of position or deformation and varies between zero and unity, where unity represents a fully activated muscle. The specific forms of the passive and active fiber energy functions are given, respectively, by

$$\bar{W}_{\text{pass}} = \frac{\mu\mu_{\text{p}}}{2}\left(\bar{I}_4 - 1\right)^2, \tag{110}$$

and

$$\bar{W}_{\text{act}} = \frac{\mu\mu_{\text{a}}}{2}\left(\hat{\bar{I}}_4 - \bar{I}_{4_0}\right)^2. \tag{111}$$

Equation (110) takes the familiar form of the standard reinforcing model with μ_p as a dimensionless parameter providing the degree of anisotropy; see Horgan and Saccomandi (2005), Merodio and Ogden (2005) and Qiu and Pence (1997). Equation (111) has a similar form; however, μ_a accounts for the change in fiber stiffness due to increasing activation and deformation. This is accounted for by specifying the dependence of μ_{a} on $\bar{I}_{4_0}$ and $\hat{\bar{I}}_4$, respectively. We consider the form

$$\mu_{\text{a}} = c_1 \mathrm{e}^{-\left(\hat{\bar{I}}_4 - \bar{I}_{4_0}\right)/c_2}, \tag{112}$$

where c_1 and c_2 are material constants to be determined from experimental data. The invariant $\bar{I}_{4_0}$, defined in $(45)_1$, represents the square of the fiber stretch during the transition from the passive to the active state with no applied mechanical loads. Since the fibers contract under stimulation, $\bar{I}_{4_0} < 1$ and is considered a material parameter.

The specific form of η for the muscle tissue, valid during the first unloading, is

$$\eta = 1 - \frac{1}{r}\tanh\left[\frac{\bar{W}_{\text{m}} - \bar{W}_0(\bar{\mathbf{F}})}{m}\right], \tag{113}$$

where r and m are material parameters and W_{m} is the maximum value of isochoric energy $\bar{W}_0$ obtained during loading. In general, the loading status of a material point is given by the change in energy. For an active material, however, we assume η is either active or inactive depending on the change in *passive* fiber energy. Similar forms of η can be defined for reloading and second unloading. However, the numerical examples given in the next section do not include reloading and second unloading and therefore these forms are not included. The interested reader is referred to Paetsch et al. (2012) for derivation and the explicit expressions.

The values of the material model parameters are determined from the uniaxial test data shown in Section 3.3. They are included in Paetsch et al. (2012) and are repeated here in Table 1, for convenience.

8 Numerical Examples

8.1 Analysis of a Drug Delivery Device

In this section we analyze a bioactive drug delivery device using the proposed theory and, in the process, validate the finite element implementation. The idea is based on the micro-pump design by Tanaka et al. (2007a),

Table 1. Summary of material parameters from Paetsch et al. (2012). The units of parameters are noted with the others being dimensionless.

	μ (Pa)	μ_p –	κ (Pa)	c_1 –	c_2 –	r –	m –
passive	755.5	86.13	1.0e9	470.0	0.95	1.05	2.5
active	755.5	86.13	1.0e9	470.0	0.95	1.55	30.0

where a thin layer of cardiomyocytes is wrapped around a spherical polymer PDMS substrate. Here, we assume a cylindrical geometry to simplify the attachment of the muscular film, which allows the use of a wrapping method similar to the one outlined by Kubo et al. (2007). The objective of this study is to investigate how the performance of a cylindrical device is influenced by the fiber orientation and the geometric dimensions of the substrate. Therefore, in the following, a baseline design is outlined, about which variations are introduced.

Figure 8 provides the geometry of the baseline micro-pump for drug delivery. Manduca muscle tissue, 15 μm thick, overlays the outer surface of the cylindrical structure. The PDMS layer has a thickness of 50 μm and two Teflon® capillary tubes provide an inlet and outlet at the distal ends of the cylinder. The dimensions have been selected to provide a length to diameter ratio of two and the same surface area as that of the spherical pump presented by Tanaka et al. (2007a).

The specific design parameters to be investigated include (i) the muscle fiber orientation, (ii) the ratio of length to diameter and (iii) the thickness of the PDMS substrate. While there are countless fiber orientations which could be evaluated, we restrict ourselves to the radial, circumferential and axial directions shown in Figure 9. Given the practical limitations of manufacturing tissues, radial fibers are included to fully elaborate the behavior of the bio-actuation, which can only be studied in a numerical simulation. The next two design variables, the length to diameter ratio and substrate thickness, relate to the geometry of the PDMS. The aspect ratio of the baseline model is varied in a range from one to ten. Finally, PDMS thicknesses are modeled over a range of 50 μm to 250 μm, where the minimum is assumed to be the manufacturing limit and is taken as the baseline. The maximum value is the design thickness used by Tanaka et al. (2007a).

A finite element model of the micro-pump was developed in the commercially available finite element code, Abaqus. Despite the large aspect ratio between the length and thickness, we discretize the body utilizing 8-noded C3D8 linear continuum elements. After a mesh refinement study, the baseline model contains approximately 24,000 elements with four elements through the thickness to accurately account for bending behavior. The large number of elements is required to capture the change in internal volume, which is calculated using the nodal coordinates at the inner surface. Both external ends of the capillaries are held fixed in the axial direction. No loads are applied, including the pressure exerted by the working fluid,

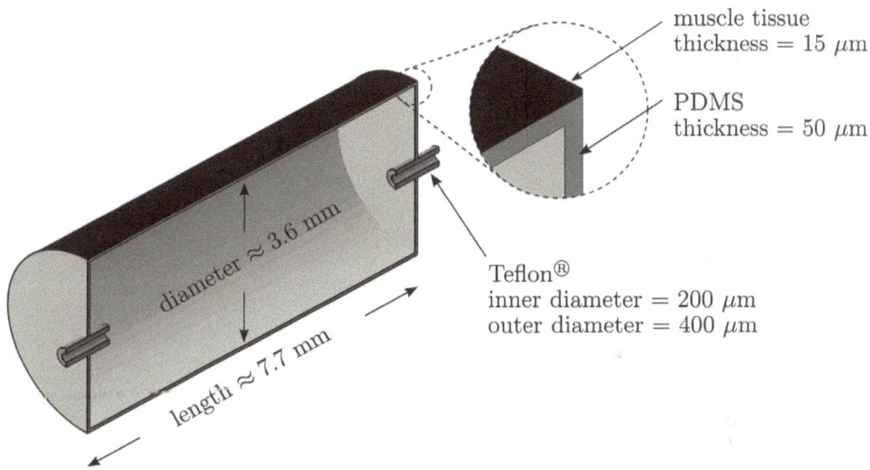

Figure 8. A half-symmetry diagram of the drug delivery micro-pump with baseline dimensions.

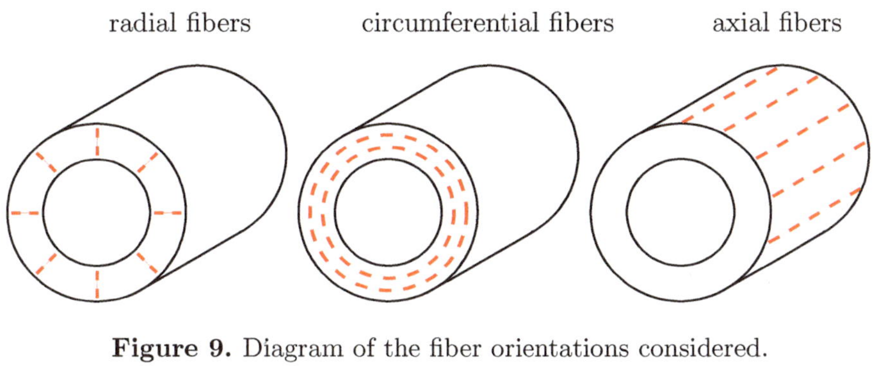

Figure 9. Diagram of the fiber orientations considered.

as the pressure-volume relation would depend on the specific design of the inlet and outlet valves and the design of such valves is well outside the scope of this work. Since the fluid–structure interaction may be significant, the results provided in the following should be referenced to the baseline and considered an indication of performance.

The change in internal volume during the activation–deactivation cycle is the primary metric used to evaluate the performance of device variations. However, additional output may be considered to help characterize the different designs, including Cauchy stress in the fiber direction, total energy and energy dissipated during the activation cycle. Change in internal volume is calculated in MATLAB using the changing nodal coordinates. MATLAB is also used in the determination of the total energy and the energy dissipated.

8.2 Fiber Orientation

We consider the three fiber orientations shown in Figure 9 for the baseline geometry described in Section 8.1. The resulting change in internal volume with the activation, α, is shown in Figure 10. Here, positive values indicate contraction and negative values expansion. While a pump may operate with either expansion or contraction, in the present study we aim for pumps which provide contraction during activation. Of the three fiber orientations considered, only the circumferential arrangement provides the desired response, with a reduction in volume of nearly 2 mm^3. Both the axial and radial fibers result in an enlarging internal volume. It can be seen from Figure 10 that the activation–deactivation paths do not coincide for the change in volume and is most pronounced for the muscle tissue with radially arranged fibers. This behavior is a result of η activating and reflects the hysteric behavior observed in uniaxial experimental data; see Section 3.1. To further understand the behavior of the micro-pump device for various fiber orientations, the total energy stored in the muscle layer with different orientations can be calculated for the activation–deactivation cycle; see Figure 11. During the activation, the muscle tissue is loading as it works against the PDMS structure and the pseudo-elastic scalar quantity η is inactive (i.e. $\eta = 1$). Therefore, the pseudo-energy function given in equation (52) coincides with the energy function of equation (106) and the maximum energy stored in the bio-actuator can be found when $\alpha = 1$. When the stimulus is removed, the muscle tissue begins to unload and $\eta \leqslant 1$. The isochoric energy, $\bar{W}_0$, is then scaled by η. The resulting total energy follows a monotonically decreasing path which differs from the corresponding path during activation. This is due to the dissipation function $\phi(\eta)$. When the muscle tissue returns to the passive natural configuration at $\alpha = 0$, we have $\bar{W}_0(\mathbf{I}) = 0$ and $U(1) = 0$. Therefore, the value of the energy upon unloading represents the energy dissipated during the activation–deactivation cycle.

From Figure 11, we see the circumferential arrangement of fibers results in the greatest energy stored during the cycle, closely followed by the radial distribution of fibers, both of which achieve a maximum energy three to four times larger than in the axial direction. The amount of energy stored during the cycle depends on the stiffness of the PDMS structure for a given direction of fiber contraction. When fully activated the muscle fiber stress-free length is shorter than the passive natural length; however, the substrate's structure resists the shortening of the fibers. The greater the resistance, the greater the energy stored in the muscle tissue. In the case of axial fibers, the micro-pump polymeric structure provides little resistance in the axial direction, resulting in the minimum energy storage. A drug delivery micro-pump with circumferential fibers, which stores the greatest energy, dissipates approximately 0.08 μJ, whereas the bio-actuator with axial fibers

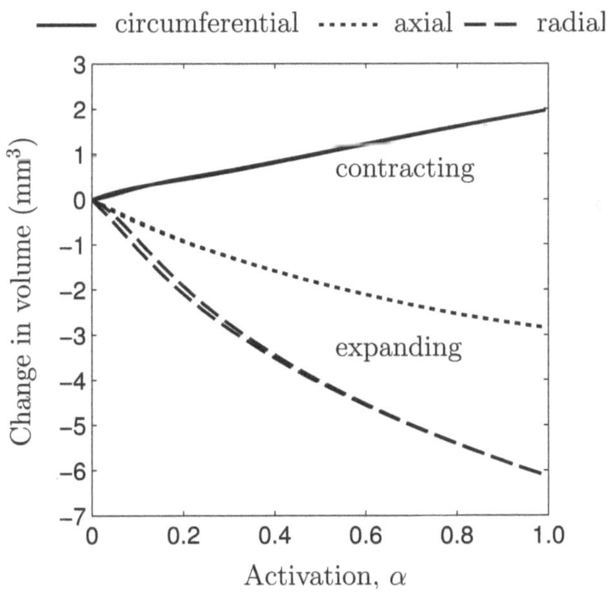

Figure 10. Change in internal volume with varying activation, α, for muscle fibers in the circumferential, axial and radial directions with the baseline geometry. Positive changes in volume represent pump contraction and negative represent expansion.

dissipates almost no energy.

The deformed shapes, scaled by a factor of five, are given for micro-pump devices with radially, circumferentially and axially oriented muscle fibers in Figure 12(a)–(c). The overlaid dashed lines provide the outline of the reference configuration. While fibers contracting in the radial direction decrease the muscle thickness, Figure 12(a) shows that the near incompressibility of the material results in the circumferential and axial expansion of the muscle tissue. Consequently, the internal volume expands, which reflects the results shown in Figure 10. For the circumferential fibers, the construct has a characteristically different response upon activation. In Figure 12(b), the circumferential fiber contraction reduces the radius of the internal volume, while the near incompressibility of the material results in a slight increase in thickness and a substantial increase in the longitudinal dimension. While the expanding thickness has little net effect on the change in internal volume, the increase in length results in a significant bulging of the cylinder's ends. This is in contrast to the concave deformed shape in the case of radial fibers. Finally, the bio-actuator with axially oriented fibers shown in

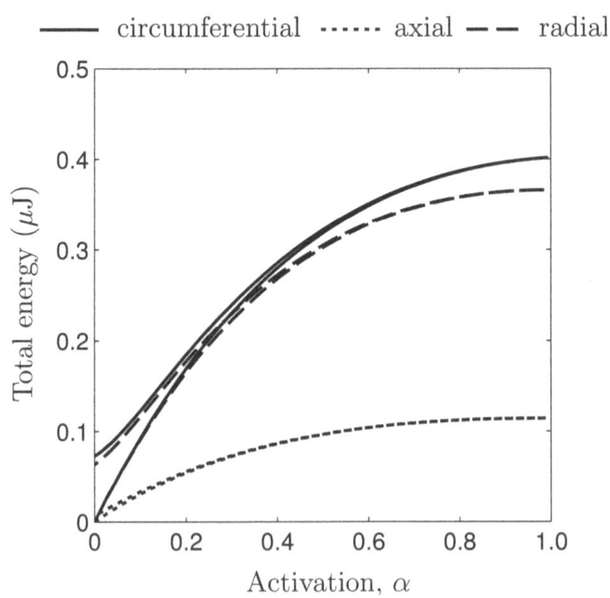

Figure 11. The *total* energy stored in the muscle tissue during the activation–deactivation cycle for muscle fibers in the circumferential, axial and radial directions with baseline geometry.

Figure 12(c) results in a reduction of the longitudinal dimension. Similar to the radial fiber arrangement, the axial fiber contraction results in the distal ends deforming into a concaved shape.

Furthermore, Figures 12(d)–(f) provide contour plots of the Cauchy stress in the same direction as the fibers. For example, Figure 12(d) gives the radial stress distribution for the radial fibers. Although the heteroge-

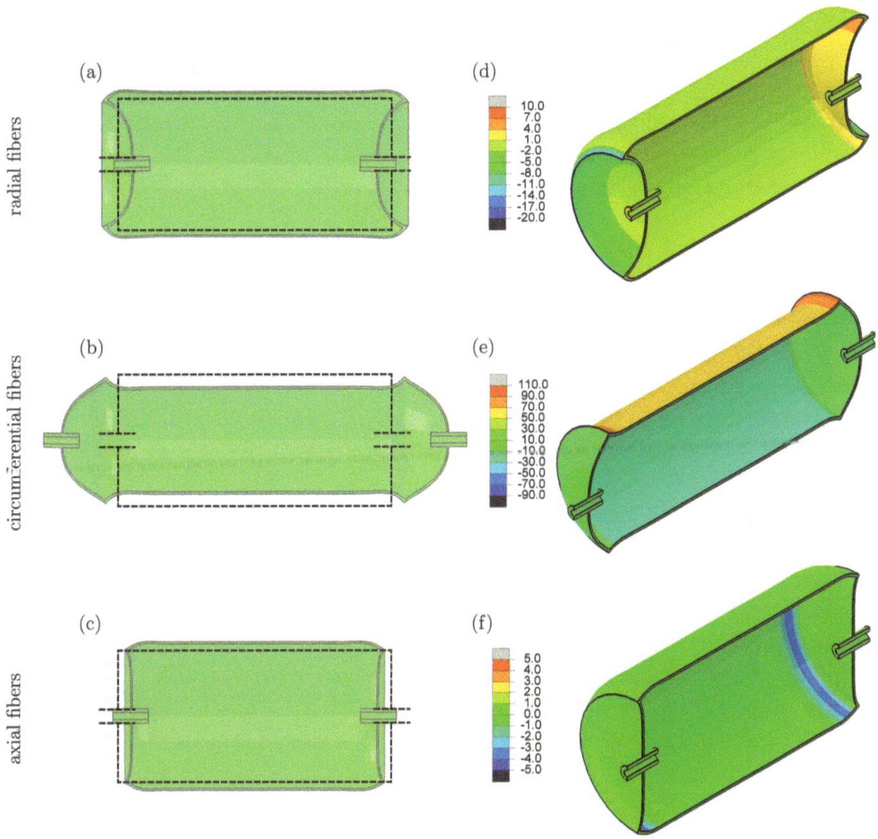

Figure 12. The deformed shapes for muscle fibers oriented in the (a) radial, (b) circumferential and (c) axial directions for $\alpha = 1$ overlaid with the undeformed configuration and contour plots of the Cauchy stress in the (d) radial, (e) circumferential and (f) axial directions for the corresponding fiber orientation on the deformed body. A deformation scale of five is used, with contour plot legends given in units of Pascals.

neous state of stress in the muscle-PDMS construct during the activation makes visualizing the stress difficult, the contours do provide some understanding of the stiffness of the structure when fibers are contracted in a particular direction. For example, the circumferentially oriented fibers result in a significantly higher state of stress than the axial fibers, similar to the results found from the maximum energy stored during contraction. Furthermore, the interaction between muscle and polymeric layers can be clearly seen in Figure 12(e) as the outer (muscle) surface is in tension while the inner (PDMS) surface is in compression, reflecting the change in the fiber's natural length.

Based on the results presented in this section, along with the assumption that the desired drug delivery device response is a contracting internal volume, it is clear the circumferential fiber orientation provides the greatest contracting change in volume, as shown in Figure 10.

Configuration of PDMS substrate. Next, we consider the design parameters of the PDMS. While the baseline geometry was selected to give the same muscle surface area as specified by Tanaka et al. (2007a), the ratio of length to diameter was arbitrarily selected as two. Here, we investigate the response of a drug delivery micro-pump with a varying aspect ratio (length to diameter). Figure 13 gives the maximum change in volume obtained during the stimulation for aspect ratios ranging from one to ten. A ratio of approximately 3.5 provides the maximum volume change with an internal displacement of 2.4 mm^3, an improvement of 0.4 mm^3 over the baseline aspect ratio of two. The peak found in Figure 13 is a result of competing factors, which can be seen from the deformed shapes. Figure 14 compares the deformed shapes of the minimum and maximum aspect ratios for $\alpha = 1$. In the case when length and diameter are equal, the distal ends bulge outward and offset the reduction of internal volume due to circumferential fiber contraction. When the length of the micro-pump is much larger than the diameter, the expansion at the ends of the cylinder is minimized. However, the smaller diameter, and thus smaller circumference, reduces the change in volume resulting from the fiber contraction. While the behavior of the device is explained with simple geometric observations, identifying the configuration for the maximum volume change is non-trivial and exemplifies the necessity of numerical modeling. In addition to the length to diameter ratio, we consider the influence of the PDMS thickness. Intuitively, the maximum change in volume obtained during stimulation decreases with an increasing thickness, as shown in Figure 15. The drug delivery micro-pump performance shows greater sensitivity to PDMS thickness for thinner layers. Figure 16 shows the deformed shape for a PDMS thickness of 0.25 mm

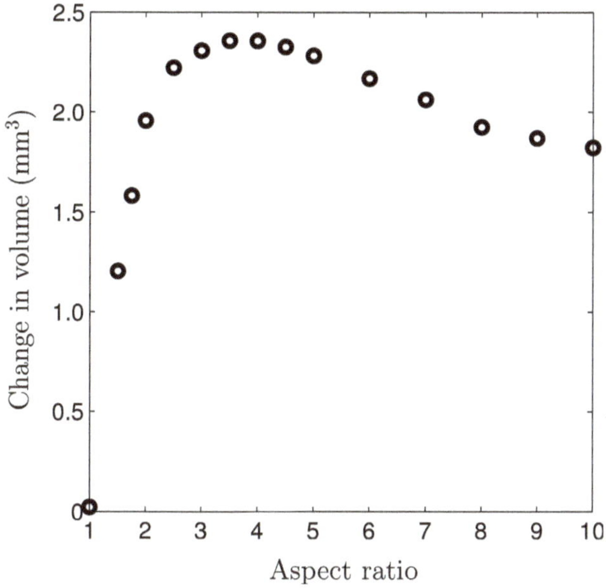

Figure 13. The maximum change in volume for various aspect ratios (length to diameter) with muscle fibers in the circumferential direction.

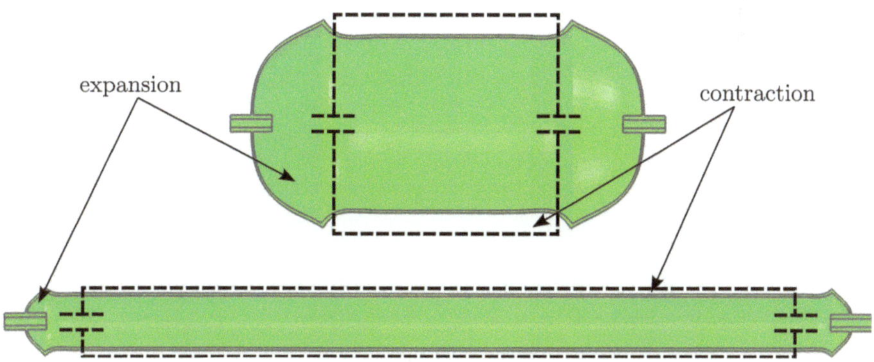

Figure 14. Comparison of undeformed–deformed shapes for a drug-delivery micro-pump with circumferentially oriented fibers, for an aspect ratio of one (top) and ten (bottom), and $\alpha = 1$. Note the deformation scale factor of five.

when $\alpha = 1$. Here, the micro-pump contracts circumferentially with axial expansion and the distal ends bulging outwards, but to a much lesser degree than that of Figure 12(b). Most noticeably, the cylindrical bulging of the thicker ends is lessened on account of the thicker section.

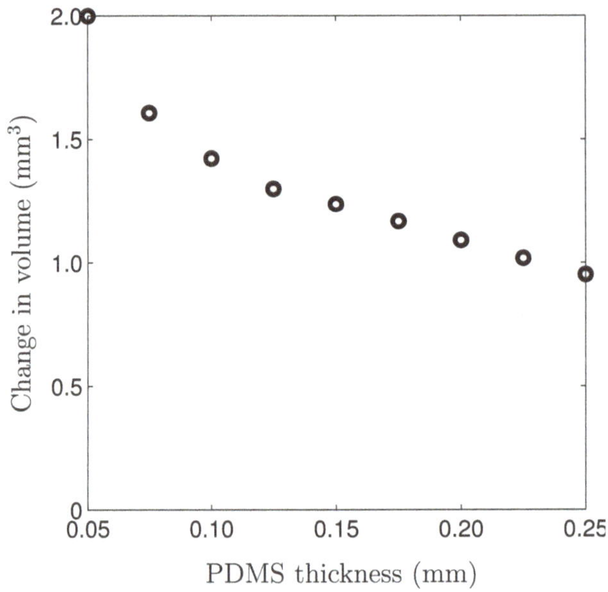

Figure 15. The maximum change in volume for various PDMS thicknesses with muscle fibers in the circumferential direction.

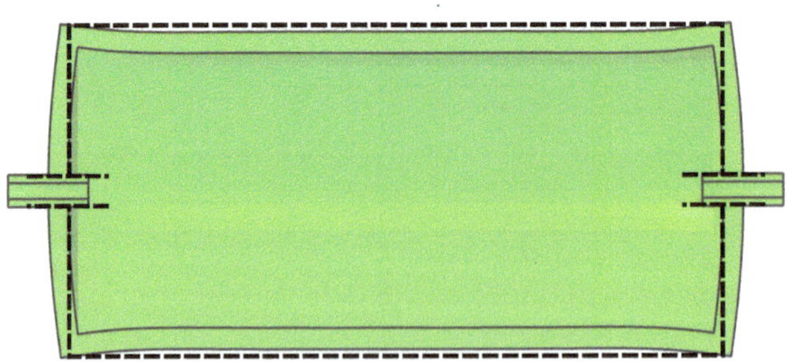

Figure 16. Undeformed-deformed shapes of a drug delivery device with circumferentially oriented fibers and a substrate thickness of 0.25 mm. Given for $\alpha = 1$ and a deformation scale factor of five.

Increasing the PDMS thickness stiffens the pump's structure, which is observed in the total energy given in Figure 17 for 0.05 mm and 0.25 mm PDMS thicknesses. When a pump with the maximum PDMS thickness is activated, the total energy of the muscle tissue is 0.95 μJ, more than twice that obtained for the baseline PDMS thickness. Furthermore, the energy increases almost linearly with activation, indicating the response is dominated by the near-linear PDMS stiffness and not the nonlinear structural behavior. The increasing stiffness associated with a thicker PDMS layer also results in a greater amount of energy being dissipated during the activation–deactivation cycle and is illustrated in Figures 17 and 18.

Optimized device configuration. With a better understanding of how design parameters influence the behavior of the device, we are now able to propose a micro-pump design with improved performance over the initial concept. Fiber orientation plays an important role in the volume change,

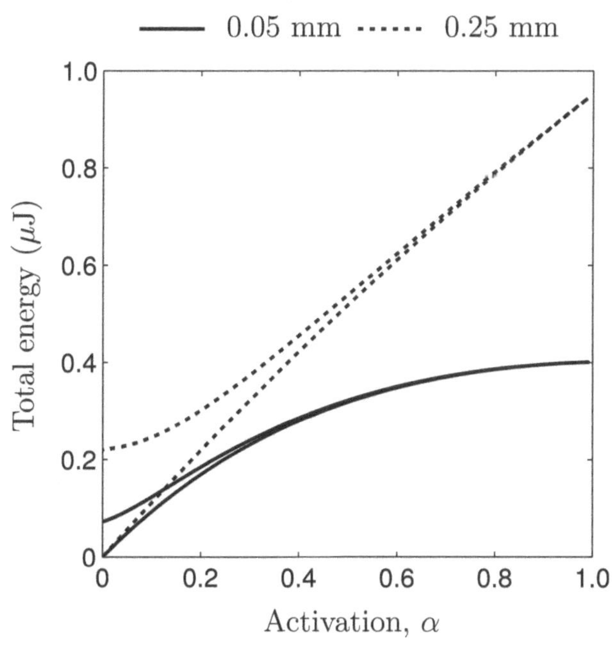

Figure 17. Comparison of total energy for two cases, (i) PDMS thickness of 0.05 mm and (ii) PDMS thickness of 0.25 mm, where the former case is repeated from Figure 11 for reference.

and bio-actuators with circumferential fibers provide the only reduction of internal volume amongst those considered. Therefore, a micro-pump for drug delivery with circumferentially arranged muscle fibers is selected for the updated model. The geometric configuration of the PDMS also plays a key role in the response of the micro-pump device. Figure 13 shows a maximum change in volume for a length to diameter ratio of approximately 3.5. Finally, we see increasing the thickness of the substrate reduces the undesirable bulging of the cylindrical ends. However, this also increases the stiffness of the overall structure, greatly restricting the contraction of the circumferential muscle fibers. Therefore, we propose utilizing a substrate with a nonuniform thickness such that the distal ends are 0.25 mm thick while the cylindrical body is 0.05 mm thick.

The behavior of the updated micro-pump device with nonuniform PDMS thickness can be seen in Figure 19. The deformed shape, overlaid with the outline of the reference configuration, at maximum activation shows significant circumferential contraction while the thicker distal ends restrict the undesirable bulging. The performance is quantified by the change in volume during the activation–deactivation cycle in Figure 20. The maximum

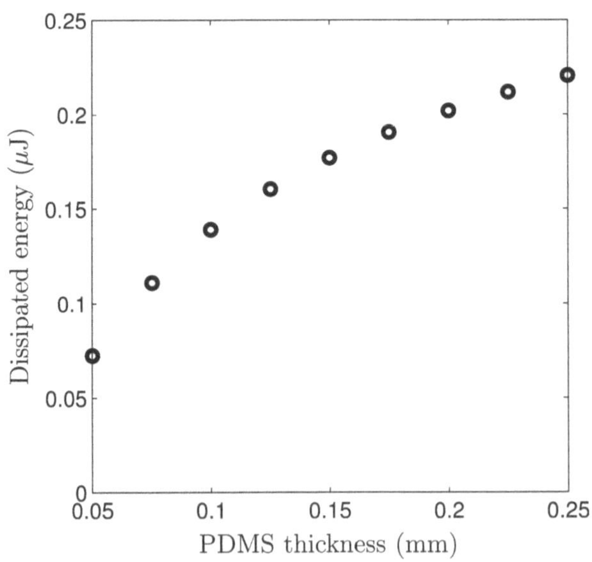

Figure 18. The energy dissipated during the activation–deactivation for various PDMS thicknesses with muscle fibers in the circumferential direction.

change in volume is nearly 4 mm^3, twice that of the baseline performance. Additionally, we see the activation–deactivation paths do not coincide due to energy dissipation.

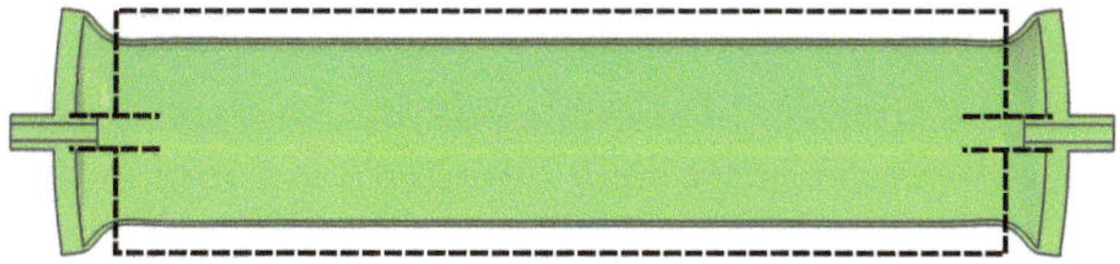

Figure 19. The deformed shape overlaid with the outline of the reference configuration for the improved micro-pump design. Note the deformation scale factor of five.

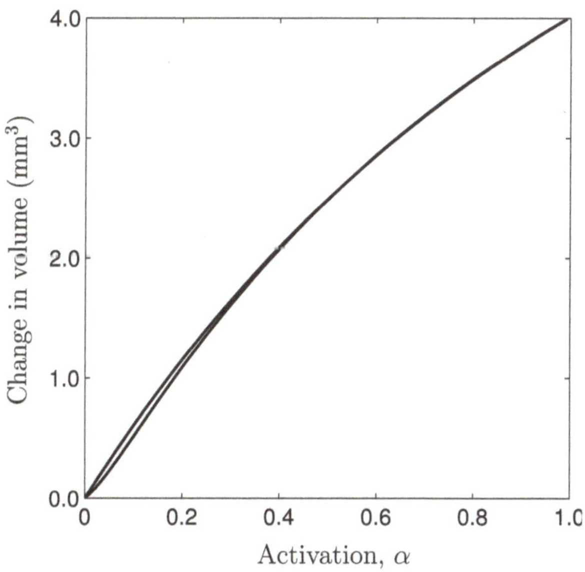

Figure 20. The volume change with activation for the improved design with circumferentially arranged muscle fibers, an aspect ratio of 3.5 and a nonuniform PDMS thickness.

Bibliography

Y. Akiyama, K. Iwabuchi, Y. Furukawa, and K. Morishima. Electrical stimulation of cultured lepidopteran dorsal vessel tissue: an experiment for development of bioactuators. *In Vitro Cell. Dev. - An.*, 46:411–415, 2010.

Y. Akiyama, T. Hoshino, K. Iwabuchi, and K. Morishima. Room temperature operable autonomously moving bio-microrobot powered by insect dorsal vessel tissue. *PLoS ONE*, 7:e38274, 2012a.

Y. Akiyama, K. Odaira, K. Sakiyama, T. Hoshino, K. Iwabuchi, and K. Morishima. Rapidly-moving insect muscle-powered microrobot and its chemical acceleration. *Biomed. Microdevices*, 7:979–986, 2012b.

D. Ambrosi, G. Arioli, F. Nobile, and A. Quarteroni. Electromechanical coupling in cardiac dynamics: The active strain approach. *SIAM J. Appl. Math.*, 71:605–621, 2011.

M. A. Bagni, B. Colombini, P. Geiger, R. Berlinguer Palmini, and G. Cecchi. Non-cross-bridge calcium-dependent stiffness in frog muscle fibers. *J. Physiol.-London*, 286:1353–1357, 2004.

A. L. Baryshyan, W. Woods, B. A. Trimmer, and D. L. Kaplan. Isolation and maintenance-free culture of contractile myotubes from *Manduca sexta* embryos. *PLoS ONE*, 7:e31598, 2012.

R. A. Bell and F. G. Joachim. Techniques for rearing laboratory colonies of tobacco hornworms and pink bollworms Lepidoptera-Sphingidae-Gelechiidae. *Ann. Entomol. Soc. Am.*, 69:365–373, 1976.

R. K. Birla, D. E. Dow, Y. C. Huang, F. Migneco, L. Khait, G. H. Borschel, V. Dhawan, and D. L. Brown. Methodology for the formation of functional, cell-based cardiac pressure generation constructs in vitro. *In Vitro Cell. Dev. Biol.-Animal*, 44:340–350, 2008.

R. K. Birla, V. Dhawan, DE Dow, YC Huang, and DL Brown. Cardiac cells implanted into a cylindrical, vascularized chamber in vivo: pressure generation and morphology. *Biotechnol. Lett.*, 31:191–201, 2009.

M. Böl, S. Reese, K. K. Parker, and E. Kuhl. Computational modeling of muscular thin films for cardiac repair. *Comput. Mech.*, 43:535–544, 2009.

M. Böl, R. Weinkert, and C. Weichert. A coupled electromechanical model for the excitation-dependent contraction of skeletal muscle. *J. Mech. Behav. Biomed.*, 4:1299–1310, 2011.

M. Böl, A. Schmitz, G. Nowak, and T. Siebert. A three-dimensional chemo-mechanical continuum model for smooth muscle contraction. *J. Mech. Behav. Biomed.*, 13:215–229, 2012.

K. S. Campbell and M. Lakie. A cross-bridge mechanism can explain the thixotropic short-range elastic component of relaxed frog skeletal muscle. *J. Physiol.-London*, 510:941–962, 1998.

M. D. Candia Carnevali. Z-line and supercontraction in the hydraulic muscular system of insect larvae. *J. Exp. Zool.*, 203:15–29, 2005.

C. Cherubini, S. Filippi, P. Nardinocchi, and L. Teresi. An electromechanical model of cardiac tissue: constitutive issues and electrophysiological effects. *Prog. Biophys. Mol. Bio.*, 97:562–573, 2008.

S. C. Cowin. Tissue growth and remodeling. *Annu. Rev. Biomed. Eng.*, 6: 77–107, 2004.

H. Demirkoparan, T. J. Pence, and A. Wineman. On dissolution and reassembly of filamentary reinforcing networks in hyperelastic materials. *Proc. R. Soc. London, Ser. A.*, 465:867–894, 2009.

A. Dorfmann and R. W. Ogden. A pseudo-elastic model for loading, partial unloading and reloading of particle-reinforced rubber. *Int. J. Solids Struct.*, 40:2699–2714, 2003.

A. Dorfmann and R. W. Ogden. A constitutive model for the Mullins effect with permanent set in particle-reinforced rubber. *Int. J. Solids Struct.*, 41:1855–1878, 2004.

A. Dorfmann, B. A. Trimmer, and W. A. Woods. A constitutive model for muscle properties in a soft-bodied arthropod. *J. Roy. Soc. Interface*, 4: 257–269, 2007.

A. L. Dorfmann, W. A. Woods, and B. A. Trimmer. Muscle performance in a soft-bodied terrestrial crawler: constitutive modeling of strain-rate dependency. *J. Roy. Soc. Interface*, 5:349–362, 2008.

C. Duch, R. J. Bayline, and R. B. Levine. Postembryonic development of the dorsal longitudinal flight muscle and its innervation in Manduca sexta. *J. Comp. Neurol.*, 422:1–17, 2000.

A. E. Ehret, M. Böl, and M. Itskov. A continuum constitutive model for the active behavior of skeletal muscle. *J. Mech. Phys. Solids*, 59:625–636, 2011.

A. W. Feinberg, A. Feigel, S. S. Shevkoplyas, S. Sheehy, G. M. Whitesides, and K. K. Parker. Muscular thin films for building actuators and powering devices. *Science*, 317:1366–1370, 2007.

G. Franceschini, D. Bigoni, P. Regitnig, and G. A. Holzapfel. Brain tissue deforms similarly to filled elastomers and follows consolidation theory. *J. Mech. Phys. Solids*, 54:2592–2620, 2006.

M. A. Goldstein and W. J. Burdette. Striated visceral muscle of Drosophila melanogaster. *J. Morphol.*, 134:315–334, 1971.

H. L. Granzier and S. Labeit. The giant protein titin: a major player in myocardial mechanics, signaling, and disease. *Circ. Res.*, 94:284–295, 2004.

H. L. Granzier and K. Wang. Interplay between passive tension and strong and weak binding cross-bridges in insect indirect flight muscle. A functional dissection by gelsolin-mediated thin filament removal. *J. Gen. Physiol.*, 101:235–270, 1993a.

H. L. Granzier and K. Wang. Passive tension and stiffness of vertebrate skeletal and insect flight muscles: the contribution of weak cross-bridges and elastic filaments. *Biophys. J.*, 65:2141–2159, 1993b.

C. M. Hai and R. A. Murphy. Cross-bridge phosphorylation and regulation of latch state in smooth muscle. *Am. J. Physiol.*, 254:C99–106, 1988.

J. Hardie. The tension/length relationship of an insect (calliphora erythrocephala) supercontracting muscle. *Experientia*, 32:714–716, 1976.

B. Hernández-Gascón, J. Grasa, B. Calvo, and J. F. Rodríguez. A 3D electro-mechanical continuum model for simulating skeletal muscle contraction. *J. Theor. Biol.*, 335:108–118, 2013.

A. Herrel, J. J. Meyers, J. P. Timmermans, and K. C. Nishikawa. Supercontracting muscle: producing tension over extreme muscle lengths. *J. Exp. Zool.*, 205:2167–2173, 2002.

J. M. Hollerbach, I. W. Hunter, and J. Ballantyne. A comparative analysis of actuator technologies for robotics. In J. Craig O. Khatib and T. Lozano-Perez, editors, *The Robotics Review*, pages 301–342. MIT Press, 1991.

K. C. Holmes and M. A. Geeves. The structural basis of muscle contraction. *Philos. T. Roy. Soc. B*, 355:419–431, 2000.

G. A. Holzapfel and R. W. Ogden. On planar biaxial tests for anisotropic nonlinearly elastic solids: A continuum mechanical framework. *Math. Mech. Solids*, 14:474–489, 2009a.

G. A. Holzapfel and R. W. Ogden. Constitutive modelling of passive myocardium: A structurally based framework for material characterization. *Phil. Trans. R. Soc. A*, 367:3445–3475, 2009b.

G. A. Holzapfel, T. C. Gasser, and R. W. Ogden. A new constitutive framework for arterial wall mechanics and a comparative study of material models. *J. Elasticity*, 61:1–48, 2000. ISSN 0374-3535.

C. O. Horgan and G. Saccomandi. A new constitutive theory for fiber-reinforced incompressible nonlinearly elastic solids. *J. Mech. Phys. Solid*, 53:1985–2015, 2005.

A. F. Huxley. Cross-bridge action: present views, prospects, and unknowns. *J. Biomech.*, 33:1189–1195, 2000.

M. Kroon. A constitutive model for smooth muscle including active tone and passive viscoelastic behaviour. *Math. Med. Biol.*, 27:129–155, 2010.

H. Kubo, T. Shimizu, M. Yamato, T. Fujimoto, and T. Okano. Creation of myocardial tubes using cardiomyocyte sheets and an in vitro cell sheet-wrapping device. *Biomaterials*, 28:3508–3516, 2007.

K. A. Lazopoulos and R. W. Ogden. Nonlinear elasticity theory with discontinuous internal variables. *Math. Mech. Solids*, 3:29–51, 1998.

R. B. Levine and J. W. Truman. Dendritic reorganization of abdominal motoneurons during metamorphosis of the moth, manduca-sexta. *J. Neurosci.*, 5:2424–2431, 1985.

R. L. Lieber, M. E. Leonard, and C. G. Brown-Maupin. Effects of muscle contraction on the load-strain properties of frog aponeurosis and tendon. *Cells Tissues Organs*, 166:48–54, 2000.

H. T. Lin, A. L. Dorfmann, and B. A. Trimmer. Soft-cuticle biomechanics: A constitutive model of anisotropy for caterpillar integument. *J. Theor. Biol.*, 256:447–457, 2009.

J. Merodio and R. W. Ogden. Mechanical response of fiber-reinforced incompressible non-linear elastic solids. *Int. J. Nonlinear Mech.*, 40:213–227, 2005.

G. Mutungi and K. W. Ranatunga. Do cross-bridges contribute to the tension during stretch of passive muscle? A response. *J. Muscle Res. Cell M.*, 21:301–302, 2000.

P. Nardinocchi and L. Teresi. On the active response of soft living tissues. *J. Elasticity*, 88:27–39, 2007.

P. Nardinocchi and L. Teresi. Electromechanical modeling of anisotropic cardiac tissues. *Math. Mech. Solids*, 18:576–591, 2013.

F. Nobile, A. Quarteroni, and R. Ruiz-Baier. An active strain electromechanical model for cardiac tissue. *Int. J. Numer. Method. Biomed. Eng.*, 28:2040–7947, 2012.

R. W. Ogden. Volume changes associated with the deformation of rubber-like solids. *J. Mech. Phys. Solids*, 24:323–338, 1976.

R. W. Ogden. Nearly isochoric elastic deformations: application to rubber-like solids. *J. Mech. Phys. Solids*, 26:37–57, 1978.

R. W. Ogden. *Non-linear Elastic Deformations*. Dover Publications, 1997.

R. W. Ogden. Elements of the theory of finite elasticity. In Y. B. Fu and R. W. Ogden, editors, *Nonlinear Elasticity: Theory and Applications*, pages 1–58. Cambridge University Press, 2001.

R. W. Ogden and D. G. Roxburgh. A pseudo-elastic model for the Mullins effect in filled rubber. *P. Roy. Soc. A-Math. Phy.*, 455:2861–2877, 1999.

M. P. Osborne. Supercontraction in the muscles of the blowfly larva: an ultrastructural study. *J. Insect. Physiol.*, 13:1472–1482, 1967.

C. Paetsch and A. Dorfmann. Non-linear modeling of active biohybrid materials. *Int. J. Nonlin. Mech.*, 56:105–114, 2013.

C. Paetsch, B. A. Trimmer, and A. Dorfmann. A constitutive model for active-passive transition of muscle fibers. *Int. J. Nonlinear Mech.*, 47: 377–387, 2012.

J. Park, J. Ryu, S. K. Choi, E. Seo, J. M. Cha, S. Ryu, J. Kim, K. Kim, and S. H. Lee. Real-time measurement of the contractile forces of self-organized cardiomyocytes on hybrid biopolymer microcantilevers. *Anal. Chem.*, 77:6571–6580, 2005.

J. Park, I. C. Kim, J. Baek, M. Cha, J. Kim, S. Park, J. Lee, and B. Kim. Micro pumping with cardiomyocye-polymer hybrid. *Lab Chip*, 7:1367–1370, 2007.

L. G. Prado, I. Makarenko, C. Andresen, M. Kruger, C. A. Opitz, and W. A. Linke. Isoform diversity of giant proteins in relation to passive and active contractile properties of rabbit skeletal muscles. *J. Gen. Physiol.*, 126: 461–480, 2005.

U. Proske and D. L. Morgan. Do cross-bridges contribute to the tension during stretch of passive muscle? *J. Muscle Res. Cell M.*, 20:433–442, 1999.

V. E. Prot and B. H. Skallerud. Nonlinear solid finite element analysis of mitral valves with heterogeneous leaflet layers. *Comput. Mech.*, 43: 353–368, 2009.

G. Y. Qiu and T. J. Pence. Remarks on the behavior of simple directionally reinforced incompressible nonlinearly elastic solids. *J. Elasticity*, 49:1–30, 1997. ISSN 0374-3535.

K. R. Rajagopal and A. S. Wineman. A constitutive equation for nonlinear solids which undergo deformation induced microstructural changes. *Int. J. Plasticity*, 8:385–395, 1992.

M. B. Rheuben and A. E. Kammer. Comparison of slow larval and fast adult muscle innervated by the same motor neuron. *J. Exp. Biol.*, 84: 103–118, 1980.

M. J. Rice. Supercontracting and non-supercontracting visceral muscles in the tsetse fly, Glossina austeni. *J. Insect Physiol.*, 16:1109–1122, 1970.

E. K. Rodriguez, A. Hoger, and A. D. McCulloch. Stress-dependent finite growth in soft elastic tissues. *J. Biomech.*, 27:455–467, 1994.

S. Rossi, R. Ruiz-Baier, L. F. Pavarino, and A. Quarteroni. Orthotropic active strain models for the numerical simulations of cardiac biomechanics. *Int. J. Numer. Method. Biomed. Eng.*, 28:761–788, 2012.

S. Rossi, T. Lassila, R. Ruiz-Baier, A. Sequeira, and A. Quarteroni. Thermodynamically consistent orthotropic activation model capturing ventricular systolic wall thickening in cardiac electromechanics. *Eur. J. Mech. A-Solid*, In press, DOI: 10.1016/j.euromechsol.2013.10.009, 2013.

M. Royuela, B. Fraile, M. I. Arenas, and R. Paniagua. Characterization of several invertebrate muscle cell types: a comparison with vertebrate muscles. *Microsc. Res. Techniq.*, 48:107–115, 2000.

R. Ruiz-Baier, A. Gizzi, S. Rossi, C. Cherubini, A. Laadhari, S. Filippi, and A. Quarteroni. Mathematical modelling of active contraction in isolated cardiomyocytes. *Math. Med. Biol.*, In press, DOI: 10.1093/imammb/dqt009, 2013.

L. M. Schwartz and R. L. Ruff. Changes in contractile properties of skeletal muscle during developmentally programmed atrophy and death. *Am. J. Psychol.-Cell Physiology*, 282:C1270–C1277, 2002.

H. Sekine, T. Shimizu, and T. Okano. Myocardial tissue engineering: toward a bioartificial pump. *Cell Tissue Res.*, 347:775–782, 2012.

B. Sharifimajd and J. Stålhand. A continuum model for skeletal muscle contraction at homogeneous finite deformations. *Biomech. Model Mechanobiol.*, 12:965–973, 2013.

J. Shim, A. Grosberg, J. C. Nawroth, K. K. Parker, and K. Bertoldi. Modeling of cardiac muscle thin films: pre-stretch, passive and active behavior. *J. Biomech.*, 45:832–841, 2012.

Dassault Systèmes Simulia. *Abaqus 6.9.* Providence RI, USA, 2009.

A. J. M. Spencer. Theory of invariants. In A. C. Eringen, editor, *Continuum Physics*, volume 1, pages 239–353. Academic Press, New York, 1971.

J. Stålhand, A. Klarbring, and G. A. Holzapfel. Smooth muscle contraction: Mechanochemical formulation for homogeneous finite strains. *Prog. Biophys. Mol. Bio.*, 96:465–481, 2008.

J. Stålhand, A. Klarbring, and G. A. Holzapfel. A mechanochemical 3D continuum model for smooth muscle contraction under finite strains. *J. Theor. Biol.*, 268:120–130, 2011.

Y. Tanaka, K. Sato, T. Shimizu, M. Yamato, T. Okano, and T. Kitamori. A micro-spherical heart pump powered by cultured cardiomyocytes. *Lab Chip*, 7:207–212, 2007a.

Y. Tanaka, K. Sato, T. Shimizu, M. Yamato, T. Okano, and T. Kitamori. Biological cells on microchips: new technologies and applications. *Biosens. Bioelectron.*, 23:449–458, 2007b.

Y. Tanaka, Y. Yanagisawa, and T. Kitamori. Fluid actuation for a bio-micropump powered by previously frozen cardiomyocytes directly seeded on a diagonally stretched thin membrane. *Sensor Actuator*, 156:494–498, 2011.

L. Tskhovrebova and J. Trinick. Titin: properties and family relationships. *Nat. Rev. Mol. Cell Bio.*, 4:679–689, 2003.

L. Tskhovrebova, J. Trinick, and J. A. Sleep. Elasticity and unfolding of single molecules of the giant muscle protein titin. *Nature*, 387:308–312, 1997.

E. Walters, P. Illich, J. Weeks, and M. Lewin. Defensive responses of larval manduca sexta and their sensitization by noxious stimuli in the laboratory and field. *J. Exp. Biol.*, 204:457–469, 2001.

J. Weickenmeier, M. Itskov, E. Mazza, and M. Jabareen. A physically motivated constitutive model for 3d numerical simulation of skeletal muscles. *Int. J. Numer. Method. Biomed. Eng.*, In press, DOI: 10.1002/cnm.2618, 2014.

W. A. Woods, Jr., S. J. Fusillo, and B. A. Trimmer. Dynamic properties of a locomotory muscle of the tobacco hornworm manduca sexta during strain cycling and simulated natural crawling. *J. Exp. Biol.*, 211:873–882, 2008.

Incremental Equations for Soft Fibrous Materials

Michel Destrade

School of Mathematics, Statistics and Applied Mathematics
National University of Ireland Galway, Ireland.*

E-mail: michel.destrade@nuigalway.ie

Abstract. The general theory of nonlinear anisotropic elasticity is extended to describe small-amplitude motions and static deformations that can be superimposed on large pre-strains of fibre-reinforced solids. The linearised governing equations of incremental motion are derived. Then they are solved for some illustrative situations which reveal a wide spectrum of possible behaviours compared to the case of initially isotropic materials. Particular attention is paid to the propagation of homogeneous waves and to the formation of static wrinkles. These objects prove useful in the investigation of the issues of material (in the bulk) and geometrical (at boundaries) stability. Attempts are also made at modelling some experimental observations made on (isotropic) silicone and (anisotropic) biological soft tissues.

1 Introduction

Consider two rectangular solid blocks, one made of silicone, the other made of mammalian skeletal muscle ('meat'), and subject them to a large shear. The first block deforms smoothly and its surface remains flat; see Figure 1(a). The second block, however, experiences a form of buckling early on, as small-amplitude wrinkles appear on its surface. From visual inspection and intuition, we can come up with an explanation for these strikingly different behaviours. If we were careful in our moulding of the silicone block, we can safely assume that it is isotropic. On the other hand, the piece of meat is clearly anisotropic, as it is 'reinforced' with visible aligned fibres. When sheared, these fibres and/or their entanglements resist compression

*I am grateful to Sophie Labat (Bordeaux), Jorge Bruno, Artur Gower and Joanne McCarthy (Galway) for their technical assistance in preparing this chapter.

L. Dorfmann, R. W. Ogden (Eds.), *Nonlinear Mechanics of Soft Fibrous Materials*,
CISM International Centre for Mechanical Sciences
DOI 10.1007/978-3-7091-1838-2_5 © CISM Udine 2015

and crumple early, lifting the surrounding tissue into a wavy pattern, with wavefronts at right-angles to the fibres; see Figure 1(b). Hence, although both blocks are soft and easily deformed, it is likely that they behave differently due to the absence or presence of fibres.

(a)  (b) 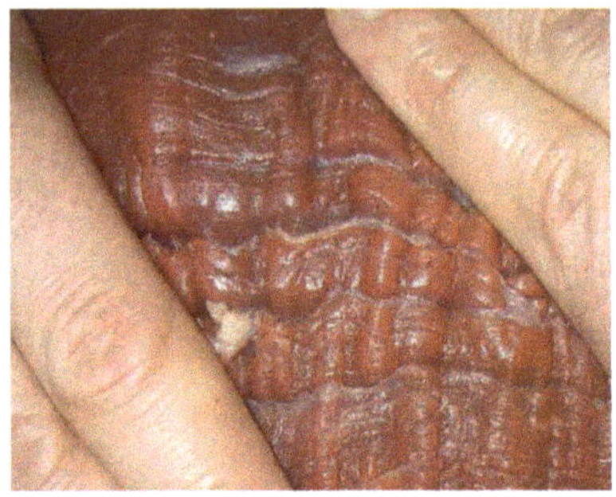

Figure 1. (a) A block of silicone subject to a large shear by hand; its surface remains flat and smooth. (b) A piece of meat sheared by hand, in a direction approximatively at 45° with respect to the fibres; its surface buckles early, with wrinkles forming at right-angle to the fibres.

From a mechanics point of view, we may now ask ourselves whether there exists a way of describing and predicting how the two blocks should behave in shear. As seen in the course of this chapter, it turns out that the simplest models of isotropic (for the silicone) and anisotropic (for the meat) *nonlinear incompressible elasticity* can indeed capture these effects. Of course the analysis itself is not easy and requires a good grasp of theoretical issues, physically-based modelling, and numerical analysis.

In order to model the small-amplitude wrinkles, we need to derive the *incremental equations of motion* of anisotropic non-linear elasticity. This procedure is described in Section 2; simply put, it relies on linearising the equations of motion in the neighbourhood of a static state of equilibrium corresponding to a large homogeneous deformation. These equations can be established in all generality, and in Section 3 we use them to study the *propagation of bulk waves* in deformed soft solids.

Indeed, wave propagation is a straightforward tool for figuring out if a solid is isotropic or not. Consider, for example, the experimental results displayed in Figure 2: they clearly show two privileged directions, along which a mechanical signal travels at different speeds than in other directions. Again, this can be captured by very simple models of nonlinear anisotropic elasticity. Deriving the speed v of a travelling wave explicitly also has an

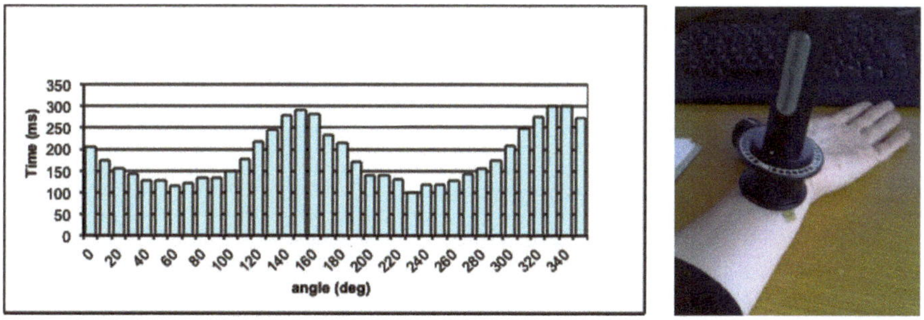

Figure 2. The Reviscometer® records the information sent by a measurement probe (placed on the skin) which can be rotated to give the variation of acoustic perturbation speed with angle. The device measures the time it takes for an impulse generated by one needle sensor to reach another sensor 2mm away by travelling on the surface of the skin. Here, measurements were made every 10° on the forearm of a 21 year old female, and averaged over 6 experiments.

advantage from the point of view of constitutive modelling, as writing down the conditions for v^2 to be positive puts physically-based restrictions on the material parameters.

Finally in Section 4, armed with all the tools and knowledge required, we can at last tackle the problem raised by our observation of Figure 1. In that section we write down the *incremental equations of static equilibrium* (i.e. we take $v = 0$) for a perturbation which is sinusoidal on the surface of a sheared half-space (to describe the wavy pattern) and decays exponentially with depth inside the substrate (to describe near-surface wrinkles). Then we set out to solve these equations. For the isotropic model of choice (the neo-Hookean model), great analytical progress and results can be found, and they square well with experimental data. For the anisotropic models, we must turn to numerics. Thankfully, a host of tools is at our disposal to help us generate a robust method of resolution, based on the Stroh formalism and its extension to include the surface impedance method.

2 Incremental Equations

In this section we derive the equations governing the propagation of small-amplitude ('incremental') motions in anisotropic hyperelastic solids subjected to large static homogeneous deformations. The method of derivation is now well established and we omit certain details which can be found elsewhere; see references at the end of the chapter, in particular the CISM chapter by Ogden (2001). For simplicity we write the components of vectors and tensors in rectangular Cartesian coordinate systems, but the analysis can generalised to other systems of coordinates at little cost.

2.1 Large Static Pre-deformation

We consider that our solid is initially at rest in the *reference configuration* $\mathcal{B}_0$ (say). It is then brought to an equilibrium configuration (the *current configuration* $\mathcal{B}$, say) by the application of a pre-stress with measure $\boldsymbol{\sigma}$ (a Cauchy stress). Hence a point initially at $\boldsymbol{X}$ in $\mathcal{B}_0$ is at $\boldsymbol{x} = \boldsymbol{\chi}(\boldsymbol{X})$ in $\mathcal{B}$, where $\boldsymbol{\chi}$ is a one-to-one mapping; see Figure 3.

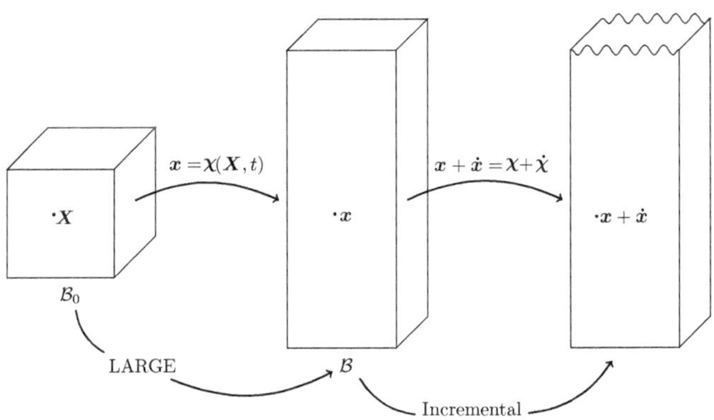

Figure 3. Sketch of the successive deformations taking place in the soft solid: first a large static homogeneous deformation, followed by an incremental motion.

The deformation $\boldsymbol{x} = \boldsymbol{\chi}(\boldsymbol{X})$ is described by *deformation gradient* $\boldsymbol{F}$:

$$\boldsymbol{F} = \frac{\partial \boldsymbol{\chi}}{\partial \boldsymbol{X}}, \qquad F_{i\alpha} = \frac{\partial \chi_i}{\partial X_\alpha}. \tag{1}$$

Here the components of $\boldsymbol{F}$ are given in the $\boldsymbol{e}_i \otimes \boldsymbol{E}_j$ system, where $\boldsymbol{E}_i$ and $\boldsymbol{e}_i$ $(i = 1, 2, 3)$ form orthonormal vector bases in $\mathcal{B}_0$ and $\mathcal{B}$, respectively. Let $J = \det \boldsymbol{F}$; this quantity measures the local volume changes, because it relates an elementary volume dV in $\mathcal{B}_0$ to its counterpart in $\mathcal{B}$ through

$$dv = J dV. \tag{2}$$

Using this identity, we may write the *conservation of mass* as follows

$$\int_{\mathcal{B}_0} \rho_0 dv = \int_{\mathcal{B}} \rho dv = \int_{\mathcal{B}_0} \rho J dV, \qquad \text{so that } \rho_0 = \rho J, \tag{3}$$

where ρ_0 and ρ are the mass densities per unit volume in $\mathcal{B}_0$ and $\mathcal{B}$, respectively.

Finally, we may write the *equations of equilibrium*, in the absence of body forces, as

$$\text{Div } \boldsymbol{S} = \boldsymbol{0}, \qquad \frac{\partial S_{\alpha i}}{\partial X_\alpha} = 0, \tag{4}$$

where $\boldsymbol{S} = J^{-1}\boldsymbol{F}\boldsymbol{\sigma}$ is the *nominal stress tensor*. We assume that the solid is hyperelastic, so that it possesses a *strain-energy density* $W = W(\boldsymbol{F})$ per unit mass. Then $\boldsymbol{S}$ is given by (Ogden, 2001),

$$\boldsymbol{S} = \frac{\partial W}{\partial \boldsymbol{F}}, \qquad S_{\alpha i} = \frac{\partial W}{\partial F_{i\alpha}}. \tag{5}$$

Notice here the convention used throughout the chapter for the ordering of the indices when differentiating tensors. For *incompressible* solids, volume changes are not permitted, so that $J = 1$ at all times. In that case $\boldsymbol{S}$ is given by

$$\boldsymbol{S} = \frac{\partial W}{\partial \boldsymbol{F}} - p\boldsymbol{F}^{-1}, \tag{6}$$

where p is a Lagrange multiplier, to be determined from boundary and/or initial conditions.

Finally, the *boundary conditions* are

$$\boldsymbol{S}^T\boldsymbol{N} = \boldsymbol{f} \text{ on } \partial\mathcal{B}_0, \tag{7}$$

where $\boldsymbol{f}$ are the applied tractions and the upper script T denotes the transpose.

2.2 Increments

Now we 'increment' the quantities and equations encountered so far. In other words, we linearise them, using a right-arrow to mean 'is incremented to', and a dot to denote an incremental quantity. Hence, for the increment of the deformation we have

$$\boldsymbol{x} = \chi(\boldsymbol{X}) \quad \longrightarrow \quad \boldsymbol{x} + \dot{\boldsymbol{x}} = \chi(\boldsymbol{X}) + \dot{\chi}(\boldsymbol{X}, t). \tag{8}$$

But $\boldsymbol{X}$ can be seen as a function of $\boldsymbol{x}$ by writing $\boldsymbol{X} = \boldsymbol{\chi}^{-1}(\boldsymbol{x})$, so that $\dot{\boldsymbol{\chi}}$ can also be seen as a function of $\boldsymbol{x}$. Then $\boldsymbol{u}(\boldsymbol{x},t) \equiv \dot{\boldsymbol{\chi}}(\boldsymbol{x},t)$ is called the *mechanical displacement*.

Next, we increment the deformation gradient:

$$\boldsymbol{F} = \frac{\partial \boldsymbol{\chi}}{\partial \boldsymbol{X}} \quad \longrightarrow \quad \boldsymbol{F} + \dot{\boldsymbol{F}}. \tag{9}$$

Here, using the chain rule, we find that

$$\dot{F}_{i\alpha} = \frac{\partial \dot{\chi}_i}{\partial X_\alpha} = \frac{\partial x_j}{\partial X_\alpha}\frac{\partial u_i}{\partial x_j} = F_{j\alpha} u_{i,j}, \qquad \text{or} \qquad \dot{\boldsymbol{F}} = \boldsymbol{\Gamma}\,\boldsymbol{F}, \tag{10}$$

where $\boldsymbol{\Gamma} = \operatorname{grad} \boldsymbol{u}$ is the Eulerian *displacement gradient*.

Moving on to the increment of the nominal stress tensor, we have

$$\boldsymbol{S} = \boldsymbol{S}(\boldsymbol{F}) \rightarrow \boldsymbol{S} = \boldsymbol{S} + \dot{\boldsymbol{S}}, \quad \text{or} \quad S_{\alpha i}\left(F_{l\beta} + \dot{F}_{l\beta}\right) = S_{\alpha i}(F_{l\beta}) + \dot{S}_{\alpha i}(F_{l\beta}). \tag{11}$$

Here, using the chain rule and (10), we find that

$$\dot{S}_{\alpha i} = \frac{\partial S_{\alpha i}}{\partial F_{l\beta}}\dot{F}_{l\beta} = \frac{\partial S_{\alpha i}}{\partial F_{l\beta}}\Gamma_{lk}F_{k\beta}, \qquad \text{or} \qquad \dot{S}_{\alpha i} = \mathcal{A}_{\alpha i\beta l}\Gamma_{lk}F_{k\beta}, \tag{12}$$

where $\mathcal{A}_{\alpha i\beta l} = \partial S_{\alpha i}/\partial F_{l\beta}$ are the *fixed-reference moduli*. By (5), they are defined as

$$\mathcal{A}_{\alpha i\beta l} = \frac{\partial^2 W}{\partial F_{i\alpha}\partial F_{l\beta}}. \tag{13}$$

We can then increment the equations of equilibrium (4), as

$$\operatorname{Div} \boldsymbol{S} = \boldsymbol{0} \quad \longrightarrow \quad \operatorname{Div} \dot{\boldsymbol{S}} = \boldsymbol{0}, \qquad \text{or} \qquad \frac{\partial S_{\alpha i}}{\partial X_\alpha} \quad \longrightarrow \quad \frac{\partial \dot{S}_{\alpha i}}{\partial X_\alpha} = 0. \tag{14}$$

Integrating over the volume of the solid, we perform the following series of operations,

$$\int_{\mathcal{B}_0} \operatorname{Div} \dot{\boldsymbol{S}}\, dV = \int_{\partial\mathcal{B}_0} \dot{\boldsymbol{S}}^T \boldsymbol{N}\, dA \qquad \text{(divergence theorem)} \tag{15}$$

$$= \int_{\partial\mathcal{B}} J^{-1} \dot{\boldsymbol{S}}^T \boldsymbol{F}^T \boldsymbol{n}\, da \qquad \text{(Nanson's formula)} \tag{16}$$

$$= \int_{\partial\mathcal{B}} \left(J^{-1}\boldsymbol{F}\,\dot{\boldsymbol{S}}\right)^T \boldsymbol{n}\, da \qquad \text{(transpose of product)} \tag{17}$$

$$= \int_{\mathcal{B}} \operatorname{div}\left(J^{-1}\boldsymbol{F}\,\dot{\boldsymbol{S}}\right) dv \qquad \text{(divergence theorem)} \tag{18}$$

so that the incremental equations of equilibrium can be put in the following equivalent (Eulerian) form:

$$\operatorname{div} \boldsymbol{\Sigma} = \mathbf{0}, \qquad \frac{\partial \Sigma_{ji}}{\partial x_j} = 0. \tag{19}$$

Here, the tensor $\boldsymbol{\Sigma}$ is defined as $\boldsymbol{\Sigma} = J^{-1}\boldsymbol{F}\,\dot{\boldsymbol{S}}$, with components

$$\Sigma_{ji} = J^{-1}F_{j\alpha}\dot{S}_{\alpha i} = J^{-1}F_{j\alpha}\mathcal{A}_{\alpha i\beta l}\Gamma_{lk}F_{k\beta}, \qquad \text{or} \qquad \Sigma_{ji} = \mathcal{A}_{0jikl}\Gamma_{lk}, \tag{20}$$

where the *instantaneous elastic moduli* are the following components of the fourth-order tensor $\boldsymbol{\mathcal{A}}_0$,

$$\mathcal{A}_{0jikl} = J^{-1}\mathcal{A}_{\alpha i\beta l}F_{j\alpha}F_{k\beta} = J^{-1}\frac{\partial^2 W}{\partial F_{i\alpha}\partial F_{l\beta}}F_{j\alpha}F_{k\beta}. \tag{21}$$

Let us increment the boundary conditions (7) where the applied tractions $\boldsymbol{f}$ are considered to correspond to a dead-load (i.e. $\boldsymbol{f}$ is constant).

$$\boldsymbol{S}^T\boldsymbol{N} = \boldsymbol{f} \quad \longrightarrow \quad \dot{\boldsymbol{S}}^T\boldsymbol{N} = \dot{\boldsymbol{f}} = \mathbf{0} \quad \text{on } \partial\mathcal{B}_0. \tag{22}$$

Using Nanson's formula, we have the equivalent Eulerian form:

$$\boldsymbol{\Sigma}^T\boldsymbol{n} = \mathbf{0} \quad \text{on } \partial\mathcal{B}. \tag{23}$$

In the case where the solid is incompressible, we need to increment the constraint of incompressibility $\det \boldsymbol{F} = 1$. Recalling that $\dfrac{\mathrm{d}}{\mathrm{d}\tau}(\det \boldsymbol{A}) = (\det \boldsymbol{A}) \operatorname{tr}\left(\dfrac{d\boldsymbol{A}}{d\tau}\boldsymbol{A}^{-1}\right)$, we find

$$J = 1 \quad \longrightarrow \quad J \operatorname{tr}\left(\dot{\boldsymbol{F}}\boldsymbol{F}^{-1}\right) = 0, \qquad \text{i.e.} \qquad \operatorname{tr}\boldsymbol{\Gamma} = u_{i,i} = 0. \tag{24}$$

Then the increment of the nominal stress for incompressible solids (6) is found by incrementing the following identity,

$$\boldsymbol{F}\,\boldsymbol{F}^{-1} = \boldsymbol{I} \quad \longrightarrow \quad \dot{\boldsymbol{F}}\,\boldsymbol{F}^{-1} + \boldsymbol{F}(\dot{\overline{\boldsymbol{F}^{-1}}}) = \mathbf{0}, \tag{25}$$

so that

$$\boldsymbol{S} \quad \longrightarrow \quad \dot{\boldsymbol{S}} = \boldsymbol{\mathcal{A}}\boldsymbol{F}\boldsymbol{\Gamma} - \dot{p}\boldsymbol{F}^{-1} + p\boldsymbol{F}^{-1}\boldsymbol{\Gamma}, \tag{26}$$

and $\boldsymbol{\Sigma} = J^{-1}\boldsymbol{F}\,\dot{\boldsymbol{S}} = \boldsymbol{F}\,\dot{\boldsymbol{S}}$ is now given by

$$\boldsymbol{\Sigma} = \boldsymbol{\mathcal{A}}_0\,\boldsymbol{\Gamma} - \dot{p}\boldsymbol{I} + p\boldsymbol{\Gamma}, \qquad \Sigma_{ji} = \mathcal{A}_{0jikl}u_{l,k} + \dot{p}\delta_{ji} + pu_{j,i}. \tag{27}$$

2.3 Elastic Moduli

Here we focus on the instantaneous elastic moduli, which are given by (21), with $J = 1$ for incompressible materials. By swapping ji with kl in (21), and taking into account that α and β are dummy indices, we obtain

$$\mathcal{A}_{0klji} = \mathcal{A}_{0jikl}, \tag{28}$$

the so-called *major symmetries*. Owing to those symmetries, we can count that in general, there are at most 45 independent instantaneous moduli.

The moduli do not possess the 'minor' symmetries, and $\mathcal{A}_{0jikl} \neq \mathcal{A}_{0ijkl}$ in general. We may find out what the difference $\mathcal{A}_{0jikl} - \mathcal{A}_{0ijkl}$ is. First, recall that the Cauchy stress $\boldsymbol{\sigma}$ is symmetric: $\boldsymbol{\sigma} = J^{-1}\boldsymbol{F}\,\boldsymbol{S} = \boldsymbol{\sigma}^T$. Hence, by incrementing,

$$\boldsymbol{F}\,\boldsymbol{S} = (\boldsymbol{F}\,\boldsymbol{S})^T \quad \longrightarrow \quad \dot{\boldsymbol{F}}\,\boldsymbol{S} + \boldsymbol{F}\,\dot{\boldsymbol{S}} = (\dot{\boldsymbol{F}}\,\boldsymbol{S} + \boldsymbol{F}\,\dot{\boldsymbol{S}})^T, \tag{29}$$

we get, using (10), (12) and (21),

$$\begin{aligned}
&\boldsymbol{\Gamma}\boldsymbol{F}\boldsymbol{S} + J\boldsymbol{\mathcal{A}}_0\boldsymbol{\Gamma} = \left(\boldsymbol{\Gamma}\boldsymbol{F}\boldsymbol{S} + J\boldsymbol{\mathcal{A}}_0\boldsymbol{\Gamma}\right)^T, \\
&\boldsymbol{\Gamma}\boldsymbol{\sigma} + \boldsymbol{\mathcal{A}}_0\boldsymbol{\Gamma} = \left(\boldsymbol{\Gamma}\boldsymbol{\sigma} + \boldsymbol{\mathcal{A}}_0\boldsymbol{\Gamma}\right)^T, \\
&\Gamma_{jk}\sigma_{ki} + \mathcal{A}_{0jikl}\Gamma_{lk} = \Gamma_{ik}\sigma_{kj} + \mathcal{A}_{0ijkl}\Gamma_{lk}, \\
&\sigma_{ki}\delta_{jl}\Gamma_{lk} + \mathcal{A}_{0jikl}\Gamma_{lk} = \sigma_{kj}\delta_{il}\Gamma_{lk} + \mathcal{A}_{0ijkl}\Gamma_{lk}.
\end{aligned} \tag{30}$$

Collecting the terms in Γ_{lk}, we reach the conclusion that (Chadwick, 1997)

$$\mathcal{A}_{0jikl} - \mathcal{A}_{0ijkl} = \sigma_{jk}\delta_{il} - \sigma_{ik}\delta_{jl}. \tag{31}$$

In particular, $\mathcal{A}_{0ijji} - \mathcal{A}_{0jiij} = \mathcal{A}_{0ijij} - \sigma_{ii}$ when $i \neq j$ (no sum).

For incompressible solids, the derivation of the difference in the moduli is similar, based on (26). We then find that (Chadwick, 1997)

$$\mathcal{A}_{0jikl} - \mathcal{A}_{0ijkl} = \left(\sigma_{jk} + p\delta_{jk}\right)\delta_{il} - \left(\sigma_{ik} + p\delta_{ik}\right)\delta_{jl}. \tag{32}$$

In particular, we have

$$\mathcal{A}_{0jiij} - \mathcal{A}_{0ijij} = -\sigma_{ii} - p. \tag{33}$$

As explained elsewhere in this book, for general nonlinear materials with two families of fibres, one oriented along $\boldsymbol{M}$, the other along $\boldsymbol{M}'$ (in $\mathcal{B}_0$), the strain energy density W can always be written as a function of the following eight invariants

$$\begin{cases} I_1 = \operatorname{tr}(\boldsymbol{C}), \\ I_2 = \frac{1}{2}\left[(\operatorname{tr}\boldsymbol{C})^2 - \operatorname{tr}\left(\boldsymbol{C}^2\right)\right] = I_3 \operatorname{tr}\left(\boldsymbol{C}^{-1}\right), \\ I_3 = \det(\boldsymbol{C}), \\ I_4 = \boldsymbol{M}\cdot\boldsymbol{C}\boldsymbol{M} = \boldsymbol{m}\cdot\boldsymbol{m}, \\ I_5 = \boldsymbol{M}\cdot\boldsymbol{C}^2\boldsymbol{M} = \boldsymbol{m}\cdot\boldsymbol{B}\boldsymbol{m}, \\ I_6 = \boldsymbol{M}'\cdot\boldsymbol{C}\boldsymbol{M}' = \boldsymbol{m}'\cdot\boldsymbol{m}', \\ I_7 = \boldsymbol{M}'\cdot\boldsymbol{C}^2\boldsymbol{M}' = \boldsymbol{m}\cdot\boldsymbol{B}\boldsymbol{m}', \\ I_8 = \boldsymbol{M}\cdot\boldsymbol{C}\boldsymbol{M}' = \boldsymbol{m}\cdot\boldsymbol{m}', \end{cases} \tag{34}$$

where $\boldsymbol{B} = \boldsymbol{F}\boldsymbol{F}^T$ is the left Cauchy–Green deformation tensor, $\boldsymbol{m} = \boldsymbol{F}\boldsymbol{M}$, $\boldsymbol{m}' = \boldsymbol{F}\boldsymbol{M}'$, and $\boldsymbol{C} = \boldsymbol{F}^T\boldsymbol{F}$ is the right Cauchy–Green deformation tensor. Hence $W = W(I_1, I_2, \ldots, I_8)$ and we can then use the chain rule to compute the moduli given by (21). For this task, we need expressions for the first derivatives of the eight invariants with respect to $\boldsymbol{F}$. They are (Ogden, 2007)

$$\begin{aligned} &\frac{\partial I_1}{\partial F_{i\alpha}} = 2F_{i\alpha}, \qquad \frac{\partial I_2}{\partial F_{i\alpha}} = 2(C_{\gamma\gamma}F_{i\alpha} - C_{\alpha\gamma}F_{i\gamma}), \qquad \frac{\partial I_3}{\partial F_{i\alpha}} = 2I_3 F_{\alpha i}^{-1}, \\ &\frac{\partial I_4}{\partial F_{i\alpha}} = 2M_\alpha F_{i\gamma}M_\gamma, \qquad \frac{\partial I_5}{\partial F_{i\alpha}} = 2(F_{i\gamma}M_\gamma C_{\alpha\gamma}M_\beta + F_{i\gamma}C_{\gamma\beta}M_\beta M_\alpha), \\ &\frac{\partial I_8}{\partial F_{i\alpha}} = F_{i\gamma}(M'_\alpha M_\gamma + M_\alpha M'_\gamma), \end{aligned} \tag{35}$$

and the derivatives of I_6 and I_7 are found from those of I_4 and I_5 by replacing $\boldsymbol{M}$ with $\boldsymbol{M}'$. We also need expressions for the second derivatives. They read

$$\begin{aligned} &\frac{\partial^2 I_1}{\partial F_{i\alpha}\partial F_{l\beta}} = 2\delta_{il}\delta_{\alpha\beta}, \\ &\frac{\partial^2 I_2}{\partial F_{i\alpha}\partial F_{l\beta}} = 2\left(2F_{i\alpha}F_{l\beta} - F_{i\beta}F_{l\alpha} + C_{\gamma\gamma}\delta_{il}\delta_{\alpha\beta} - B_{il}\delta_{\alpha\beta} - C_{\alpha\beta}\delta_{il}\right), \\ &\frac{\partial^2 I_3}{\partial F_{i\alpha}\partial F_{l\beta}} = 2I_3\left(2F_{\alpha i}^{-1}F_{\beta l}^{-1} - F_{\alpha l}^{-1}F_{\beta i}^{-1}\right), \\ &\frac{\partial^2 I_4}{\partial F_{i\alpha}\partial F_{l\beta}} = 2\delta_{il}M_\alpha M_\beta, \\ &\frac{\partial^2 I_5}{\partial F_{i\alpha}\partial F_{l\beta}} = 2\left[\delta_{il}(C_{\alpha\gamma}M_\gamma M_\beta + C_{\beta\gamma}M_\gamma M_\alpha) + \delta_{\alpha\beta}F_{i\gamma}M_\gamma F_{l\delta}M_\delta \right. \\ &\qquad\qquad \left. + F_{i\gamma}M_\gamma F_{l\alpha}M_\beta + F_{l\gamma}M_\gamma F_{i\beta}M_\alpha + B_{il}M_\alpha M_\beta\right]. \end{aligned} \tag{36}$$

Again, the expressions for the derivatives of I_6 and I_7 are found from those of I_4 and I_5 by replacing $\boldsymbol{M}$ with $\boldsymbol{M}'$. Finally,

$$\frac{\partial^2 I_8}{\partial F_{i\alpha} F_{l\beta}} = \delta_{il}(M'_\alpha M_\beta + M_\alpha M'_\beta). \tag{37}$$

Computing the instantaneous moduli from (21) and (35)–(37) is not that difficult in practice, and the process can be automated with a computer algebra system if needs be. Take the case of *incompressible solids reinforced with one family of parallel fibres.* Their strain-energy density is of the form

$$W = W(I_1, I_2, I_4, I_5), \tag{38}$$

only, because $I_3 = 1$ at all times. Then we find that

$$\begin{aligned}
\mathcal{A}_{0jikl} &= 2W_1\delta_{il}B_{jk} \\
&+ 2W_2\left[2B_{ij}B_{kl} - B_{ik}B_{jl} - B_{il}B_{jk} + I_1\delta_{il}B_{jk} - \delta_{il}(\boldsymbol{B}^2)_{jk}\right] \\
&+ 2W_4\delta_{il}m_j m_k \\
&+ 2W_5\left[\delta_{il}(\boldsymbol{Bm})_j m_k + \delta_{il}(\boldsymbol{Bm})_k m_j\right. \\
&\qquad\left. + B_{jk}m_i m_l + B_{jl}m_i m_k + B_{ik}m_j m_l + B_{il}m_j m_k\right] \\
&+ 4W_{11}B_{ij}B_{kl} \\
&+ 4W_{22}(I_1\boldsymbol{B} - \boldsymbol{B}^2)_{ij}(I_1\boldsymbol{B} - \boldsymbol{B}^2)_{kl} \\
&+ 4W_{12}\left[B_{ij}(I_1\boldsymbol{B} - \boldsymbol{B}^2)_{kl} + B_{kl}(I_1\boldsymbol{B} - \boldsymbol{B}^2)_{ij}\right] \\
&+ 4W_{14}\left(B_{ij}m_k m_l + B_{kl}m_i m_j\right) \\
&+ 4W_{24}\left[(I_1\boldsymbol{B} - \boldsymbol{B}^2)_{kl}m_i m_j + (I_1\boldsymbol{B} - \boldsymbol{B}^2)_{ij}m_k m_l\right] \\
&+ 4W_{44}m_i m_j m_k m_l \\
&+ 4W_{55}\left[(\boldsymbol{Bm})_i m_j + (\boldsymbol{Bm})_j m_i\right]\left[(\boldsymbol{Bm})_l m_k + (\boldsymbol{Bm})_k m_l\right] \\
&+ 4W_{15}\left[B_{ij}m_k(\boldsymbol{Bm})_l + B_{ij}m_l(\boldsymbol{Bm})_k\right. \\
&\qquad\left. + B_{kl}m_j(\boldsymbol{Bm})_i + B_{kl}m_i(\boldsymbol{Bm})_j\right] \\
&+ 4W_{25}\left\{(I_1\boldsymbol{B} - \boldsymbol{B}^2)_{ij}\left[(\boldsymbol{Bm})_l m_k + (\boldsymbol{Bm})_k m_l\right]\right. \\
&\qquad\left. + (I_1\boldsymbol{B} - \boldsymbol{B}^2)_{kl}\left[(\boldsymbol{Bm})_i m_j + (\boldsymbol{Bm})_j m_i\right]\right\} \\
&+ 4W_{45}\left[m_i m_j m_k(\boldsymbol{Bm})_l + m_i m_j m_l(\boldsymbol{Bm})_k\right. \\
&\qquad\left. + m_j m_k m_l(\boldsymbol{Bm})_i + m_i m_k m_l(\boldsymbol{Bm})_j\right].
\end{aligned} \tag{39}$$

Consider for example the following strain energy density of the *neo-Hookean reinforcing model* (Merodio and Ogden, 2002),

$$W = \mu(I_1 - 3)/2 + F(I_4), \tag{40}$$

for an incompressible solid with one family of fibres, where F is a function of I_4 only and $\mu > 0$ is the initial shear modulus of the solid when $F \equiv 0$. Then the expressions above for the moduli reduce to

$$\mathcal{A}_{0jikl} = \mu\delta_{il}B_{jk} + 2F'(I_4)\delta_{il}m_j m_k + 4F''(I_4)m_i m_j m_k m_l. \tag{41}$$

Here, in addition to the major symmetries (28), we also find that $\mathcal{A}_{0jilk} = \mathcal{A}_{0jkli}$. It follows that we end up with 23 non-zero components, several of which are equal to one another. *In toto* there are 13 independent moduli.

Or take the case of a *plane pre-strain* in the 1, 2 plane say, for an incompressible solid with only one family of fibres, lying in that plane. Then, in the coordinate system of the Eulerian principal axes,

$$\boldsymbol{B} = \begin{bmatrix} \lambda^2 & 0 & 0 \\ 0 & \lambda^{-2} & 0 \\ 0 & 0 & 1 \end{bmatrix}, \quad \boldsymbol{B}^{-1} = \begin{bmatrix} \lambda^{-2} & 0 & 0 \\ 0 & \lambda^2 & 0 \\ 0 & 0 & 1 \end{bmatrix} \tag{42}$$

and clearly

$$I_1 = I_2 = \lambda^2 + \lambda^{-2} + 1, \qquad I_3 = 1. \tag{43}$$

Also, $\boldsymbol{M} = (M_1, M_2, 0)$, say, with $M_1^2 + M_2^2 = 1$, so that

$$I_4 = \lambda^2 M_1^2 + \lambda^{-2} M_2^2 = m_1^2 + m_2^2, \qquad I_5 = \lambda^4 M_1^2 + \lambda^{-4} M_2^2 = \lambda^2 m_1^2 + \lambda^{-2} m_2^2. \tag{44}$$

Then we find the connection: $I_5 = (I_1 - 1)I_4 - 1$. Here there are thus only 2 independent invariants, and we can introduce the following function of two variables:

$$\hat{W}(I_1, I_4) = W\left(I_1, I_1, 1, I_4, (I_1 - 1)I_4 - 1\right). \tag{45}$$

That function allows for a reduced form of the moduli (Merodio and Ogden, 2002):

$$\begin{aligned} \mathcal{A}_{0jikl} = {} & 2\hat{W}_1\delta_{il}B_{jk} + 2\hat{W}_4\delta_{il}m_j m_k + 4\hat{W}_{11}B_{ij}B_{kl} \\ & + 4\hat{W}_{44}m_j m_i m_k m_l + 4\hat{W}_{14}(B_{ij}m_k m_l + B_{kl}m_i m_j). \end{aligned} \tag{46}$$

For an example in that category, take the *Mooney-Rivlin standard reinforcing model*,

$$W = C(I_1 - 3)/2 + D(I_2 - 3)/2 + E(I_4 - 1)^2/4, \tag{47}$$

where C, D and E are positive constants. Then $\hat{W}(I_1, I_4) = (C + D)(I_1 - 3) + E(I_4 - 1)^2/4$, and

$$\mathcal{A}_{0jikl} = (C + D)\delta_{il}B_{jk} + E(m_1^2 + m_2^2 - 1)\delta_{il}m_j m_k + 2Em_i m_j m_k m_l. \tag{48}$$

Finally, consider the following strain energy density used by Ciarletta et al. (2013) for modelling *skin tissue*, when seen as reinforced with a single family of fibres, oriented in $\mathcal{B}_0$ along the unit vector $\boldsymbol{M}$. It is given by

$$W = \frac{\mu}{2}(I_1 - 3) + \beta I_\alpha, \tag{49}$$

where $\mu > 0$ and $\beta > 0$ are constants, and the structural anisotropic invariant I_α is defined as

$$I_\alpha := \boldsymbol{M} \cdot \left[\boldsymbol{C} + \boldsymbol{C}^{-1} - 2\boldsymbol{I}\right] \boldsymbol{M} = (\lambda_\alpha - \lambda_\alpha^{-1})^2. \tag{50}$$

Here $\lambda_\alpha := (\boldsymbol{M} \cdot \boldsymbol{C}\boldsymbol{M})^{\frac{1}{2}}$ represents the fibre stretch. For this model, the instantaneous moduli can again be computed from the formula (39), by noticing the connection

$$I_\alpha = (1 - I_1)I_4 + I_5 + I_2 - 2. \tag{51}$$

This is quite a long exercise however, and it is simpler to rely on the following formula:

$$\frac{\partial F^{-1}_{\beta k}}{\partial F_{i\alpha}} = -F^{-1}_{\beta i}F^{-1}_{\alpha k}, \tag{52}$$

which can be established by differentiating the identity $\boldsymbol{F}\boldsymbol{F}^{-1} = \boldsymbol{I}$. The end result is that the corresponding moduli are

$$\begin{aligned} \mathcal{A}_{0jikl} = \mu\delta_{il}B_{jk} + 2\beta(&\delta_{il}m_j m_k + \delta_{ik}F^{-1}_{\alpha l}M_\alpha F^{-1}_{\beta j}M_\beta \\ &+ \delta_{jl}F^{-1}_{\alpha i}M_\alpha F^{-1}_{\beta k}M_\beta + \delta_{jk}F^{-1}_{\alpha i}M_\alpha F^{-1}_{\beta l}M_\beta). \end{aligned} \tag{53}$$

3 Wave Propagation

In this section, we study small-amplitude wave propagation in infinite, homogeneous, incompressible, hyperelastic solids subject to large, static, homogeneous deformations. We look at the effect of anisotropy on the wave speed in initially isotropic solids and initially anisotropic (fibre-reinforced) tissues.

We take the axes of the Cartesian coordinate system in $\mathcal{B}$ to be aligned with the principal axes of deformation, that is we describe the pre-strain by

$$x_1 = \lambda_1 X_1, \qquad x_2 = \lambda_2 X_2, \qquad x_3 = \lambda_3 X_3, \tag{54}$$

where the λ's are the constant principal stretch ratios. Then the constraint of incompressibility dictates that a unit cube keeps its volume, so that $\lambda_1\lambda_2\lambda_3 = 1$.

Now, deriving the equations of *incremental motion* is a process similar to deriving those of incremental equilibrium (19); see Ogden (2007) for example. The end result is that they read

$$\Sigma_{ji,j} = \rho u_{i,tt}, \qquad u_{j,j} = 0, \tag{55}$$

where the components Σ_{ji} are given by (27) and ρ is the mass density per unit volume (ρ is constant in homogeneous bodies because of the internal constraint of incompressibility, see (2)). Let us solve the equations for a motion in the form of a *plane homogeneous wave.* For these, the mechanical displacement is written in all generality as

$$\boldsymbol{u} = \boldsymbol{a} f\left(\boldsymbol{n}\cdot\boldsymbol{x} - vt\right), \qquad \dot{p} = g\left(\boldsymbol{n}\cdot\boldsymbol{x} - vt\right), \tag{56}$$

where f and g are arbitrary single-variable functions, $\boldsymbol{n}$ is a unit vector in direction of *propagation*, $\boldsymbol{a}$ is a unit vector in direction of *polarisation*, and the real scalar v is the *speed.*

Then, the incremental incompressibility $(55)_2$ gives $a_j n_j f' = 0$, or

$$\boldsymbol{a}\cdot\boldsymbol{n} = 0. \tag{57}$$

In other words, the motion is *transverse.*

Next we note that the moduli $\mathcal{A}_{0jikl}$ and the Lagrange multiplier p are constants because the components $F_{i\alpha}$ of the deformation gradient $\boldsymbol{F}$ derived from (54) are constants themselves. It follows that the incremental equations of motion $(55)_1$ read

$$\mathcal{A}_{0jikl} a_l n_k n_j f'' - n_i g' = \rho v^2 f'' a_i \tag{58}$$

(note that the term involving p is $p u_{j,ij}$, which is zero by incremental incompressibility). Taking the product of this equation with a_i and using $\boldsymbol{a}\cdot\boldsymbol{n} = 0$, we arrive at the *secular equation,*

$$\rho v^2 = \mathcal{A}_{0jikl} a_i a_l n_j n_k, \tag{59}$$

which gives the wave speed in terms of the wave characteristics, the material properties and the pre-strain.

3.1 Strong Ellipticity for the Isotropic Matrix

From (59), we see that the speed is *real* when

$$\mathcal{A}_{0jikl} n_j n_k a_i a_l > 0, \qquad \text{for all } \boldsymbol{a},\boldsymbol{n} \text{ such that } \boldsymbol{a}\cdot\boldsymbol{a} = \boldsymbol{n}\cdot\boldsymbol{n} = 1,\ \boldsymbol{a}\cdot\boldsymbol{n} = 0. \tag{60}$$

This is the so-called *Strong Ellipticity* (SE) condition.

From here on we focus on wave propagation in a principal plane, by taking $\boldsymbol{n}$ in the (x_1, x_2) plane. Hence we have

$$\boldsymbol{n} = (n_1, n_2, 0), \quad \text{and either (i) } \boldsymbol{a} = (0,0,1) \text{ or (ii) } \boldsymbol{a} = (-n_2, n_1, 0), \tag{61}$$

where $n_1^2 + n_2^2 = 1$.

Recall that in *isotropic* materials, the principal axes of stress are aligned with the principal axes of strain. Then, in the (x_1, x_2, x_3) coordinate system, there are only 15 non-zero components of $\boldsymbol{\mathcal{A}}_0$ (Ogden, 1997). They are

$$\mathcal{A}_{0iijj}, \qquad \mathcal{A}_{0ijij}, \qquad \mathcal{A}_{0ijji}, \tag{62}$$

where there is no sum on the repeated indices.

Case (i): $\boldsymbol{a} = (0,0,1)$. Here SE reads $\mathcal{A}_{0j3k3} n_j n_k > 0$, i.e. $\mathcal{A}_{01313} n_1^2 + \mathcal{A}_{02323} n_2^2 > 0$ for all n_1, n_2. It follows that Strong Ellipticity imposes restrictions on the sign of some moduli, here

$$\mathcal{A}_{01313} > 0, \qquad \mathcal{A}_{02323} > 0. \tag{63}$$

Case (ii): $\boldsymbol{a} = (-n_2, n_1, 0)$. Here SE can be written in the compact form

$$\alpha\, n_1^4 + 2\beta\, n_1^2 n_2^2 + \gamma\, n_2^4 > 0, \tag{64}$$

for all n_1, n_2, where

$$\alpha = \mathcal{A}_{01212}, \quad 2\beta = \mathcal{A}_{01111} + \mathcal{A}_{02222} - 2\mathcal{A}_{01122} - 2\mathcal{A}_{01221}, \quad \gamma = \mathcal{A}_{02121}. \tag{65}$$

Taking in turn $n_2 = 0$ and $n_1 = 0$ imposes $\alpha > 0$ and $\gamma > 0$. Then, taking $(n_2/n_1)^4 = \alpha/\gamma$ gives $\beta + \sqrt{\alpha\gamma} > 0$. Conversely, if the inequalities $\alpha > 0$, $\gamma > 0$, $\beta + \sqrt{\alpha\gamma} > 0$ are assumed, then

$$\alpha n_1^4 + 2\beta n_1^2 n_2^2 + \gamma n_2^4 > \alpha n_1^4 - 2\beta n_1^2 n_2^2 + \gamma n_2^4 = \left(\sqrt{\alpha} n_1^2 - \sqrt{\gamma} n_2^2\right) > 0, \tag{66}$$

and SE is satisfied. In conclusion, the SE condition is equivalent to

$$\alpha > 0, \qquad \gamma > 0, \qquad \beta + \sqrt{\alpha\gamma} > 0. \tag{67}$$

From now on, we assume that the *SE condition holds for the isotropic matrix.*

If for example the isotropic matrix is modelled by the Mooney–Rivlin material: $W = C(I_1 - 3)/2 + D(I_2 - 3)/2$, as in (47), then we find that

$$\alpha = \lambda_1^2 \left(C + D\lambda_3^2\right), \qquad \gamma = \lambda_2^2 \left(C + D\lambda_3^2\right), \qquad 2\beta = \alpha + \gamma. \tag{68}$$

Hence, SE reads: $C + D\lambda_3^2 > 0$, for all λ_3, so that we must have $C > 0$, $D > 0$. Similarly, if the matrix is modelled by the neo-Hookean model: $W = \mu(I_1 - 3)/2$, as in (40) and (49), then SE leads to $\mu > 0$.

3.2 Acoustic Tensor

Now we go back to the incremental equations of motion (58) for plane homogeneous waves. Multiplying the first equation by n_i and using (57) we obtain $g' = \mathcal{A}_{0jikl}n_i n_j n_k m_l f''$. Then we substitute back this expression for g' and drop the common factor f'' to arrive at

$$(\mathcal{A}_{0jikl}n_j n_k - \mathcal{A}_{0jpkl}n_j n_p n_k n_i)a_l = \rho v^2 a_i. \tag{69}$$

By introducing the symmetric *acoustic tensor* $\boldsymbol{Q}(\boldsymbol{n})$, with components

$$Q_{il}(\boldsymbol{n}) = \mathcal{A}_{0jikl}n_j n_k, \tag{70}$$

the equations above can be written in the compact form

$$[\boldsymbol{I} - \boldsymbol{n}\otimes\boldsymbol{n}]\boldsymbol{Q}(\boldsymbol{n})\boldsymbol{a} = \rho v^2\boldsymbol{a}, \qquad \boldsymbol{a}\cdot\boldsymbol{n} = 0. \tag{71}$$

Note that the tensor $[\boldsymbol{I} - \boldsymbol{n}\otimes\boldsymbol{n}]\boldsymbol{Q}(\boldsymbol{n})$ is *not* symmetric. However, because the motion is transverse, we may write the equation above in the following equivalent form

$$[\boldsymbol{I} - \boldsymbol{n}\otimes\boldsymbol{n}]\boldsymbol{Q}(\boldsymbol{n})[\boldsymbol{I} - \boldsymbol{n}\otimes\boldsymbol{n}]\boldsymbol{a} = \rho v^2\boldsymbol{a}, \qquad \boldsymbol{a}\cdot\boldsymbol{n} = 0. \tag{72}$$

By introducing $\bar{\boldsymbol{I}} = \boldsymbol{I} - \boldsymbol{n}\otimes\boldsymbol{n}$, the projection tensor onto the plane normal to $\boldsymbol{n}$, this reads as

$$\bar{\boldsymbol{Q}}(\boldsymbol{n})\boldsymbol{a} = \rho v^2\boldsymbol{a}, \qquad \boldsymbol{a}\cdot\boldsymbol{n} = 0, \qquad \text{where} \quad \bar{\boldsymbol{Q}}(\boldsymbol{n}) = \bar{\boldsymbol{I}}\boldsymbol{Q}(\boldsymbol{n})\bar{\boldsymbol{I}} \tag{73}$$

is a *symmetric tensor*. Note that this symmetrization of the acoustic tensor in constrained materials (including incompressible materials as here) is attributed to M.A. Hayes in the PhD thesis of N.H. Scott (see Scott and Hayes, 1985).

For any given direction of propagation $\boldsymbol{n}$, we now have (73), a 2D symmetric algebraic eigenvalue problem for determining ρv^2 and $\boldsymbol{a}$. Thanks to the symmetry of $\bar{\boldsymbol{Q}}$, we know that there are two mutually orthogonal eigenvectors $\boldsymbol{a}$ and $\boldsymbol{b}$, say, and that the two corresponding eigenvalues ρv_a^2 and ρv_b^2, say, must be real. They are obtained by solving the characteristic equation

$$\det[\bar{\boldsymbol{Q}}(\boldsymbol{n}) - \rho v^2\boldsymbol{I}] = 0, \tag{74}$$

which factorises into the product of ρv^2 by a quadratic. The eigenvalue $\rho v^2 = 0$ corresponds to eigenvector $\boldsymbol{n}$ and is discarded; the two other eigenvalues correspond to two purely transverse waves. When the eigenvalues coincide, we can superpose the two corresponding waves to form a *circularly-polarized wave*, which propagates along one of the so-called *acoustic axes*.

3.3 Example: Deformed Isotropic Material

First we treat the case of wave propagation when the anisotropy is *strain-induced* only. We take the deformed body to be made of an incompressible *isotropic Mooney–Rivlin material*, with strain-energy density $W = C(I_1 - 3)/2 + D(I_2 - 3)/2$, which is (47) when $E \equiv 0$.

First we consider waves travelling in a *principal plane* of pre-deformation, so that $\boldsymbol{n} = (\cos\theta, \sin\theta, 0)$ say, where θ is the angle of propagation in the principal plane with respect to the x_1-direction. Then the two transverse eigenvectors are $\boldsymbol{a} = (-\sin\theta, \cos\theta, 0)$ and $\boldsymbol{b} = (0, 0, 1)$, with corresponding eigenvalues

$$\rho v_a^2 = \lambda_1^2(C + D\lambda_3^2)\cos^2\theta + \lambda_2^2(C + D\lambda_3^2)\sin^2\theta,$$
$$\rho v_b^2 = \lambda_1^2(C + D\lambda_2^2)\cos^2\theta + \lambda_2^2(C + D\lambda_1^2)\sin^2\theta, \tag{75}$$

respectively. To find these compact expressions we used the connections (68).

For our numerical calculations, we pick $\lambda_1 = 2$, $\lambda_2 = 0.45$, $\lambda_3 = (\lambda_1\lambda_2)^{-1}$ for the pre-stretch, and $C = 0.6\mu$, $D = 0.4\mu$ for the material parameters, where μ is the initial shear modulus (in Pa). In this example, the sinusoidal variations of the speeds according to (75) are found to coincide twice over a half-period, indicating the presence of two acoustic axes in the deformed solid. To see this, we can either solve (75) for θ when $v_a = v_b$, or plot the variations of both speeds with θ and look for intersections, see Figure 4(a).

In wave acoustics, it is common to plot the *slowness surface*, because its shape is related to vibrational wavefronts which can be visualised experimentally (Musgrave, 1970; Wolfe, 1995). This is a spherical plot of the slowness $s := 1/v$ for each propagation direction $\boldsymbol{n}$. Hence for our waves propagating in the x_3–principal plane, the propagation vector is of the form $\boldsymbol{n} = (\cos\theta, \sin\theta, 0)$. For each angle of propagation θ, there are two slownesses $s_a := 1/v_a$ and $s_b := 1/v_b$ giving 2 points on the slowness surface: $(s_a, \theta, 0)$ and $(s_b, \theta, 0)$. Clearly, the two slowness sheets intersect in the directions of acoustic axes, see Figure 4(b) for the intersection of the slowness surface with the plane orthogonal to $(0, 0, 1)$.

Now for waves propagating *in any direction* $\boldsymbol{n}$, not necessarily in a principal plane, we specialise (70) and (73) to the Mooney–Rivlin strain energy function. We find that the reduced acoustic tensor reads

$$\bar{\boldsymbol{Q}}(\boldsymbol{n}) = C\,(\boldsymbol{n}\cdot\boldsymbol{B}\boldsymbol{n})\,\bar{\boldsymbol{I}} + D\bar{\boldsymbol{I}}\boldsymbol{B}^{-1}\bar{\boldsymbol{I}}. \tag{76}$$

Hence here, it is perfectly possible to find explicit expressions for the wave speeds and polarisation, simply by solving the eigenvalue problem for this tensor. In particular we can see straightaway that the polarisation vectors

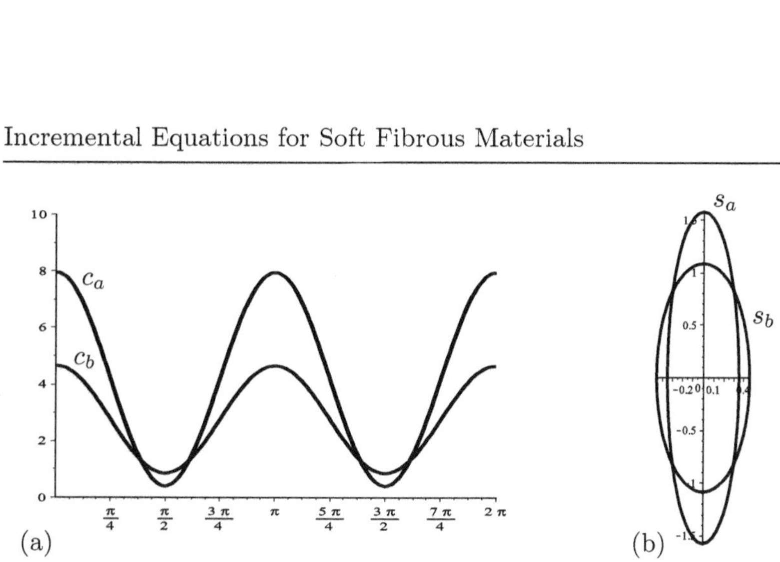

Figure 4. Homogeneous plane wave travelling in a principal plane in a deformed (initially isotropic) Mooney–Rivlin material. (a) Variations of the non-dimensional speeds $c_a := \sqrt{\rho v_a^2/\mu}$ and $c_b := \sqrt{\rho v_b^2/\mu}$ with the angle of propagation θ measured with respect to the direction of greatest stretch. (b) Polar plot of the variations of the non-dimensional slownesses $s_a := 1/c_a$ and $s_b := 1/c_b$ with θ. The intersections of the plots point to the directions of 2 acoustic axes, along which circularly-polarised waves propagate.

$\boldsymbol{a}$ and $\boldsymbol{b}$ are along the axes of the ellipsoidal section of the $\boldsymbol{x} \cdot \boldsymbol{B}^{-1}\boldsymbol{x} = 1$ ellipsoid by the plane normal to $\boldsymbol{n}$, as they are eigenvectors of both the $\bar{\boldsymbol{I}}\boldsymbol{B}^{-1}\bar{\boldsymbol{I}}$ and $\bar{\boldsymbol{I}}$ tensors. The corresponding speeds are then

$$\rho v_a^2 = C\,(\boldsymbol{n} \cdot \boldsymbol{B}\boldsymbol{n}) + D\left(\boldsymbol{a} \cdot \boldsymbol{B}^{-1}\boldsymbol{a}\right), \qquad \rho v_b^2 = C\,(\boldsymbol{n} \cdot \boldsymbol{B}\boldsymbol{n}) + D\left(\boldsymbol{b} \cdot \boldsymbol{B}^{-1}\boldsymbol{b}\right); \tag{77}$$

see Boulanger and Hayes (1992) for a full analysis of this problem.

To plot the entire slowness surface, we can take advantage of the graphical abilities of any mathematical/numerical package using a simple algorithm: Pick a propagation direction $\boldsymbol{n}$ with Cartesian components $\boldsymbol{n} = (\cos\theta\sin\phi, \sin\theta\sin\phi, \cos\phi)$, where θ and ϕ are its azimuthal and polar angles, respectively; Compute the corresponding reduced acoustic tensor $\bar{\boldsymbol{Q}}$ and its two eigenvalues ρv_a^2 and ρv_b^2; Plot the two points with spherical coordinates $(1/v_a, \theta, \phi)$ and $(1/v_b, \theta, \phi)$. Figure 5 illustrates the results of this algorithm for the same material and pre-strain parameter values used in Figure 4.

3.4 Example: Deformed Anisotropic Material

To present a representative example, we focus here on a material reinforced by one family of parallel fibres behaving according to the Mooney–

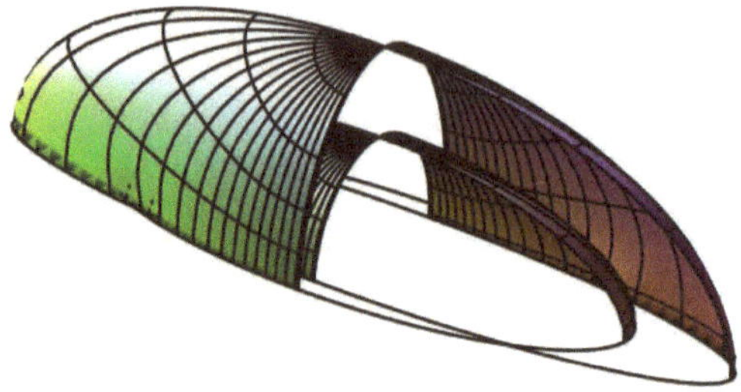

Figure 5. One quarter of a slowness surface composed of two sheets, each corresponding to the non-dimensional slownesses $s_a := \sqrt{\mu/(\rho v_a^2)}$ and $s_b := \sqrt{\mu/(\rho v_b^2)}$, using the same material and pre-strain parameter values used in Figure 4. The sheets are represented by a collection of points (s, θ, ϕ) in the spherical coordinate system, where θ and ϕ are the azimuthal and polar angles of propagation, respectively. The solid lines correspond to the intersection of the slowness surface sheets with the plane orthogonal to $(0, 0, 1)$.

Rivlin standard reinforcing model (47).

Putting together Equations (70) and (73) for a general propagation direction $\boldsymbol{n}$, we arrive in that case at the following expression for the reduced acoustic tensor,

$$\bar{\boldsymbol{Q}}(\boldsymbol{n}) = \Big[C\,(\boldsymbol{n}\cdot\boldsymbol{B}\boldsymbol{n}) + E\,(\boldsymbol{m}\cdot\boldsymbol{m}-1)\,(\boldsymbol{m}\cdot\boldsymbol{n})^2\Big]\,\bar{\boldsymbol{I}} \\ + D\bar{\boldsymbol{I}}\boldsymbol{B}^{-1}\bar{\boldsymbol{I}} + 2E\,(\boldsymbol{m}\cdot\boldsymbol{n})^2\,\bar{\boldsymbol{I}}\boldsymbol{m}\otimes\bar{\boldsymbol{I}}\boldsymbol{m}. \quad (78)$$

This expression makes it clear that it not obvious at all now to find polarisation vectors as analytical eigenvectors for $\bar{\boldsymbol{Q}}$ in general, except in the special case where the direction of the fibres lies in a principal plane of deformation. Then, the two eigenvectors are clearly $\boldsymbol{b} := \boldsymbol{n}\times\boldsymbol{m}$ and $\boldsymbol{a} := \boldsymbol{n}\times\boldsymbol{b}$.

For an illustration, we take $C = 0.6\mu$, $D = 0.4\mu$, $E = 40\mu$ in (47), where μ (in Pa) can be seen as the initial shear modulus of the soft matrix. We let the fibres be originally at 45° between the X_1- and X_2-axes, i.e. $\boldsymbol{M} = (1/\sqrt{2}, 1/\sqrt{2}, 0)$. Then we subject the solid to the pre-strain (54), with $\lambda_1 = 1.4$, $\lambda_2 = 0.8$, $\lambda_3 = (\lambda_1\lambda_2)^{-1}$, so that $\boldsymbol{m} = \boldsymbol{F}\boldsymbol{M}$ is at angle $\tan^{-1}[\lambda_2\sin\alpha/(\lambda_1\cos\alpha)] \simeq 27.4°$ with the x_1-axis, which gives the direction of the fibres in the deformed configuration.

By taking $\boldsymbol{n} = (\cos\theta, \sin\theta, 0)$ and varying θ, we obtain the dispersion curves and slowness surface principal sections of Figure 6. Clearly, the

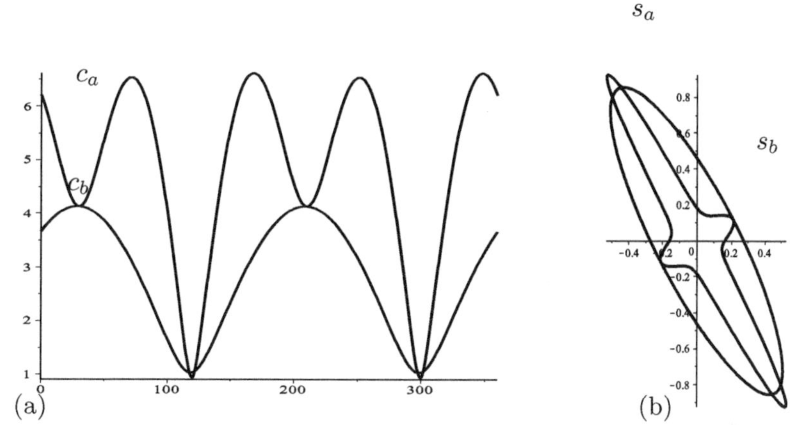

Figure 6. Homogeneous plane wave travelling in a principal plane in a deformed (initially anisotropic) Mooney–Rivlin material reinforced with one family of parallel fibres. (a) Variations of the non-dimensional speeds $c_a := \sqrt{\rho v_a^2/\mu}$ and $c_b := \sqrt{\rho v_b^2/\mu}$ with the angle of propagation θ measured with respect to the direction of greatest stretch. Here the fibres are in the principal plane, originally aligned in the direction $\theta = 45°$. (b) Polar plot of the variations of the non-dimensional slownesses $s_a := 1/c_a$, $s_b := 1/c_b$ with θ.

introduction of fibres brings in a more colourful behaviour, as their presence is strongly felt in the variations of c_a and s_a for the wave polarised in the principal plane. In particular we see that this transverse wave travels at its slowest in the direction perpendicular to the fibres. The speed c_a is at a local minimum when the corresponding wave travels in the direction of the fibres, and reaches its maximum at an intermediate angle. Meanwhile the speed c_b of the wave polarised along the x_3−direction has more regular (sinusoidal) variations. Although the variation of these speeds could be studied analytically, we do not pursue this avenue here, as they are attached to the special case of principal plane propagation. In general, it is better to turn to a numerical treatment of the eigenvalues, following the algorithm presented in the previous subsection. For the present case study, it leads to the slowness surface depicted in Figure 7.

4 Surface Stability

Now we move on from the study of infinitesimal waves in unbounded (infinite) deformed solids to the study of small-amplitude perturbations in a

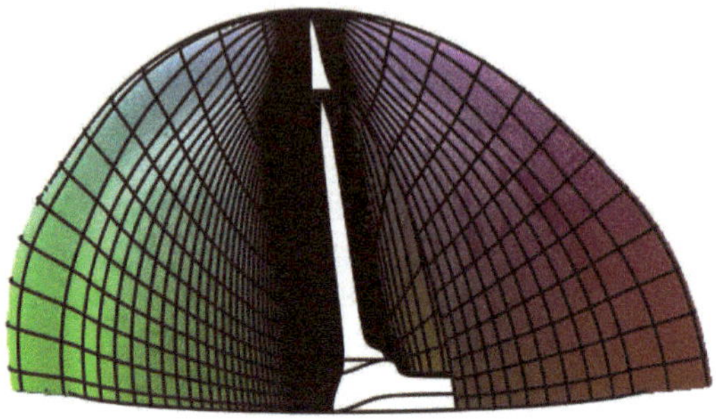

Figure 7. One quarter of a slowness surface composed of two sheets, each corresponding to the non-dimensional slownesses $s_a := \sqrt{\mu/(\rho v_a^2)}$ and $s_b := \sqrt{\mu/(\rho v_b^2)}$, using the same material and pre-strain parameter values used in Figure 6. The sheets are represented by a collection of points (s, θ, ϕ) in the spherical coordinate system, where θ and ϕ are the azimuthal and polar angles of propagation, respectively. The solid lines correspond to the intersection of the slowness surface sheets with the plane orthogonal to $(0, 0, 1)$.

deformed *bounded* media reinforced with fibres. The complexity of the analysis increases dramatically and we must be careful to take a step-by-step approach to tackling this complexity. Hopefully then, each result we establish along the way can serve as a benchmark or limiting case for the next degree of difficulty.

Hence we look at the simplest boundary problem possible, that of a *semi-infinite solid* limited by a plane. Since there is no characteristic length in a half-space geometry, the results ought to be *non-dispersive* (independent of a wavelength). Then we focus on small-amplitude *static wrinkles* instead of waves, again to decrease the number of parameters. To fix the ideas, we select *simple shear* as the base large pre-strain, as it described by a single kinematic parameter, the amount of shear K, say, and we imagine that it takes place in planes parallel to the free surface. Thus we take the solid to fill the half-space $X_2 \geq 0$ in the reference configuration. We subject it to the simple shear

$$x_1 = X_1 + KX_2, \qquad x_2 = X_2, \qquad x_3 = X_3, \tag{79}$$

so that it fills the region $x_2 \geq 0$ in the deformed configuration, and the planes $X_2 = 0$ and $x_2 = 0$ are assumed to be free of traction. We call $\boldsymbol{E}_i$ and $\boldsymbol{e}_i$ $(i = 1, 2, 3)$ the unit vectors along the X_i- and x_i-axes, respectively.

Then, in the $\boldsymbol{E}_i \otimes \boldsymbol{e}_j$ and $\boldsymbol{e}_i \otimes \boldsymbol{e}_j$ bases, respectively, the deformation gradient $\boldsymbol{F} = \partial \boldsymbol{x}/\partial \boldsymbol{X}$ and the left Cauchy–Green deformation gradient $\boldsymbol{B} = \boldsymbol{F}\boldsymbol{F}^T$ have components

$$\boldsymbol{F} = \begin{bmatrix} 1 & K & 0 \\ 0 & 1 & 0 \\ 0 & 0 & 1 \end{bmatrix}, \qquad \boldsymbol{B} = \begin{bmatrix} 1+K^2 & K & 0 \\ K & 1 & 0 \\ 0 & 0 & 1 \end{bmatrix}. \tag{80}$$

We expect that as the magnitudes of the strains increase during a large simple shear, there will be directions of severe compression developing for some line elements and eventually, at a *critical amount of shear* K_{cr}, say, the surface will buckle and let wrinkles appear. For the solid, we consider in turn an isotropic neo-Hookean model, and then a soft neo-Hookean matrix reinforced with a single family of stiffer parallel fibres lying in the planes parallel to the free surface.

In other words, our ultimate goal in this section is to model the physical behaviours reported in Figure 1.

4.1 Surface Instability for a Sheared Isotropic Material

We start off by studying the onset of wrinkles in a sheared isotropic neo-Hookean material, with strain-energy density

$$W = \mu(I_1 - 3), \tag{81}$$

where $\mu > 0$ is the shear modulus.

Biot (1963) showed that surface instability occurs for this material when the following *wrinkling condition* is met

$$\lambda_1^2 \lambda_3 = \sigma_0, \tag{82}$$

where $\sigma_0 \simeq 0.296$ is the real root of the cubic $\sigma^3 + \sigma^2 + 3\sigma - 1 = 0$. Here it is assumed that the free surface is parallel to a principal plane of pre-strain. In that plane, $\lambda_1 < 1$ and $\lambda_3 \geq 1$ are the principal stretches and $\lambda_2 = (\lambda_1 \lambda_3)^{-1}$ is the principal stretch in the direction normal to the surface (the depth). The wrinkles decay exponentially with depth and have sinusoidal variations along the surface, with their front running parallel to the principal Eulerian axis where λ_3 occurs. That is, they are parallel to the direction of greatest tension and thus, orthogonal to the direction of greatest compression.

The simple shear described by (79) is a *plane strain deformation*, such that

$$\lambda_1 = \lambda, \qquad \lambda_2 = 1, \qquad \lambda_3 = \lambda^{-1}, \tag{83}$$

say. Then Biot's wrinkling condition clearly gives the *critical stretch of compression* as $\lambda_{\text{cr}} = \sigma_0 \simeq 0.296$, and thus the critical stretch of extension as $\lambda_3 = \sigma_0^{-1} \simeq 3.38$. We conclude that the wrinkles appear when the solid is compressed by 71%, or equivalently, when it is stretched by 238%.

This is quite an extreme deformation for a soft solid. To connect it with the kinematics of simple shear, we recall the links between the principal stretches and the amount of shear (Ogden, 1997)

$$K = \lambda_3 - \lambda_1, \qquad \lambda_3 = K/2 + \sqrt{1 + K^2/4}, \tag{84}$$

and that the *angle of shear* is $\tan^{-1} K$ (when the solid is not sheared, this angle is zero; when it is infinitely sheared, this angle tends to 90° as a limit).

So here, surface instability in a sheared isotropic neo-Hookean material occurs when

$$K = K_{\text{cr}} = \sigma_0^{-1} - \sigma_0 \simeq 3.09, \qquad \text{and} \qquad \tan^{-1} K = \tan^{-1} K_{\text{cr}} \simeq 72.0°. \tag{85}$$

This is a very large shear, as seen Figure 8. These values go a long way to explaining why the surface of the sheared silicone in Figure 1(a) remains stable, as the angle of shear there is clearly less than 72°.

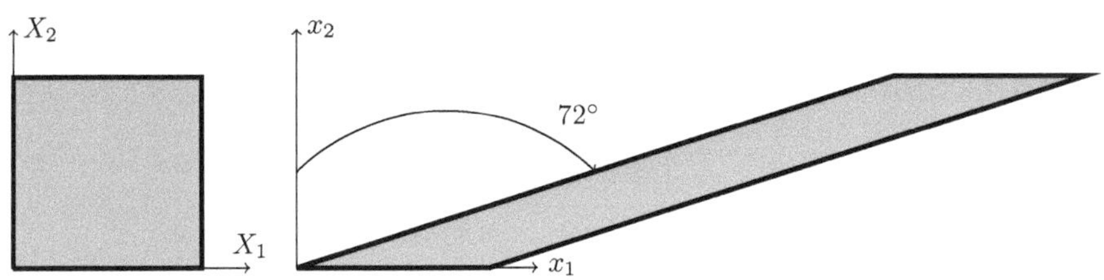

Figure 8. The amount of shear required to reach surface instability for an isotropic neo-Hookean half-space is $K_{\text{cr}} = 3.09$, which represents an enormous strain, as shown here.

However large, it is nonetheless possible to create a homogeneous pre-strain of that magnitude in a soft solid, and to observe the appearance of surface instability when a critical threshold is reached. To produce the experiment photographed in Figure 9, we filled up the third of a four-sided plexiglass box with commercial gelatine. The four connecting edges of the box are hinged so that a large homogeneous 'shear-box' deformation is possible, which can be decomposed into the combination of a simple shear and a tri-axial stretch (Stolz, 2010). At a very large amount of pre-strain, we see small-amplitude wrinkles form on the surface, and they are aligned with the long diagonal of the deformed shear-box, which is in the direction of largest stretch.

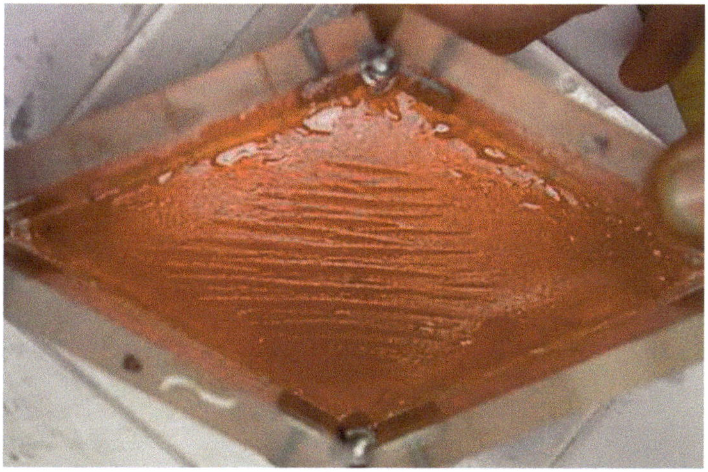

Figure 9. Large homogenous strain of soft gelatine in a shear-box apparatus, up to the critical threshold of surface instability, as indicated by the formation of small-amplitude wrinkles aligned with the long diagonal.

To complete the picture, we can ask ourselves an additional question: Do the wrinkles always occur along the direction of greatest stretch? Indeed, Biot (1963) only studied wrinkles appearing along a principal direction, at a certain degree of compression. But what if 'non-principal' wrinkles had appeared prior to reaching his critical state of deformation? Intuition and the observation of Figure 9 tell us that 'principal' wrinkles should be the general case, but we must keep in mind that wrinkling in a principal direction is not necessarily the sole outcome of a surface instability analysis (Gower and Destrade, submitted) nor of experiments: for instance, 'oblique' buckling wrinkles have been observed in polymeric strips (Wang et al., 2011) or in meta-sedimentary rocks (Meere et al., 2013).

Thankfully, answering this question for the neo-Hookean surface instability problem is easy, as we can rely on the works of Flavin (1963). He showed that wrinkles develop parallel to the direction making an angle θ with the principal direction of strain associated with the largest stretch λ_3 when the following *wrinkling condition* is met

$$\lambda_1^2\lambda_3^2(\lambda_1^2\cos^2\theta + \lambda_3^2\sin^2\theta) = \sigma_0^2. \tag{86}$$

In the plane strain situation above, this condition reads as

$$\lambda^4\cos^2\theta - \lambda^2\sigma_0^2 + \sin^2\theta = 0. \tag{87}$$

This quadratic in $\lambda^2 < 1$ has real roots provided θ is in the range $-\theta_0 \leq \theta \leq \theta_0$, where $\theta_0 = (1/2)\sin^{-1}\sigma_0^2 \simeq 2.51°$. In that narrow range, λ reaches a maximum of 0.296 when $\theta = 0°$, indicating that wrinkles take place at that level of compression and are aligned with the direction of greatest stretch. We expect this situation to be completely different for anisotropic materials, because the fibres and their orientation should play a major role in the onset of wrinkles, see Figure 1(b).

4.2 Surface Instability for a Sheared Anisotropic Material

Now we introduce one family of parallel fibres, orientated along the unit vector $\boldsymbol{M}$ with components

$$\boldsymbol{M} = \cos\Phi \boldsymbol{E}_1 + \sin\Phi \boldsymbol{E}_3 \tag{88}$$

in the reference configuration, where Φ is the angle between the direction of shear and the fibres. Simple shear is a homogeneous deformation and so, using (80), we see that $\boldsymbol{M}$ is transformed into $\boldsymbol{m} = \boldsymbol{F}\boldsymbol{M}$ in the current configuration,

$$\boldsymbol{m} = (\cos\Phi + K\sin\Phi)\boldsymbol{e}_1 + \sin\Phi\, \boldsymbol{e}_3. \tag{89}$$

It follows that the anisotropic invariant $I_4 \equiv \boldsymbol{m} \cdot \boldsymbol{m}$ is given by

$$I_4 = 1 + K\sin 2\Phi + K^2 \sin^2\Phi. \tag{90}$$

Recall that I_4 is the squared stretch in the fibre direction: if $I_4 \geq 1$ then the fibres are in extension, if $I_4 \leq 1$ then they are in compression. For our constitutive assumptions, we move on from the isotropic neo-Hookean material to the *standard reinforcing neo-Hookean material*, a subcase of (47):

$$W = \mu(I_1 - 3)/2 + E(I_4 - 1)^2/4. \tag{91}$$

In that case, it is a simple exercise to show that the corresponding Cauchy stress tensor $\boldsymbol{\sigma}$ is

$$\boldsymbol{\sigma} = -p\boldsymbol{I} + \mu\boldsymbol{B} + E(I_4 - 1)\boldsymbol{m} \otimes \boldsymbol{m}, \tag{92}$$

where p is the Lagrange multiplier introduced by the constraint of incompressibility. Its value is found by imposing that the surface $x_2 = 0$ be free of traction: $\sigma_{22} = 0$, which gives $p = \mu$.

Thus, the pre-stress required to maintain the large static simple shear is

$$\boldsymbol{\sigma} = \mu(\boldsymbol{B} - \boldsymbol{I}) + E(I_4 - 1)\boldsymbol{m} \otimes \boldsymbol{m}, \tag{93}$$

showing clearly that the directions of principal stress and of principal strain do not coincide in general (except of course in the special cases where $\boldsymbol{m}$ is aligned with principal directions of strain).

Finally, we note that according to the Appendix, the quantity $2E/3\mu$ is a measure of the stiffness of the fibres compared to that of the soft matrix: if it is greater than 1, then the fibres are stiffer than the matrix, at least in a tensile, infinitesimal, tensile test.

Now we seek a perturbation solution in the following form,

$$\{\boldsymbol{u}, \dot{p}\} = \{\boldsymbol{U}(kx_2), \mathrm{i}kP(kx_2)\}\mathrm{e}^{\mathrm{i}k(\cos\theta x_1 + \sin\theta x_3)}, \tag{94}$$

where k is the "wave"-number and $\boldsymbol{U}$, P are functions of kx_2 alone. This form describes wrinkles in the deformed configuration with sinusoidal variations in the direction of $\boldsymbol{n} = (\cos\theta, \sin\theta, 0)$, and depth variations according to the amplitude functions U and P, to be determined later. Then, according to (27), the components Σ_{ji} for the incremental traction are of a similar form,

$$\Sigma_{ji} = \mathrm{i}kS_{ji}(kx_2)\mathrm{e}^{\mathrm{i}k(\cos\theta x_1 + \sin\theta x_3)}, \tag{95}$$

say, where the amplitude functions S_{ij} are to be determined. All the kinematical quantities of interest are represented in Figure 10.

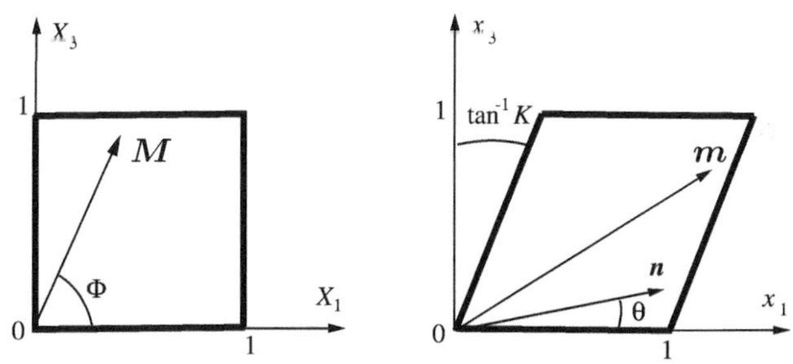

Figure 10. Simple shear of a unit cuboid in the boundary of a half-space by amount K. The face shown is parallel to the free surface. The solid is reinforced with a family of parallel fibres aligned in the direction of $\boldsymbol{M} = (\cos\Phi, \sin\Phi, 0)$ in the reference configuration and of $\boldsymbol{m} = \boldsymbol{F}\boldsymbol{M}$ in the current configuration. The unit vector $\boldsymbol{n} = (\cos\theta, \sin\theta, 0)$ in the current configuration is orthogonal to the wavefront of wrinkles which may develop if the shear is severe enough.

By relying on a systematic procedure devised by Chadwick (1997), we can eliminate P and write the governing equations (19) as a first-order linear ordinary differential system. This is known as the *Stroh formulation* of the problem (Stroh, 1962),

$$\begin{bmatrix} \boldsymbol{U}' \\ \boldsymbol{S}' \end{bmatrix} = \mathrm{i}\boldsymbol{N} \begin{bmatrix} \boldsymbol{U} \\ \boldsymbol{S} \end{bmatrix}, \text{ where } \boldsymbol{U} := \begin{bmatrix} U_1 \\ U_2 \\ U_3 \end{bmatrix}, \quad \boldsymbol{S} := \begin{bmatrix} S_{21} \\ S_{22} \\ S_{23} \end{bmatrix}, \quad \boldsymbol{N} := \begin{bmatrix} \boldsymbol{N}_1 & \boldsymbol{N}_2 \\ \boldsymbol{N}_3 & \boldsymbol{N}_1 \end{bmatrix}, \tag{96}$$

and the prime denotes differentiation with respect to the argument kx_2. Here the symmetric 3×3 matrices $\boldsymbol{N}_1$, $\boldsymbol{N}_2$, $\boldsymbol{N}_3$ are given by

$$-\boldsymbol{N}_1 = \begin{bmatrix} 0 & \cos\theta & 0 \\ \cos\theta & 0 & \sin\theta \\ 0 & \sin\theta & 0 \end{bmatrix}, \quad \boldsymbol{N}_2 = \begin{bmatrix} 1/\mu & 0 & 0 \\ 0 & 0 & 0 \\ 0 & 0 & 1/\mu \end{bmatrix}, \quad -\boldsymbol{N}_3 = \begin{bmatrix} \eta & 0 & \kappa \\ 0 & \nu & 0 \\ \kappa & 0 & \chi \end{bmatrix},$$

with

$$\begin{aligned} \eta &= (\mathcal{A}_{01111} + 3\mu)\cos^2\theta + 2\mathcal{A}_{01131}\cos\theta\sin\theta + \mathcal{A}_{03131}\sin^2\theta, \\ \nu &= \mathcal{A}_{01212}\cos^2\theta + 2\mathcal{A}_{01232}\cos\theta\sin\theta + \mathcal{A}_{03232}\sin^2\theta - \mu, \\ \chi &= \mathcal{A}_{01313}\cos^2\theta + 2\mathcal{A}_{01333}\cos\theta\sin\theta + (\mathcal{A}_{03333} + 3\mu)\sin^2\theta, \\ \kappa &= \mathcal{A}_{01113}\cos^2\theta + (2\mathcal{A}_{01133} + 3\mu)\cos\theta\sin\theta + \mathcal{A}_{03133}\sin^2\theta, \end{aligned}$$

and the components of $\boldsymbol{\mathcal{A}}_0$ are given explicitly by (41), with $F' = 2E(I_4 - 1)$ and $F'' = 2E$. Notice how all the "anisotropy" information is located in the $\boldsymbol{N}_3$ sub-matrix. For details of this derivation, see Destrade et al. (2008).

The system (96) can be written in compact form as

$$\boldsymbol{\xi}' = \mathrm{i}\boldsymbol{N}\boldsymbol{\xi}, \qquad \text{where} \qquad \boldsymbol{\xi} := [\boldsymbol{U}, \boldsymbol{S}]^T, \tag{97}$$

is the Stroh displacement–traction vector, a function of the dimensionless quantity kx_2. Its solution is clearly an exponential:

$$\boldsymbol{\xi}(kx_2) = \left[\boldsymbol{U}^0, \boldsymbol{S}^0\right]^T \mathrm{e}^{\mathrm{i}kqx_2}, \tag{98}$$

say, where $\boldsymbol{U}^0, \boldsymbol{S}^0$ are constant vectors and q is an eigenvalue of $\boldsymbol{N}$. In effect we now have to solve an *eigenvalue problem* for the Stroh matrix $\boldsymbol{N}$:

$$(\boldsymbol{N} - q\boldsymbol{I})\boldsymbol{\xi}^0 = \boldsymbol{0}. \tag{99}$$

The associated characteristic equation $\det(\boldsymbol{N} - q\boldsymbol{I}) = 0$ is a *cubic* in q^2,

$$q^6 - \left(2 - \frac{\chi + \eta}{\mu}\right) q^4 + \left(1 + \frac{\nu - 2\epsilon}{\mu} + \frac{\chi\eta - \kappa^2}{\mu^2}\right) q^2 + \frac{\epsilon(\mu + \nu)}{\mu^2} = 0, \tag{100}$$

where the quantity ϵ is defined by

$$\epsilon := \chi \cos^2\theta - 2\kappa \cos\theta \sin\theta + \eta \sin^2\theta. \tag{101}$$

Here we focus on complex roots to the bicubic, because we are looking for solutions with amplitudes decaying with depth–real roots would lead to inadequate sinusoidal non-decaying depth variations. Since the polynomial characteristic equation has real coefficients, the complex roots come in pairs of complex conjugate numbers. Specifically, out of the six possible complex roots, we can only keep those three satisfying Im $q > 0$; see the form of the solution in (98). These three adequate roots q_1, q_2, q_3 allow us to find three independent solutions to the differential system, which we combine to form its general solution as

$$\boldsymbol{\xi}(kx_2) = c_1\boldsymbol{\xi}^{(1)}\mathrm{e}^{\mathrm{i}q_1kx_2} + c_2\boldsymbol{\xi}^{(2)}\mathrm{e}^{\mathrm{i}q_2kx_2} + c_3\boldsymbol{\xi}^{(3)}\mathrm{e}^{\mathrm{i}q_3kx_2}, \tag{102}$$

where the c_i are scalar constants and the $\boldsymbol{\xi}^{(i)}$ are the eigenvectors. We write this general solution as

$$\boldsymbol{\xi}(kx_2) = \left[\boldsymbol{\xi}^{(1)}|\boldsymbol{\xi}^{(2)}|\boldsymbol{\xi}^{(3)}\right]\left\langle \mathrm{e}^{\mathrm{i}qkx_2}\right\rangle \begin{bmatrix} c_1 \\ c_2 \\ c_3 \end{bmatrix}, \tag{103}$$

where we adopted the notation $\left\langle \mathrm{e}^{\mathrm{i}qkx_2}\right\rangle := \mathrm{diag}\left(\mathrm{e}^{\mathrm{i}q_1kx_2}, \mathrm{e}^{\mathrm{i}q_2kx_2}, \mathrm{e}^{\mathrm{i}q_3kx_2}\right)$. We re-write it even more compactly as

$$\boldsymbol{\xi}(kx_2) = \begin{bmatrix} \boldsymbol{A} \\ \boldsymbol{B} \end{bmatrix} \left\langle \mathrm{e}^{\mathrm{i}qkx_2}\right\rangle \boldsymbol{c}, \tag{104}$$

where $\boldsymbol{A}$ ($\boldsymbol{B}$) is the 3×3 matrix made with the top (bottom) three rows of $\left[\boldsymbol{\xi}^{(1)}|\boldsymbol{\xi}^{(2)}|\boldsymbol{\xi}^{(3)}\right]$. Hence now

$$\boldsymbol{U}(kx_2) = \boldsymbol{A}\left\langle \mathrm{e}^{\mathrm{i}qkx_2}\right\rangle \boldsymbol{c}, \qquad \boldsymbol{S}(kx_2) = \boldsymbol{B}\left\langle \mathrm{e}^{\mathrm{i}qkx_2}\right\rangle \boldsymbol{c}. \tag{105}$$

Now that we have the general solution, we can solve the boundary value problem, which states that there should be no incremental traction on the free surface, that is

$$\boldsymbol{S}(0) = \boldsymbol{0}. \tag{106}$$

According to $(105)_2$, this is possible for non-trivial $\boldsymbol{c}$ only when $\det \boldsymbol{B} = 0$, which is the *bifurcation criterion*. However this is not expressed in the most optimal way. In particular, computing numerically the eigenvalues and corresponding linearly independent eigenvectors of $\boldsymbol{N}$ is quite demanding from a computational point of view. For certain cases, the complex matrix

$\boldsymbol{B}$ becomes ill-conditioned and finding numerically for which values of K it becomes singular is not an easy task at all. So we turn to a much more efficient method, based on the notion of mechanical impedance.

We remark that it follows from (105) that $\boldsymbol{c} = \left\langle \mathrm{e}^{-\mathrm{i}qkx_2} \right\rangle \boldsymbol{A}^{-1}\boldsymbol{U}(kx_2)$, and that

$$\boldsymbol{S}(kx_2) = \mathrm{i}\boldsymbol{Z}\boldsymbol{U}(kx_2), \qquad \text{where} \quad \boldsymbol{Z} := -\mathrm{i}\boldsymbol{B}\boldsymbol{A}^{-1} \tag{107}$$

is the so-called *surface impedance matrix*. Clearly, the matrix $\boldsymbol{Z}$ is constant (independent of x_2). Also, it is singular, because the boundary condition (106) can only be satisfied for a nontrivial surface displacement $\boldsymbol{U}(0)$ when

$$\det \boldsymbol{Z} = 0. \tag{108}$$

Finally, it can be shown (e.g., Shuvalov, 2000; Mielke and Fu, 2004) that $\boldsymbol{Z}$ is Hermitian so that it is of the form

$$\boldsymbol{Z} = \begin{bmatrix} Z_1 & Z_4 + \mathrm{i}Z_5 & Z_6 + \mathrm{i}Z_7 \\ Z_4 - \mathrm{i}Z_5 & Z_2 & Z_8 + \mathrm{i}Z_9 \\ Z_6 - \mathrm{i}Z_7 & Z_8 - \mathrm{i}Z_9 & Z_3 \end{bmatrix}, \tag{109}$$

where the Z_i are real constants.

Then, using (107), the governing equations (97) read as

$$\boldsymbol{U}' = \mathrm{i}\boldsymbol{N}_1\boldsymbol{U} + \mathrm{i}\boldsymbol{N}_2\boldsymbol{S} = \mathrm{i}\boldsymbol{N}_1\boldsymbol{U} + \boldsymbol{N}_2\boldsymbol{Z}\boldsymbol{U}, \tag{110}$$
$$\boldsymbol{S}' = \mathrm{i}\boldsymbol{N}_3\boldsymbol{U} + \mathrm{i}\boldsymbol{N}_1\boldsymbol{S} = \mathrm{i}\boldsymbol{N}_3\boldsymbol{U} + \boldsymbol{N}_1\boldsymbol{Z}\boldsymbol{U} = -\mathrm{i}\boldsymbol{Z}\boldsymbol{U}'. \tag{111}$$

Substitute the first line into the second to end up with

$$\boldsymbol{Z}\boldsymbol{N}_2\boldsymbol{Z} + \mathrm{i}\boldsymbol{Z}\boldsymbol{N}_1 - \mathrm{i}\boldsymbol{N}_1\boldsymbol{Z} + \boldsymbol{N}_3 = \boldsymbol{0}, \tag{112}$$

an algebraic *Riccati matrix equation*. Because the $\boldsymbol{N}_i$ are real symmetric matrices and $\boldsymbol{Z}$ is Hermitian, writing separately the real and imaginary parts of this equation yields nine real quadratic equations for the Z_i of (109). Together with the traction-free boundary condition (106), we have an algebraic system of 10 equations for the 10 unknowns $Z_1, Z_2, \dots, Z_9$ and

K. It goes as follows

$$
\begin{aligned}
&2\mu\cos\theta Z_5 - Z_1^2 - Z_6^2 - Z_7^2 + \mu\eta = 0,\\
&\mu\sin\theta Z_7 - Z_1Z_4 - Z_6Z_8 - Z_7Z_9 = 0,\\
&\mu\sin\theta Z_5 - \mu\cos\theta Z_9 - Z_1Z_6 - Z_3Z_6 + \mu\kappa = 0,\\
&\mu\cos\theta(Z_1 - Z_2) + \mu\sin\theta Z_6 + Z_1Z_5 - Z_6Z_9 + Z_7Z_8 = 0,\\
&2\mu\cos\theta Z_5 - 2\mu\sin\theta Z_9 + Z_5^2 + Z_9^2 + Z_4^2 + Z_8^2 - \mu\nu = 0,\\
&\mu\cos\theta Z_7 + Z_5Z_7 + Z_3Z_8 + Z_4Z_6 = 0,\\
&\mu\cos\theta Z_8 - \mu\sin\theta Z_4 - Z_1Z_7 - Z_3Z_7 = 0,\\
&\mu\cos\theta Z_6 - \mu\sin\theta(Z_2 - Z_3) - Z_3Z_9 - Z_4Z_7 + Z_5Z_6 = 0,\\
&2\mu\sin\theta Z_9 + Z_7^2 + Z_3^2 + Z_6^2 - \mu\chi = 0,\\
&Z_1Z_2Z_3 - Z_1Z_8^2 - Z_1Z_9^2 - Z_2Z_6^2 - Z_2Z_7^2 - Z_3Z_4^2 - Z_3Z_5^2\\
&\qquad + 2Z_4Z_6Z_8 + 2Z_4Z_7Z_9 - 2Z_5Z_6Z_9 + 2Z_5Z_7Z_8 = 0. \qquad (113)
\end{aligned}
$$

The system is non-linear and may thus have several solutions, but there is only one for which $\boldsymbol{Z}$ is positive semi-definite, corresponding to the decaying solution (Fu and Mielke, 2002). This system provides us with the following algorithm for solving the bifurcation criterion.

We prescribe the material parameters: μ the shear modulus, E the fibre modulus, and Φ the angle of the fibres. Then we fix θ, the wrinkles angle. Then, the only remaining unknowns are the solutions, if they exist, to the system above.

We start at $K = 0$ (no shear) and increment slowly this quantity, until it reaches a value K_{cr} at which the system has a solution. If it exists, then that is the point at which the bifurcation criterion is met. Of course, it could be that wrinkles have appeared earlier, at a lower amount of shear, for another θ. We must thus vary θ from $0°$ to $180°$ and find the angle θ^* for which the amount of shear K_{cr} is minimal, $K_{\mathrm{cr}} = K^*$, say, which corresponds to the earliest onset of wrinkling in an oblique direction.

At the end of this process, we have access to the critical amount of shear K^* for a given material. It tells us how much the body can be sheared before instability, and gives us the deformation gradient and the current orientation of the fibres. With the value of θ^* we know how the wrinkles are oriented with respect to the fibres. With the value of $\boldsymbol{Z}$, we can solve $\boldsymbol{ZU}(0) = \boldsymbol{0}$ for $\boldsymbol{U}(0)$ to find the shape of the wrinkles on the free surface (up to an arbitrary multiplicative parameter: recall that this is a linearised stability analysis). In fact, by solving numerically the differential equation (110) we can find the variations of the full incremental field displacements with depth, and then use (107) to obtain those of the traction field.

Now if we want to investigate the effect of the orientation of the fibres with respect to the direction of shear, we can vary Φ and eventually obtain a $K^* - \Phi$ stability plot. Then we can vary the value of E/μ to investigate the effect of the relative strength of the fibres compared to the soft matrix. It transpires that those later curves are quite tricky to acquire due the following reason.

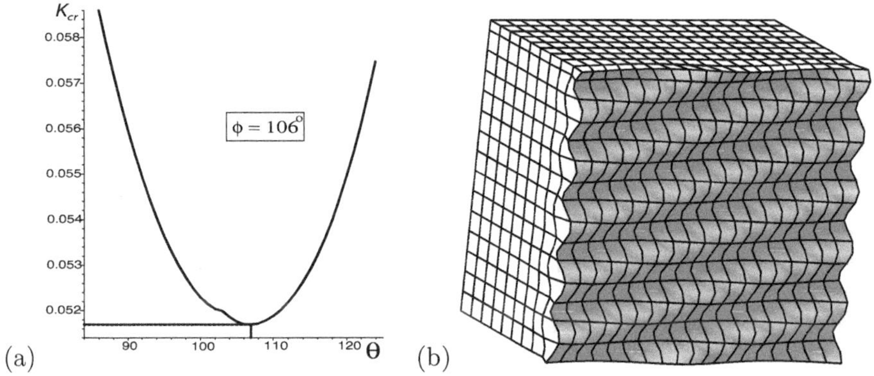

Figure 11. A half-space made of neo-Hookean soft matrix with standard reinforcement can buckle very early in simple shear. Here the E/μ ratio is equal to 10.0. When the fibres are originally at $\Phi = 106°$ to the shear direction, wrinkles appear at $K^* \simeq 0.052$, with the normal to the wrinkle front at $\theta^* \simeq 107°$ from the wrinkles. (a) The $K_{\text{cr}} - \theta$ curve, leading to the determination of the critical values K^* and θ^* (intersection of the dashed lines). (b) The corresponding surface buckling pattern, with wrinkles forming at right angle to the fibres.

At $\Phi = 0$, the critical amount of shear is very close to the value $\sigma_0^{-1} - \sigma_0 \simeq 3.09$ found in (85) for the isotropic neo-Hookean half-space ($E \equiv 0$). Then as Φ is increased, a smooth increase in K^* follows. Hence for each increase in Φ, the natural inclination is to look for the next K^* in the neighborhood of the previous one. However, the search should always start from $K = 0$ instead, because as it happens, the $K^* - \Phi$ stability plot is discontinuous: at a certain angle Φ_0, the material stops being made more stable by the fibres and instead, becomes unstable at a very low value of K^*. This is illustrated by the example in Figure 11, where the situation is summarised in the caption. In that example it is found that when $E/\mu = 10.0$, and the fibres are originally at an angle $\Phi = 106°$, then the surface of the half-space buckles as soon as K reaches $K^* \simeq 0.052$, a tiny amount of shear. In the current configuration, the direction of the fibres is found from (89) as $103°$.

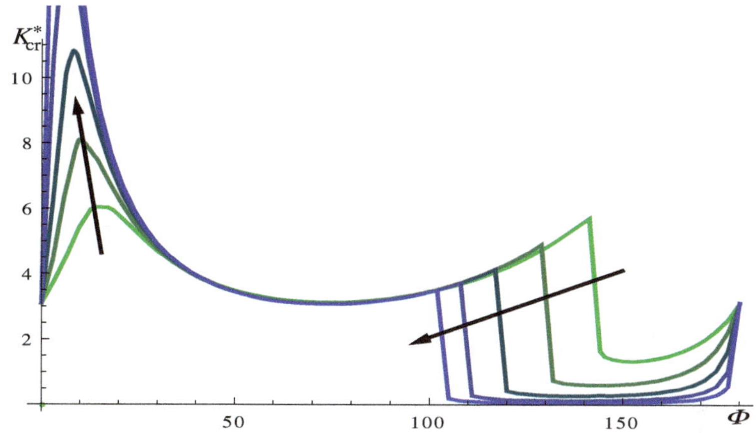

Figure 12. Bifurcation plots for surface instability of a neo-Hookean soft matrix with standard reinforcement in simple shear: Critical amount of shear K^* vs orientation Φ of the fibres in the reference configuration, for $E/\mu = 1.0, 2.0, 4.0, 8.0, 16.0$ (increasing values indicated by arrows).

The normal to the wrinkles-front is oriented at an angle $\theta^* \simeq 107°$, which means that the wrinkles are almost at right-angle with the fibres, a nice agreement with the observation in the simple experiment of Figure 1(b).

In Figure 12 we collect the $K^* - \Phi$ plots for different values of E/μ. The Appendix shows that in the linear regime of a tensile test along the fibres, the Young modulus of the material is $3\mu + 2E$, so that in a sense, $2E/3\mu$ represents the fibre-to-matrix stiffness ratio. For the figure we take $E/\mu = 1.0$ (fibres softer than the matrix), 2.0, 4.0, 8.0, 16.0 (fibres stiffer than the matrix). We find that when the fibres are originally more or less oriented along the direction of shear, the stability is enhanced compared to the non-reinforced case and $K^* > 3.09$. Then, as mentioned earlier for the $E/\mu = 10.0$ case, a dramatic drop in the value of K^* occurs at some angle Φ, whose value decreases as E/μ increases. In other words, the stiffer the fibres, the wider the range of early surface instability in shear.

4.3 Surface Instability for Another Sheared Anisotropic Material

Finally, we consider materials with a neo-Hookean soft matrix, also reinforced by one family of fibres, but this time modelled according to the strain energy density (49). According to Ciarletta et al. (2011, 2013), this constitutive modelling ensures *strong convexity* of the whole tissue (matrix and fibres) in planar deformations, in contrast with standard models of fibre reinforcement.

Again, it is easy to find the Lagrange multiplier p such that the surface $x_2 = 0$ is free of traction as $p = \mu$. We then find that the Cauchy pre-stress required to maintain the large static simple shear is

$$\boldsymbol{\sigma} = \mu(\boldsymbol{B} - \boldsymbol{I}) + 2\beta(\boldsymbol{FM} \otimes \boldsymbol{FM} - \boldsymbol{F}^{-T}\boldsymbol{M} \otimes \boldsymbol{F}^{-T}\boldsymbol{M}), \tag{114}$$

where $\boldsymbol{F}$ and $\boldsymbol{B}$ are given by (80).

Now we can conduct a surface stability analysis similar to the previous one, by using the moduli of (53). For the details see Ciarletta et al. (2013). Figure 13 displays the end result in the form of the $K^* - \Phi$ plots obtained for several values of the ratio β/μ. According to the Appendix, the quantity $8\beta/3\mu$ measures the infinitesimal matrix-to-fibres stiffness ratio. Here we consider in turn the cases $\beta/\mu = 0.0$ (no fibres), 0.8 (fibres softer than matrix), 2.0, and 4.0 (fibres stiffer than matrix.)

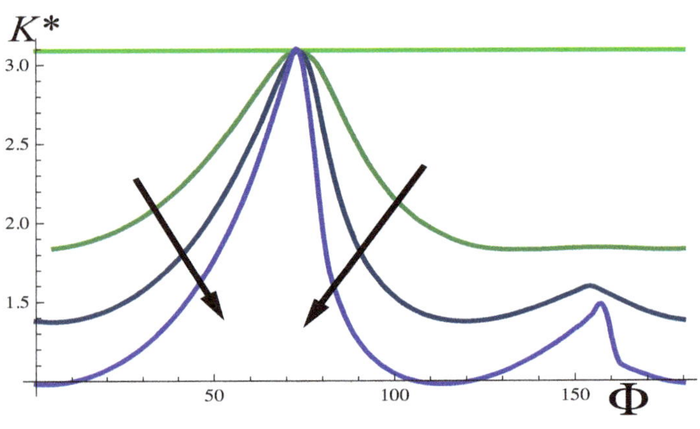

Figure 13. Bifurcation plots for surface instability of a neo-Hookean soft matrix with convex reinforcement in simple shear: Critical amount of shear K^* vs orientation Φ of the fibres in the reference configuration, for $\beta/\mu = 0.0, 0.8, 2.0, 4.0$ (increasing values indicated by arrows).

We remark that all the curves have a common point at $(K^*, \Phi) = (3.09, 73.3°)$, but this is an artefact due to the model. It can be explained by computing the components of the Stroh matrix in the coordinate system aligned with the Eulerian principal axes of deformation; see Ciarletta et al. (2013). Otherwise, the inclusion of fibres clearly makes the half-space more unstable in shear, as all the curves are below the isotropic $K^* = 3.09$ horizontal line. In the process of determining K^* numerically we also find that the wrinkle wavefronts are almost orthogonal to the fibres, in line with experimental observations, and also in line with the results of the previous subsection.

The main difference between the predictions of the standard reinforcing model (91) and of the anisotropic model of Ciarletta et al. (2011) is the disappearance of the discontinuities in the dispersion curves. This is to be expected, because of the strong convexity associated with the latter model. Whether one model is more accurate than the other (or accurate at all) is of course a matter for experimentalists to determine. This comment leads us naturally to the conclusion of this chapter, for which we invoke the wise recommendations of Golomb (1970).

5 Epilogue

The five DONT's of Modelling (Golomb, 1970)

1. Don't believe that the model is the reality;
2. Don't extrapolate beyond the region of fit;
3. Don't distort reality to fit the model;
4. Don't retain a discredited model;
5. Don't fall in love with your model.

A Appendix: Fibre to Matrix Stiffness Ratio

To compare the stiffness of the isotropic matrix to that of the fibres (in the case of a material reinforced with a single family of parallel fibres), we perform a tensile test in the direction of the fibres. For this test, the lateral contractions are equal by symmetry and the deformation gradient has components $\boldsymbol{F} = \text{diag}\left(\lambda, \lambda^{-1/2}, \lambda^{-1/2}\right)$, where λ is the stretch ratio in the direction of uniaxial stretch. Then the left Cauchy–Green deformation tensor is $\boldsymbol{B} = \text{diag}\left(\lambda^2, \lambda^{-1}, \lambda^{-1}\right)$, and $I_4 = \lambda^2$, $I_5 = \lambda^4$. It follows that the Cauchy stress has the following components

$$\begin{aligned} \sigma_1 &= -p + 2W_1\lambda^2 - 2W_2\lambda^{-2} + 2W_4\lambda^2 + 4W_5\lambda^4, \\ \sigma_2 &= \sigma_3 = 0 = -p + 2W_1\lambda^{-1} - 2W_2\lambda. \end{aligned} \tag{115}$$

Subtracting one equation from the other to eliminate p, we find that

$$\sigma_1 = \lambda \widehat{W}'(\lambda), \quad \text{where} \quad \widehat{W}(\lambda) := W(\lambda^2 + 2\lambda^{-1}, \lambda^{-2} + 2\lambda, \lambda^2, \lambda^4). \tag{116}$$

When there is no stress ($\sigma_1 = 0$), there is no stretch ($\lambda = 1$), so that $\widehat{W}'(1) \equiv 0$. Then, for a small elongation, $\lambda = 1 + \epsilon$, say, we have by expansion of (116),

$$\sigma_1 = E_{11}\epsilon, \quad \text{where} \quad E_{11} = \widehat{W}''(1) \tag{117}$$

is the Young modulus in the fibre direction. Hence, in the case of the neo-Hookean standard reinforcing model (91), where

$$\widehat{W}(\lambda) = \mu(\lambda^2 + 2\lambda^{-1} - 3)/2 + E(\lambda^2 - 1)^2/4, \tag{118}$$

we find that

$$E_{11} = 3\mu + 2E, \tag{119}$$

In the case of the strain energy density (49), where

$$\widehat{W}(\lambda) = \mu(\lambda^2 + 2\lambda^{-1} - 3)/2 + \beta(\lambda - \lambda^{-1})^2, \tag{120}$$

we find that

$$E_{11} = 3\mu + 8\beta. \tag{121}$$

Bibliography

M. A. Biot. Surface instability of rubber in compression. *Applied Science Research, Series A*, 12:168–182, 1963.

Ph. Boulanger and M. Hayes. Finite-amplitude waves in deformed Mooney-Rivlin material. *Quarterly Journal of Mechanics and Applied Mathematics*, 45:575–593, 1992.

P. Chadwick. The application of the Stroh formalism to prestressed elastic media. *Mathematics and Mechanics of Solids*, 2:379–403, 1997.

P. Ciarletta, I. Izzo, S. Micera, and F. Tendick. Stiffening by fibre reinforcement in soft materials: A hyperelastic theory at large strains and its application. *Journal of the Biomechanical Behavior of Biomedical Materials*, 4:1359–1368, 2011.

P. Ciarletta, M. Destrade, and A. L. Gower. Shear instability in skin tissue. *Quarterly Journal of Mechanics and Applied Mathematics*, 66:273–288, 2013.

M. Destrade, M. D. Gilchrist, D. A. Prikazchikov, and G. Saccomandi. Surface instability of sheared soft tissues. *Journal of Biomechanical Engineering*, 130:061007, 2008.

J. N. Flavin. Surface waves in pre-stressed Mooney material. *Quarterly Journal of Mechanics and Applied Mathematics*, 16:441–449, 1963.

Y. B. Fu and A. Mielke. A new identity for the surface impedance matrix and its application to the determination of surface-wave speeds. *Proceedings of the Royal Society of London, Series A*, 458:2523–2543, 2002.

S. W. Golomb. Mathematical models–uses and limitations. *Simulation*, 4: 197–198, 1970.

A. L. Gower and M. Destrade. Biot surface instability: Theory vs observation. submitted.

P. A. Meere, K. F. Mulchrone, and M. Timmerman. Shear folding in low-grade metasedimentary rocks: Reverse shear along cleavage at a high angle to the maximum compressive stress. *Geology*, 41:879–882, 2013.

J. Merodio and R. W. Ogden. Material instabilities in fiber-reinforced nonlinearly elastic solids under plane deformation. *Archives of Mechanics*, 54:525–552, 2002.

A. Mielke and Y. B. Fu. Uniqueness of the surface-wave speed: A proof that is independent of the stroh formalism. *Mathematics and Mechanics of Solids*, 9:5–15, 2004.

M. J. P. Musgrave. *Crystal Acoustics*. Holden-Day, 1970.

R. W. Ogden. *Non-linear Elastic Deformations*. Dover, 1997.

R. W. Ogden. Elements of the theory of finite elasticity. In Y. B. Fu and R. W. Ogden, editors, *Nonlinear Elasticity: Theory and Applications*, pages 1–58. Cambridge University Press, 2001.

R. W. Ogden. Incremental statics and dynamics of pre-stressed elastic materials. In M. Destrade and G. Saccomandi, editors, *Waves in Nonlinear Pre-Stressed Materials, CISM Lecture Notes*, volume 495, pages 1–26. Springer, 2007.

N. H. Scott and M. Hayes. A note on wave propagation in internally constrained hyperelastic materials. *Wave Motion*, 7:601–605, 1985.

A. L. Shuvalov. On the theory of wave propagation in anisotropic plates. *Proceeding of the Royal Society of London, Series A*, 456:2197–2222, 2000.

C. Stolz. *Milieux Continus en Transformations Finies*. Editions de l'Ecole Polytechnique, 2010.

A. N. Stroh. Some analytic solutions for Rayleigh waves in cubic crystals. *Journal of Mathematics and Physics*, 41:77–103, 1962.

Q. Wang, Y. Yin, H. Xie, J. Liu, W. Yang, P. Chen, and Q. Zhang. Buckling modes of polymer membranes restricted by metal wires. *Soft Matter*, 7: 2888–2894, 2011.

J. P. Wolfe. Acoustic wavefronts in crystalline solids. *Physics Today*, September:34–40, 1995.

Effects of Fiber Bending and Twisting Resistance on the Mechanics of Fiber-reinforced Elastomers

David J. Steigmann

Department of Mechanical Engineering, University of California
Berkeley, CA 94720, USA

E-mail: dsteigmann@berkeley.edu

Abstract. A model of nonlinearly elastic solids reinforced by continuously distributed embedded fibers is formulated in which elastic resistance of the fibers to extension, bending and twist is taken into account explicitly.

1 Introduction

The mechanics of fiber-reinforced solids has generated a substantial literature (Pipkin, 1979; Spencer, 1972). It has been based almost entirely on the concept of a simple anisotropic material in which the response functions depend on the conventional deformation gradient, possibly subject to constraints such as bulk incompressibility or fiber inextensibility. The continuum theory presumes the fibers to be so densely distributed as to render meaningful the idealization of a continuous distribution, and purports to describe homogenized fiber-matrix composites.

A major generalization of the continuum theory, accommodating the intrinsic flexural stiffness of the fibers, is given in Spencer and Soldatos (2007). This is framed in the setting of the nonlinear strain-gradient theory (Toupin, 1964; Mindlin and Tiersten, 1962; Koiter, 1964) of anisotropic elasticity in which elastic resistance is assigned to changes in curvature (flexure) of the fibers. The latter is calculated from the second gradient of the continuum deformation in which the fibers are regarded as convected curves. The model also accounts for additional effects associated with the gradients of the fiber stretches.

In the present work we develop a model in which the fibers offer elastic resistance to twist in addition to flexure and stretch. Thus the fibers are

L. Dorfmann, R. W. Ogden (Eds.), *Nonlinear Mechanics of Soft Fibrous Materials*,
CISM International Centre for Mechanical Sciences
DOI 10.1007/978-3-7091-1838-2_6 © CISM Udine 2015

regarded as continuously distributed spatial rods of the Kirchhoff type in which the kinematics are based on a position field and an orthonormal triad field (Landau and Lifshitz, 1986; Dill, 1992; Antman, 2005). Variation of the triad along the length of a fiber accounts for flexure and twist, while the position field generates the fiber stretch. We seek a model that accounts for these effects but which does not attribute elastic resistance to the gradient of fiber stretch. This restriction is in accord with established rod theories. We emphasize the fact that the theory advanced in Spencer and Soldatos (2007) also models fiber twist, but does so in a manner that we believe to be unduly restrictive, at least insofar as the three-dimensional theory is concerned. This view is substantiated in Section 3. The present model is a special case of the Cosserat theory of nonlinear elasticity (Reissner, 1987; Pietraszkiewicz and Eremeyev, 2009; Neff, 2006).

The elements of Kirchhoff rod theory are summarized in Section 2, and serve to motivate the three-dimensional theory discussed in the sequel. This is followed, in Section 3, by a survey of nonlinear Cosserat elasticity, specialized to model the effects of a single family of embedded fibers interacting with an elastic matrix. The resulting model is similar in structure to the Kirchhoff theory, with the effects of fiber-matrix interaction manifesting themselves as forces and couples distributed along the lengths of the embedded fibers.

Sections 2 and 3 draw largely on Steigmann (1996, 2012), together with an Appendix from the latter which is reproduced here for the sake of convenience. The main additional contributions of the present chapter are the derivation of the relevant Legendre-Hadamard necessary conditions for energy-minimizing states (Section 4) and an elaboration of the theory of material symmetry (Section 5). The chapter concludes with a few simple illustrative examples based on Ericksen's so-called controllable deformations (Ericksen, 1954). A few of the sections culminate in open questions and suggestions for further work.

We use standard notation such as $\boldsymbol{A}^t$, $\boldsymbol{A}^{-1}$, $\boldsymbol{A}^*$, $\mathrm{Sym}\boldsymbol{A}$, $\mathrm{Skw}\boldsymbol{A}$ and $\mathrm{tr}\,\boldsymbol{A}$. These are respectively the transpose, the inverse, the cofactor, the symmetric part, the skew part and the trace of a tensor $\boldsymbol{A}$, regarded as a linear transformation from a three-dimensional vector space to itself. We also use Sym and Skw to denote the linear subspaces of symmetric and skew tensors, and Orth^+ the group of rotation tensors. The tensor product of three-vectors is indicated by interposing the symbol $\otimes$, and the Euclidean inner product of tensors $\boldsymbol{A}, \boldsymbol{B}$ is denoted and defined by $\boldsymbol{A} \cdot \boldsymbol{B} = \mathrm{tr}(\boldsymbol{A}\boldsymbol{B}^t)$; the associated norm is $|\boldsymbol{A}| = \sqrt{\boldsymbol{A} \cdot \boldsymbol{A}}$. The symbol $|\cdot|$ is also used to denote the usual Euclidean norm of three-vectors. Latin and Greek indices take values in $\{1, 2, 3\}$ and $\{2, 3\}$ respectively, and, when repeated, are summed over their ranges. Finally, the notation $F_{\boldsymbol{A}}$ stands for the tensor-valued derivative of a scalar-valued function $F(\boldsymbol{A})$.

2 Kirchhoff's Theory for Spatial Rods

We present a brief modern survey of the Kirchhoff theory for spatial rods upon which the subsequent development is based. In this model the rod is regarded as a spatial curve endowed with an elastic energy density, per unit initial length, that depends on the curvature and twist of the rod. According to the derivation from conventional three-dimensional nonlinear elasticity given in Dill (1992), this theory also accommodates small axial strain along the rod, although this effect is suppressed in derivations based on asymptotic analysis or the method of gamma convergence. We include it here, however. Further, we forego any discussion of the connection between Kirchhoff theory and three-dimensional elasticity and simply regard the rod as a directed curve (Antman, 2005) in which certain *a priori* constraints are imposed. It is possible, of course, to base a rod theory on the general Cosserat theory of directed curves, but this requires additional constitutive data and so we do not pursue such a development here. Detailed background material on the Kirchhoff theory may be found in Landau and Lifshitz (1986).

2.1 Kinematics

The basic kinematic variables in the theory are a (deformed) position field $\boldsymbol{r}(S)$, where $S \in [0, L]$ and L is the length of the rod in a reference configuration, and a right-handed, orthonormal triad $\{\boldsymbol{d}_i(S)\}$ in which $\boldsymbol{d}_1 = \boldsymbol{d}$, where $\boldsymbol{d}$ is the unit vector defined by

$$\boldsymbol{r}'(S) = \lambda \boldsymbol{d}, \quad \text{and} \quad \lambda = |\boldsymbol{r}'(S)| \tag{1}$$

is the *stretch* of the rod. Thus $\boldsymbol{d}_1$ is the unit tangent to the rod in a deformed configuration and $\boldsymbol{d}_\alpha$ $(\alpha = 2, 3)$ span its cross-sectional plane at arclength station S.

A central aspect of the Kirchhoff theory is the stipulation that each cross section deforms as a rigid disc, the rigid deformation varying from one section to another. Accordingly, there is a rotation field $\boldsymbol{R}(S)$ such that $\boldsymbol{d}_i = \boldsymbol{R}\boldsymbol{D}_i$, where $\boldsymbol{D}_i(S)$ are the values of $\boldsymbol{d}_i(S)$ in the reference configuration. Using the representation $\boldsymbol{D}_j \otimes \boldsymbol{D}_j$ for the three-dimensional identity, we infer that

$$\boldsymbol{R} = \boldsymbol{d}_i \otimes \boldsymbol{D}_i. \tag{2}$$

The curvature and twist of the rod are embodied in the derivatives $\boldsymbol{d}_i'(S)$, where

$$\boldsymbol{d}_i' = \boldsymbol{R}'\boldsymbol{D}_i + \boldsymbol{R}\boldsymbol{D}_i'. \tag{3}$$

Let $\{\boldsymbol{E}_i\}$ be a fixed right-handed background frame. Then $\boldsymbol{D}_i(S) = \boldsymbol{A}(S)\boldsymbol{E}_i$ for some rotation field $\boldsymbol{A}$, yielding

$$\boldsymbol{d}_i' = \boldsymbol{W}\boldsymbol{d}_i = \boldsymbol{w} \times \boldsymbol{d}_i, \tag{4}$$

where

$$\boldsymbol{W} = \boldsymbol{R}'\boldsymbol{R}^t + \boldsymbol{R}\boldsymbol{A}'\boldsymbol{A}^t\boldsymbol{R}^t \tag{5}$$

is a skew tensor and $\boldsymbol{w} = \mathrm{ax}\boldsymbol{W}$ is its axial vector. If the rod is straight and untwisted in its reference configuration, i.e. if $\boldsymbol{D}_i' = \boldsymbol{0}$, then we have simply $\boldsymbol{W} = \boldsymbol{R}'\boldsymbol{R}^t$.

2.2 Strain-energy Function

We assume the strain energy $\mathcal{S}$ stored in the rod to be expressible in the form

$$\mathcal{S} = \int_0^L W dS, \tag{6}$$

where W, the strain energy per unit length, is a function of the list $\{\boldsymbol{R}, \boldsymbol{R}', \boldsymbol{r}'\}$, possibly depending explicitly on S. Explicit S dependence may arise from the initial curvature or twist of the rod, or from possibly non-uniform material properties.

We require W to be Galilean invariant and thus stipulate that its values be unaffected by the substitution $\{\boldsymbol{R}, \boldsymbol{R}', \boldsymbol{r}'\} \to \{\boldsymbol{Q}\boldsymbol{R}, \boldsymbol{Q}\boldsymbol{R}', \boldsymbol{Q}\boldsymbol{r}'\}$, where $\boldsymbol{Q}$ is an arbitrary *uniform* rotation. Because W is defined pointwise, to derive a necessary condition we may choose the particular rotation $\boldsymbol{Q} = \boldsymbol{R}^t_{|S}$ to conclude that W is determined by the list $\{\boldsymbol{R}^t\boldsymbol{R}', \boldsymbol{R}^t\boldsymbol{r}'\}$. This list is easily shown to be Galilean invariant and so our necessary condition is also sufficient. It is equivalent to $\{\boldsymbol{R}^t\boldsymbol{W}\boldsymbol{R} - \boldsymbol{A}'\boldsymbol{A}^t, \lambda\boldsymbol{D}\}$, where $\boldsymbol{D} = \boldsymbol{D}_1$. But $\boldsymbol{D}$ and $\boldsymbol{A}'\boldsymbol{A}^t$ are independent of the deformation and serve merely to confer an explicit S dependence. Accordingly, we write $W = W(\boldsymbol{R}^t\boldsymbol{W}\boldsymbol{R}, \lambda; S)$. Henceforth the S dependence plays a passive role and so will be suppressed in the notation.

We observe that

$$\boldsymbol{R}^t\boldsymbol{W}\boldsymbol{R} = W_{ij}\boldsymbol{D}_i \otimes \boldsymbol{D}_j, \quad \text{with} \quad W_{ij} = \boldsymbol{d}_i \cdot \boldsymbol{W}\boldsymbol{d}_j = \boldsymbol{d}_i \cdot \boldsymbol{d}_j', \tag{7}$$

and thus conclude that

$$W = w(\boldsymbol{\kappa}, \lambda), \tag{8}$$

where $\boldsymbol{\kappa}\,(=\kappa_i \boldsymbol{D}_i)$ is the axial vector of $\boldsymbol{R}^t\boldsymbol{W}\boldsymbol{R}$, i.e.

$$\kappa_i = \tfrac{1}{2}e_{ijk}W_{kj}, \tag{9}$$

in which e_{ijk} is the Levi-Civita permutation symbol ($e_{123} = +1$, etc.). Here κ_1 is the twist of the rod and κ_α are the curvatures.

2.3 Variational Theory

Although the equilibrium equations of the Kirchhoff theory are well known and easily derived from elementary considerations, it is instructive – and also helpful in advancing our further objectives – to derive them from a variational argument. Thus let $\epsilon \in (-\epsilon_0, \epsilon_0)$ be a parameter and consider the configuration defined by $\boldsymbol{r}^*(S;\epsilon)$ and $\boldsymbol{d}_i^*(S;\epsilon)$, with (cf. (1))

$$(\boldsymbol{r}^*)' = \lambda^* \boldsymbol{d}^*, \quad \lambda^* = |(\boldsymbol{r}^*)'|, \tag{10}$$

where $\boldsymbol{r}^*(S;0) = \boldsymbol{r}(S)$ and $\boldsymbol{d}_i^*(S;0) = \boldsymbol{d}_i(S)$. We take the state defined by $\epsilon = 0$ to be an equilibrium configuration.

We assume, for ϵ close to zero, that

$$\begin{aligned} \boldsymbol{r}^*(S;\epsilon) &= \boldsymbol{r}(S) + \epsilon\boldsymbol{u}(S) + \tfrac{1}{2}\epsilon^2\boldsymbol{v}(S) + o(\epsilon^2), \\ \boldsymbol{d}_i^*(S;\epsilon) &= \boldsymbol{d}_i(S) + \epsilon\dot{\boldsymbol{d}}_i(S) + \tfrac{1}{2}\epsilon^2\ddot{\boldsymbol{d}}_i(S) + o(\epsilon^2), \end{aligned} \tag{11}$$

where superposed dots identify ϵ-derivatives, evaluated at $\epsilon = 0$; thus, $\boldsymbol{u} = \dot{\boldsymbol{r}}$ and $\boldsymbol{v} = \ddot{\boldsymbol{r}}$. The associated strain energy is

$$\mathcal{S}^* = \int_0^L w(\boldsymbol{\kappa}^*, \lambda^*)dS = \mathcal{S} + \epsilon\dot{\mathcal{S}} + \tfrac{1}{2}\epsilon^2\ddot{\mathcal{S}} + o(\epsilon^2), \tag{12}$$

where

$$\dot{\mathcal{S}} = \int_0^L [(\partial w/\partial\lambda)\dot{\lambda} + m_i\dot{\kappa}_i]dS, \tag{13}$$

with

$$m_i = \partial w/\partial\kappa_i, \tag{14}$$

is the first variation of the energy, and

$$\ddot{\mathcal{S}} = \int_0^L [(\partial^2 w/\partial\lambda^2)(\dot{\lambda})^2 + (\partial w/\partial\lambda)\ddot{\lambda} + 2\delta_i\dot{\kappa}_i\dot{\lambda} + C_{ij}\dot{\kappa}_i\dot{\kappa}_j + m_i\ddot{\kappa}_i]dS, \tag{15}$$

with

$$\delta_i = \partial^2 w/\partial\lambda\partial\kappa_i \quad \text{and} \quad C_{ij} = \partial^2 w/\partial\kappa_i\partial\kappa_j, \tag{16}$$

is the second variation.

Let

$$\boldsymbol{R}^*(S;\epsilon) = \boldsymbol{d}_i^*(S;\epsilon) \otimes \boldsymbol{D}_i(S). \tag{17}$$

Then,

$$\dot{\boldsymbol{d}}_i = \boldsymbol{a} \times \boldsymbol{d}_i, \tag{18}$$

where $\boldsymbol{a}(S)$ is the axial vector of the skew tensor $\dot{\boldsymbol{R}}\boldsymbol{R}^t$. Similarly,

$$\ddot{\boldsymbol{d}}_i = \boldsymbol{a} \times (\boldsymbol{a} \times \boldsymbol{d}_i) + \boldsymbol{b} \times \boldsymbol{d}_i, \tag{19}$$

where $\boldsymbol{b}(S)$ is the axial vector of the skew tensor $(\dot{\boldsymbol{R}}\boldsymbol{R}^t)^{\cdot}$. Finally, let $a(S) = \dot{\lambda}$ and $b(S) = \ddot{\lambda}$, i.e.

$$\lambda^*(S;\epsilon) = \lambda(S) + \epsilon a(S) + \tfrac{1}{2}\epsilon^2 b(S) + o(\epsilon^2). \tag{20}$$

On combining these formulae we have

$$\dot{\mathcal{S}} = I[a, \boldsymbol{a}] \quad \text{and} \quad \ddot{\mathcal{S}} = I[b, \boldsymbol{b}] + J[a, \boldsymbol{a}], \tag{21}$$

where

$$I[a, \boldsymbol{a}] = \int_0^L [(\partial w/\partial\lambda)a + \boldsymbol{m} \cdot \boldsymbol{a}']dS, \tag{22}$$

with

$$\boldsymbol{m} = m_i \boldsymbol{d}_i, \tag{23}$$

and

$$J[a, \boldsymbol{a}] = \int_0^L [(\partial^2 w/\partial\lambda^2)a^2 + 2a\boldsymbol{\delta} \cdot \boldsymbol{a}' + \boldsymbol{m} \cdot \boldsymbol{a}' \times \boldsymbol{a} + \boldsymbol{a}' \cdot \boldsymbol{C}\boldsymbol{a}']dS, \tag{24}$$

with

$$\boldsymbol{\delta} = \delta_i \boldsymbol{d}_i \quad \text{and} \quad \boldsymbol{C} = C_{ij} \boldsymbol{d}_i \otimes \boldsymbol{d}_j. \tag{25}$$

Differentiation of (10) with respect to ϵ, using (18)–(20), furnishes the connections

$$\begin{aligned} \boldsymbol{u}'(S) &= a\boldsymbol{d} + \boldsymbol{a} \times \boldsymbol{r}' \quad \text{and} \\ \boldsymbol{v}'(S) &= 2a(\boldsymbol{a} \times \boldsymbol{d}) + \boldsymbol{a} \times (\boldsymbol{a} \times \boldsymbol{r}') + b\boldsymbol{d} + \boldsymbol{b} \times \boldsymbol{r}', \end{aligned} \tag{26}$$

whereas differentiation of $(\boldsymbol{r}^*)' \cdot \boldsymbol{d}_\alpha^* = 0$ furnishes

$$\boldsymbol{u}' \cdot \boldsymbol{d}_\alpha + \boldsymbol{r}' \cdot \boldsymbol{a} \times \boldsymbol{d}_\alpha = 0, \quad \alpha = 2, 3. \tag{27}$$

Further, using (7) we obtain (Steigmann, 1996)

$$\dot{\kappa}_i = \boldsymbol{d}_i \cdot \boldsymbol{a}' \quad \text{and} \quad \ddot{\kappa}_i = \boldsymbol{d}_i \cdot \boldsymbol{a}' \times \boldsymbol{a} + \boldsymbol{d}_i \cdot \boldsymbol{b}'. \tag{28}$$

To accommodate these constraints in the variational formulation we follow standard practice and introduce the energy

$$\mathcal{T} = \mathcal{S} + \int_0^L f_\alpha \boldsymbol{r}' \cdot \boldsymbol{d}_\alpha dS, \tag{29}$$

where $f_\alpha(S)$ are Lagrange multipliers. This is an extension of the actual energy from the class of deformations defined by the constraints to arbitrary deformations. First, we note that variations with respect to the Lagrange multipliers simply return the constraints as the relevant Euler–Lagrange equations; thus we do not make these explicit. Further, the stationarity of $\mathcal{T}$ with respect to arbitrary variations implies stationarity with respect to variations satisfying the constraints in particular; this in turn ensures the stationarity of $\mathcal{S}$ – and hence equilibrium – for the actual constrained problem. Conversely, if the Euler–Lagrange equations and natural boundary conditions for $\mathcal{T}$ are satisfied, which are necessary and sufficient for its stationarity against arbitrary variations, then $\mathcal{T}$ is stationary against constrained variations in particular, and this ensures the stationarity of $\mathcal{S}$. Accordingly, the Euler–Lagrange equations for $\mathcal{T}$, augmented by the constraints, are the equilibrium equations for the actual constrained problem. This train of logic is well known in principle. We belabor it here because it is rarely discussed in the literature.

Of course the foregoing statements about stationarity do not extend to criteria for a minimum; that is, we cannot use the extended energy to extract necessary conditions for minimizers of the actual energy. This follows from relaxation of constraints, i.e. $\inf \mathcal{T} \leq \inf \mathcal{S}$.

Proceeding, then, to stationarity conditions ($\dot{\mathcal{T}} = 0$), we have

$$I[a, \boldsymbol{a}] + \int_0^L f_\alpha (\boldsymbol{u}' \cdot \boldsymbol{d}_\alpha + \boldsymbol{r}' \cdot \boldsymbol{a} \times \boldsymbol{d}_\alpha) dS = 0, \tag{30}$$

which is equivalent to

$$\int_0^L (\boldsymbol{f} \cdot \boldsymbol{u}' + \boldsymbol{m} \cdot \boldsymbol{a}' + \boldsymbol{a} \cdot \boldsymbol{f} \times \boldsymbol{r}') dS = 0, \tag{31}$$

where

$$\boldsymbol{f} = (\partial w / \partial \lambda) \boldsymbol{d} + f_\alpha \boldsymbol{d}_\alpha. \tag{32}$$

This in turn is equivalent to

$$[\boldsymbol{f} \cdot \boldsymbol{u} + \boldsymbol{m} \cdot \boldsymbol{a}]_0^L - \int_0^L [\boldsymbol{u} \cdot \boldsymbol{f}' + \boldsymbol{a} \cdot (\boldsymbol{m}' - \boldsymbol{f} \times \boldsymbol{r}')] dS = 0, \tag{33}$$

and the Euler–Lagrange equations are given simply by

$$\boldsymbol{f}' = \mathbf{0} \quad \text{and} \quad \boldsymbol{m}' + \boldsymbol{r}' \times \boldsymbol{f} = \mathbf{0}. \tag{34}$$

These of course are the equilibrium equations of the Kirchhoff theory (Landau and Lifshitz, 1986) in the absence of distributed forces or moments. If forces and moments per unit initial length, $\boldsymbol{g}$ and $\boldsymbol{\pi}$ respectively, are acting, then

$$\boldsymbol{f}' + \boldsymbol{g} = \mathbf{0} \quad \text{and} \quad \boldsymbol{m}' + \boldsymbol{\pi} + \mathbf{r}' \times \boldsymbol{f} = \mathbf{0}. \tag{35}$$

Boundary conditions may be deduced from the vanishing of the bracketed term in (33) (Steigmann, 1996), adjusted to account for applied forces or moments, if any.

2.4 Energy Minimizers and the Legendre Necessary Condition

Let $\mathcal{S}$ be the equilibrium energy. We require (cf. (12))

$$0 \leq \mathcal{S}^* - \mathcal{S} = \tfrac{1}{2}\epsilon^2[\ddot{\mathcal{S}} + o(\epsilon^2)/\epsilon^2], \tag{36}$$

where $\ddot{\mathcal{S}}$ is given by $(21)_2$. Dividing by ϵ^2 and passing to the limit then yields $\ddot{\mathcal{S}} \geq 0$. On substitution of the equilibrium equations (34) and boundary conditions, after some effort we obtain the equilibrium value (Steigmann, 1996)

$$\ddot{\mathcal{S}} = F + G, \tag{37}$$

where

$$F = \int_0^L F^*(a, \boldsymbol{a}')dS \quad \text{and} \quad G = \int_0^L G^*(a, \boldsymbol{a}, \boldsymbol{a}')dS, \tag{38}$$

with

$$\begin{aligned} F^*(a, \boldsymbol{a}') &= (\partial^2 w/\partial\lambda^2)a^2 + 2a\boldsymbol{\delta} \cdot \mathbf{a}' + \boldsymbol{a}' \cdot \boldsymbol{C}\boldsymbol{a}', \\ G^*(a, \boldsymbol{a}, \boldsymbol{a}') &= \boldsymbol{m} \cdot \boldsymbol{a}' \times \boldsymbol{a} - 2a\boldsymbol{f} \cdot \boldsymbol{a} \times \boldsymbol{d} - \boldsymbol{f} \cdot \boldsymbol{a} \times (\boldsymbol{a} \times \boldsymbol{r}'), \end{aligned} \tag{39}$$

use having been made of $(26)_2$.

Consider the variation defined by $a = 0$ and $\boldsymbol{a} = f(S)\boldsymbol{e}$, with $\boldsymbol{e}$ fixed and

$$f(S) = \begin{cases} S - S_1, & S_1 \leq S \leq S_2 \\ -\dfrac{\theta}{1-\theta}(S - S_3), & S_2 < S \leq S_3 \\ 0, & S \in [0, L] \setminus (S_1, S_3), \end{cases} \tag{40}$$

where $S_1 < S_2 < S_3$, $(S_1, S_2) \subset (0, L)$ and

$$\theta = \frac{S_2 - S_1}{S_3 - S_1} \in (0, 1). \tag{41}$$

The induced variation $\boldsymbol{u}(S)$, if desired, is obtained by integrating the constraint equation $\boldsymbol{u}' = \boldsymbol{a} \times \boldsymbol{r}'$ (cf. $(26)_1$).

Let

$$\Delta = S_3 - S_1, \quad \Delta_1 = S_2 - S_1 = \theta\Delta \quad \text{and} \quad \Delta_2 = S_3 - S_2 = (1-\theta)\Delta. \tag{42}$$

Then,

$$\Delta^{-1}F = \frac{\theta}{\Delta_1}\int_{S_1}^{S_2} F^*(0,\boldsymbol{e})dS + \frac{1-\theta}{\Delta_2}\int_{S_2}^{S_3} F^*\left(0,\frac{-\theta}{1-\theta}\boldsymbol{e}\right)dS, \tag{43}$$

where (cf. $(39)_1$) $F^*(0,\frac{-\theta}{1-\theta}\boldsymbol{e}) = (\frac{\theta}{1-\theta})^2 F^*(0,\boldsymbol{e})$, yielding

$$\Delta^{-1}F = \left(\frac{\theta}{1-\theta}\right)F^*(0,\boldsymbol{e}) \tag{44}$$

and hence $F = O(\Delta)$. Similarly,

$$\Delta^{-1}G = \frac{\theta}{\Delta_1}\int_{S_1}^{S_2} G^*(0,\boldsymbol{a},\boldsymbol{e})dS + \frac{1-\theta}{\Delta_2}\int_{S_2}^{S_3} G^*\left(0,\boldsymbol{a},\frac{-\theta}{1-\theta}\boldsymbol{e}\right)dS. \tag{45}$$

Passing to the limit $\Delta \to 0$ and using the fact that $\boldsymbol{a}(S)$ then approaches zero pointwise, we find, from the Mean-Value theorem, that

$$\Delta^{-1}G \to \theta G^*(0,\boldsymbol{0},\boldsymbol{e}) + (1-\theta)G^*\left(0,\boldsymbol{0},\frac{-\theta}{1-\theta}\boldsymbol{e}\right). \tag{46}$$

But $(39)_2$ implies that $G^*(0,\boldsymbol{a},\cdot) = 0$, yielding $G = o(\Delta)$.

The restriction on the second variation then implies that

$$0 \le \Delta^{-1}(F+G) = \left(\frac{\theta}{1-\theta}\right)F^*(0,\boldsymbol{e}) + \Delta^{-1}o(\Delta), \tag{47}$$

and passage to the limit yields $F^*(0,\boldsymbol{e}) \ge 0$, i.e.

$$\boldsymbol{e}\cdot\boldsymbol{C}\boldsymbol{e} \ge 0 \quad \text{for all} \quad \boldsymbol{e}. \tag{48}$$

This is the Legendre necessary condition in the present context. Further necessary conditions for a minimum of the energy are given in Steigmann (1996).

In the classical theory for initially straight rods composed of isotropic materials, the strain energy at unit stretch is (Landau and Lifshitz, 1986)

$$w(1,\boldsymbol{\kappa}) = \tfrac{1}{2}GJ\kappa_1^2 + \tfrac{1}{2}EI\kappa_\alpha\kappa_\alpha, \tag{49}$$

where G and E are the (positive) shear and Young moduli; and J and I are the polar- and second-moments of area of the rod cross section. In this case $(C_{ij}) = \text{diag}(GJ, EI, EI)$, and (48) is automatically satisfied.

We will use a similar procedure to extract necessary conditions for minimizers in the much more complicated theory of elastic solids with embedded fibers.

3 Materials with Embedded Fibers: Cosserat Elasticity

Cosserat elasticity emerges as the natural setting for elastic solids with embedded fibers that transmit bending and twisting moments. We present a development of the associated theory in the present section, paralleling that discussed in the previous section for single fibers. The model is intended for application to fiber-reinforced solids in which the constituent fibers are prismatic cylinders and much stiffer than the matrix material in which they are embedded. Accordingly, the fibers are modeled as continuously distributed Kirchhoff rods.

To motivate the kinematical hypotheses underpinning the theory, consider the case in which the fibers and matrix are perfectly bonded. The interface between the matrix and a typical fiber is then convected by the deformation as a material surface. Hadamard's compatibility condition requires that $\boldsymbol{F}^+ - \boldsymbol{F}^- = \boldsymbol{f} \otimes \boldsymbol{N}$ for some vector $\boldsymbol{f}$, where $\boldsymbol{N}$ is a unit normal to the interface and $\boldsymbol{F}^\pm$ are the values of the deformation gradients in the fiber and matrix at the interface. In particular, if $\boldsymbol{D}\,(= \boldsymbol{D}_1)$ is the unit tangent to the centerline of a fiber, then $\boldsymbol{N} \in \text{Span}\{\boldsymbol{D}_\alpha\}$ because the fiber is a prismatic cylinder; here $\boldsymbol{D}_\alpha$, $\alpha = 2, 3$, are orthonormal unit vectors in the fiber cross section. It follows that

$$\boldsymbol{F}^+\boldsymbol{D} = \boldsymbol{F}^-\boldsymbol{D}, \quad \text{but} \quad \boldsymbol{F}^+\boldsymbol{D}_\alpha \neq \boldsymbol{F}^-\boldsymbol{D}_\alpha, \tag{50}$$

and hence that the deformation gradients in the fiber and matrix are generally unequal. This stands in contrast to the model proposed in Spencer and Soldatos (2007), based on a single deformation gradient.

If a fiber is sufficiently stiff relative to the matrix, its deformation gradient is given approximately by a rotation field $\boldsymbol{R}$, say, combined with axial stretch, in accordance with Dill's interpretation of the Kirchhoff theory. The situation is illustrated in Figure 1 and is consistent with the micromechanical analysis of deBotton and Hariton (2006). Thus we interpret $(50)_1$ in the form

$$\boldsymbol{F}\boldsymbol{D} = \lambda\boldsymbol{R}\boldsymbol{D}, \tag{51}$$

where $\lambda\,(= |\boldsymbol{FD}|)$ is the fiber stretch and $\boldsymbol{F}$ is the matrix deformation gradient. The fields $\boldsymbol{F}$ and $\boldsymbol{R}$ are otherwise independent, in accordance with $(50)_2$.

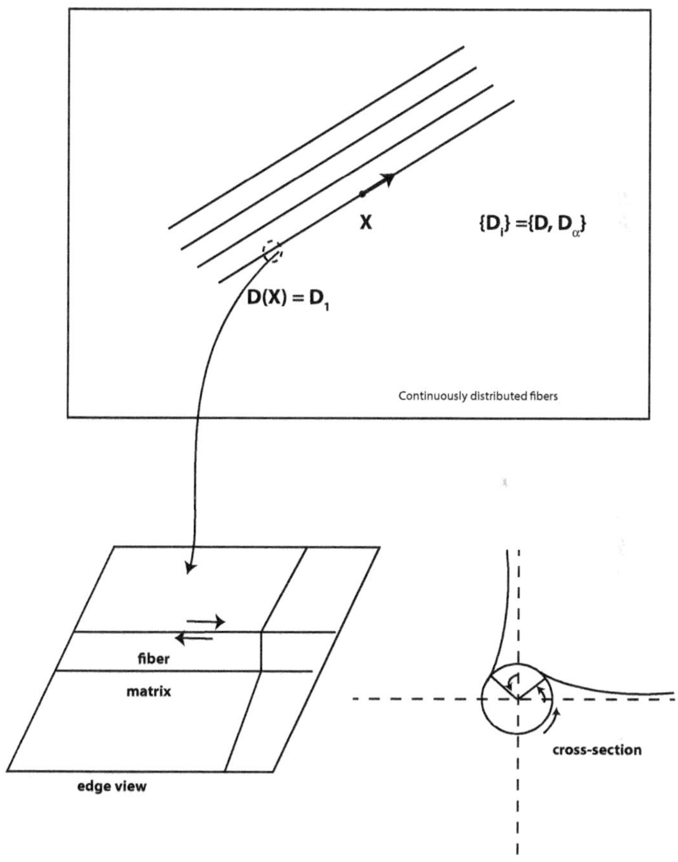

Figure 1. Fibers and matrix are kinematically independent; their interface convects as a material surface.

3.1 Kinematics

To model the kinematics of the embedded fibers, we assume the body, regarded as a homogenized continuum consisting of matrix material and fibers together, to be endowed with a rotation field $\boldsymbol{R}(\boldsymbol{X})$ in addition to the usual deformation $\boldsymbol{\chi}(\boldsymbol{X})$. To exhibit the main ideas as simply and clearly as possible, we confine attention here to materials reinforced by a single family of fibers.

Drawing on the structure of rod theory with axial extension, we assume the existence of a referential energy density $U(\boldsymbol{F}, \boldsymbol{R}, \boldsymbol{S}; \boldsymbol{X})$, where $\boldsymbol{F}$ is the usual deformation gradient and $\boldsymbol{S}$ is the rotation gradient; thus,

$$\boldsymbol{F} = F_{iA}\boldsymbol{e}_i \otimes \boldsymbol{E}_A, \quad \boldsymbol{R} = R_{iA}\boldsymbol{e}_i \otimes \boldsymbol{E}_A \quad \text{and} \quad \boldsymbol{S} = S_{iAB}\boldsymbol{e}_i \otimes \boldsymbol{E}_A \otimes \boldsymbol{E}_B \tag{52}$$

with

$$F_{iA} = \chi_{i,A} \quad \text{and} \quad S_{iAB} = R_{iA,B}, \tag{53}$$

where $(\cdot)_{,A} = \partial(\cdot)/\partial X_A$ and we use a Cartesian index notation that emphasizes the two-point character of the deformation gradient and rotation fields. Here $\{\boldsymbol{e}_i\}$ and $\{\boldsymbol{E}_A\}$ are fixed orthonormal bases associated with the coordinates x_i and X_A, where $x_i = \chi_i(X_A)$.

The rotation field acts on the orthonormal triad field $\{\boldsymbol{D}_i(\boldsymbol{X})\}$ associated with the unit tangents and cross sections of embedded fibers. To make the roles of these vectors explicit, we write $\{\boldsymbol{D}_i\} = \{\boldsymbol{D}, \boldsymbol{D}_\alpha\}$, $\alpha = 2, 3$, where $\boldsymbol{D}\,(= \boldsymbol{D}_1)$ is the unit tangent to a fiber in the reference configuration, denoted by κ, and $\boldsymbol{D}_\alpha$ are cross-sectional vectors embedded in the fiber, but *not* in the matrix (cf. (50)). Thus,

$$\boldsymbol{d}_i = \boldsymbol{R}\boldsymbol{D}_i \tag{54}$$

is the (orthonormal) fiber triad in the current configuration, where $\boldsymbol{d}\,(= \boldsymbol{d}_1)$ is the unit tangent to a fiber.

We regard the fiber as an embedded curve, and hence the tangent field $\boldsymbol{D}(\boldsymbol{X})$ as being convected by the deformation $\boldsymbol{\chi}(\boldsymbol{X})$. This generates the connection (cf. (51))

$$\boldsymbol{F}\boldsymbol{D} = \lambda \boldsymbol{d}, \quad \text{where} \quad \boldsymbol{d} = \boldsymbol{R}\boldsymbol{D} \quad \text{and} \quad \lambda = |\boldsymbol{F}\boldsymbol{D}|. \tag{55}$$

The cross-sectional vectors $\boldsymbol{D}_\alpha$ are not embedded in the matrix, and so in general their images $\boldsymbol{d}_\alpha$ in the current configuration are not directly connected to the deformation of the matrix. Rather, they are free to shear relative to the matrix while remaining mutually unsheared. This effectively extends the kinematics of the ideal theory of elastic materials with embedded inextensible fibers (Pipkin, 1979; Spencer, 1972) to allow for fiber extension, flexure and twist. Equations (54) and (55) generate the two constraints

$$\boldsymbol{R}\boldsymbol{D}_\alpha \cdot \boldsymbol{F}\boldsymbol{D} = 0, \quad \alpha = 2,3, \tag{56}$$

between the fiber rotation and matrix deformation, and thus yield the interpretation of the present model as a constrained variant of the Cosserat theory of nonlinear elasticity (Reissner, 1987).

3.2 Strain-energy Function

We observe that the constraints (56) are invariant under the transformations $\boldsymbol{F} \to \boldsymbol{Q}\boldsymbol{F}$ and $\boldsymbol{R} \to \boldsymbol{Q}\boldsymbol{R}$, where $\boldsymbol{Q}$ is the spatially uniform rotation associated with an arbitrary superposed rigid-body motion. We assume the energy density function to be likewise, and thus impose the requirement

$$U(\boldsymbol{F}, \boldsymbol{R}, \boldsymbol{S}; \boldsymbol{X}) = U(\boldsymbol{Q}\boldsymbol{F}, \boldsymbol{Q}\boldsymbol{R}, \boldsymbol{Q}\boldsymbol{S}; \boldsymbol{X}), \tag{57}$$

where $(\boldsymbol{Q}\boldsymbol{S})_{iAB} = (Q_{ij}R_{jA})_{,B} = Q_{ij}S_{jAB}$. The restriction

$$U(\boldsymbol{F}, \boldsymbol{R}, \boldsymbol{S}; \boldsymbol{X}) = W(\boldsymbol{E}, \boldsymbol{\Gamma}; \boldsymbol{X}), \tag{58}$$

where (Pietraszkiewicz and Eremeyev, 2009; Steigmann, 2012)

$$\boldsymbol{E} = \boldsymbol{R}^t\boldsymbol{F} = E_{AB}\boldsymbol{E}_A \otimes \boldsymbol{E}_B, \quad E_{AB} = R_{iA}F_{iB}, \tag{59}$$

$$\boldsymbol{\Gamma} = \Gamma_{DC}\boldsymbol{E}_D \otimes \boldsymbol{E}_C, \quad \Gamma_{DC} = \tfrac{1}{2}e_{BAD}R_{iA}R_{iB,C}, \tag{60}$$

with W a suitable function and e_{ABC} the permutation symbol, is both necessary and sufficient for the stated invariance. Sufficiency is obvious, while necessity follows by choosing $\boldsymbol{Q} = \boldsymbol{R}^t_{|\boldsymbol{X}}$, where $\boldsymbol{X}$ is the material point in question, and making use of the fact that for each fixed $C \in \{1,2,3\}$ the matrix $R_{iA}R_{iB,C}$ is skew; this following on differentiation of $R_{iA}R_{iB} = \delta_{AB}$ (the Kronecker delta). The associated axial vectors $\boldsymbol{\gamma}_C$ have components

$$\gamma_{D(C)} = \tfrac{1}{2}e_{BAD}R_{iA}R_{iB,C}, \tag{61}$$

yielding (Pietraszkiewicz and Eremeyev, 2009)

$$\boldsymbol{\Gamma} = \boldsymbol{\gamma}_C \otimes \boldsymbol{E}_C, \tag{62}$$

and so $\boldsymbol{\Gamma}$ is equivalent to $\boldsymbol{R}^t\boldsymbol{S}$.

3.3 Stationary Energy and Equilibrium

In the case of conservative loading equilibria may be interpreted as states that render stationary the potential energy

$$E = \int_{\kappa} W dv - L, \tag{63}$$

where L is a suitable load potential. For example, the extension of the classical dead-load problem under negligible body forces and couples entails the assignment of fixed traction $\boldsymbol{t}$ on a part $\partial\kappa_t$ of the boundary and fixed couples $\boldsymbol{m}_i$ on a part $\partial\kappa_c$, such that

$$L = \int_{\partial\kappa_t} \mathbf{t} \cdot \boldsymbol{\chi} da + \int_{\partial\kappa_c} \boldsymbol{m}_i \cdot \boldsymbol{d}_i da. \tag{64}$$

The virtual work of the force and couples is

$$\dot{L} = \int_{\partial\kappa_t} \mathbf{t} \cdot \dot{\boldsymbol{\chi}} da + \int_{\partial\kappa_c} \mathbf{c} \cdot \boldsymbol{\omega} da, \tag{65}$$

where the superposed dots refer to derivatives with respect to a parameter of a one-parameter family $\{\boldsymbol{F}(\boldsymbol{X};\epsilon), \boldsymbol{R}(\boldsymbol{X};\epsilon)\}$ of deformation and rotation fields, $\boldsymbol{\omega} = \text{ax}\boldsymbol{\Omega}$ is the axial vector of the skew tensor $\boldsymbol{\Omega} = \dot{\boldsymbol{R}}^t \boldsymbol{R}$ ($\boldsymbol{\Omega v} = \boldsymbol{\omega} \times \boldsymbol{v}$ for all $\boldsymbol{v}$) and (Steigmann, 2012)

$$\mathbf{c} = \text{ax}[(\boldsymbol{D}_i \otimes \boldsymbol{m}_i)\boldsymbol{R} - \boldsymbol{R}^t(\boldsymbol{m}_i \otimes \boldsymbol{D}_i)] \tag{66}$$

is the (configuration dependent) couple traction.

Stationarity of the energy is subsumed under the general virtual-work statement (Steigmann, 2012)

$$\int_{\kappa} \dot{W} dv = \int_{\partial\kappa_t} \mathbf{t} \cdot \dot{\boldsymbol{\chi}} da + \int_{\partial\kappa_c} \mathbf{c} \cdot \boldsymbol{\omega} da, \tag{67}$$

where the derivatives are evaluated at equilibrium, corresponding to $\epsilon = 0$, say. We regard the values of $\boldsymbol{\chi}$ as being assigned on $\partial\kappa \setminus \partial\kappa_t$, and those of $\boldsymbol{R}$ as being assigned on $\partial\kappa \setminus \partial\kappa_c$. We emphasize the fact that this holds whenever the virtual work of the loads is expressible as a linear form in $\dot{\boldsymbol{\chi}}$ and $\boldsymbol{\omega}$, including the case of non-conservative loads or conservative loads besides those discussed above.

Global balance statements may be derived from (63) by assuming that $\partial\kappa_t = \partial\kappa_c = \partial\kappa$ and imposing (67) for arbitrary rigid-body motions. This is shown in Steigmann (2012) to yield

$$\int_{\partial\kappa} \boldsymbol{t} da = \mathbf{0} \quad \text{and} \quad \int_{\partial\kappa} (\boldsymbol{\chi} \times \mathbf{t} - \boldsymbol{R}\mathbf{c}) da = \mathbf{0}, \tag{68}$$

and hence the interpretation of $-\boldsymbol{Rc}$ as a couple density.

In the course of deriving further consequences of (67) we proceed as in the case of single fibers and consider, as in Section 2, the extended energy

$$\bar{E} = \int_{\kappa} \bar{W} dv - L \tag{69}$$

in which

$$\bar{W} = W + \Lambda_\alpha \boldsymbol{D}_\alpha \cdot \boldsymbol{E}\boldsymbol{D}, \tag{70}$$

where Λ_α are Lagrange multipliers associated with (56). Here $\bar{E}$ is to be regarded as an unconstrained functional of $\boldsymbol{\chi}$, $\boldsymbol{R}$ and Λ_α. Thus E and $\bar{E}$ coincide when the constraints are in effect, whereas the latter effectively extends the former to states in which the constraints are relaxed. Variation of the multipliers simply returns the constraints, ensuring that states which render $\bar{E}$ stationary are also stationary states for E. This follows from the fact that stationarity in the class of unrestricted variations of $\boldsymbol{\chi}$ and $\boldsymbol{R}$ implies stationarity in the restricted class defined by the constraints. In this way equilibrium equations may be derived by requiring that $\bar{E}$ be stationary for unconstrained variations.

The results of the Appendix may be used to reduce the statement $\dot{\bar{E}} = 0$ to

$$\begin{aligned} \int_{\kappa} \{ \dot{\Lambda}_\alpha \boldsymbol{D}_\alpha \cdot \boldsymbol{E}\boldsymbol{D} + \boldsymbol{\omega} \cdot (\mathrm{Div}\, \boldsymbol{\mu} + 2\mathrm{ax}\, \mathrm{Skw}[(\boldsymbol{\sigma} + \boldsymbol{\Lambda} \otimes \boldsymbol{D})\boldsymbol{E}^t + \boldsymbol{\mu}\boldsymbol{\Gamma}^t]) \\ - \dot{\boldsymbol{\chi}} \cdot \mathrm{Div}(\boldsymbol{R}\boldsymbol{\sigma} + \boldsymbol{\lambda} \otimes \boldsymbol{D}) \} dv \\ = \int_{\partial\kappa_t} \dot{\boldsymbol{\chi}} \cdot [\boldsymbol{t} - (\boldsymbol{R}\boldsymbol{\sigma} + \boldsymbol{\lambda} \otimes \boldsymbol{D})\boldsymbol{n}] da + \int_{\partial\kappa_c} \boldsymbol{\omega} \cdot (\boldsymbol{c} + \boldsymbol{\mu}\boldsymbol{n}) da, \end{aligned} \tag{71}$$

where $\boldsymbol{n}$ is the exterior unit normal to $\partial\kappa$,

$$\boldsymbol{\Lambda} = \Lambda_\alpha \boldsymbol{D}_\alpha, \quad \boldsymbol{\lambda} = \boldsymbol{R}\boldsymbol{\Lambda}, \quad \boldsymbol{\sigma} = W_{\boldsymbol{E}} \quad \text{and} \quad \boldsymbol{\mu} = W_{\boldsymbol{\Gamma}} \tag{72}$$

and the variations of the multipliers have been made explicit. Hence the equilibrium equations

$$\mathrm{Div}(\boldsymbol{R}\boldsymbol{\sigma} + \boldsymbol{\lambda} \otimes \boldsymbol{D}) = \boldsymbol{0}, \quad \mathrm{Div}\, \boldsymbol{\mu} + \mathrm{ax}\{2\, \mathrm{Skw}[(\boldsymbol{\sigma} + \boldsymbol{\Lambda} \otimes \boldsymbol{D})\boldsymbol{E}^t + \boldsymbol{\mu}\boldsymbol{\Gamma}^t]\} = \boldsymbol{0}, \tag{73}$$

and boundary conditions

$$\boldsymbol{t} = (\boldsymbol{R}\boldsymbol{\sigma} + \boldsymbol{\lambda} \otimes \boldsymbol{D})\boldsymbol{n} \quad \text{on} \quad \partial\kappa_t \quad \text{and} \quad \boldsymbol{c} + \boldsymbol{\mu}\boldsymbol{n} = \boldsymbol{0} \quad \text{on} \quad \partial\kappa_c. \tag{74}$$

The traction condition, with $\boldsymbol{n} = \boldsymbol{D}$, yields the interpretation of $\boldsymbol{\lambda} = \Lambda_\alpha \boldsymbol{d}_\alpha$ as a density of kinematically undetermined transverse shear force acting on the fiber cross sections.

Remarks.

1. Fiber inextensibility is accommodated by appending the constraint $\boldsymbol{D}\cdot\boldsymbol{ED}=1$; $\boldsymbol{\Lambda}$ and $\boldsymbol{\lambda}$ are now 3-vectors given respectively by $\Lambda_i\boldsymbol{D}_i$ and $\Lambda_i\boldsymbol{d}_i$ in which Λ_1 is a kinematically undetermined density of axial force exerted on the fibers.
2. Incompressibility entails the constraint $\det\boldsymbol{F}\,(=\det\boldsymbol{E})=1$, which may be accommodated by using

$$\bar{W}=W+\Lambda_\alpha\boldsymbol{D}_\alpha\cdot\boldsymbol{ED}-p(\det\boldsymbol{E}-1) \tag{75}$$

in place of (70), where p is the associated Lagrange multiplier. With reference to the Appendix, this affects only equations $(73)_1$ and $(74)_1$, which are replaced by

$$\mathrm{Div}(\boldsymbol{R\sigma}-p\boldsymbol{F}^*+\boldsymbol{\lambda}\otimes\boldsymbol{D})=\boldsymbol{0}\quad\text{and}\quad\boldsymbol{t}=(\boldsymbol{R\sigma}-p\boldsymbol{F}^*+\boldsymbol{\lambda}\otimes\boldsymbol{D})\boldsymbol{n}, \tag{76}$$

respectively, augmented by the Piola identity $\mathrm{Div}\,\boldsymbol{F}^*=\boldsymbol{0}$.

3.4 A Simple Model for Fiber-reinforced Material

The kinematics of embedded fibers may be described in the present framework by using (54) to write (cf. (7), (9))

$$\kappa_i=\tfrac{1}{2}e_{ijk}\boldsymbol{D}_k\cdot\boldsymbol{R}^t\boldsymbol{R}'\boldsymbol{D}_j, \tag{77}$$

where $(\cdot)'$ is the directional, or *fiber* derivative along the fiber axis $\boldsymbol{D}$ and we have assumed, with minor loss of generality, that the fibers are straight and untwisted in κ, i.e. that $\boldsymbol{D}_j'=\boldsymbol{0}$. Here we use $R_{iA}'=R_{iA,B}D_B$ to derive (cf. (A.9))

$$\boldsymbol{R}^t\boldsymbol{R}'=R_{iC}S_{iAB}D_B\boldsymbol{E}_C\otimes\boldsymbol{E}_A=e_{ACD}\Gamma_{DB}D_B\boldsymbol{E}_C\otimes\boldsymbol{E}_A, \tag{78}$$

which implies that $\boldsymbol{\kappa}=\kappa_i\boldsymbol{D}_i$ is determined by $\boldsymbol{\Gamma}$.

In view of the structure of the rod theory discussed in Section 2, we assume the constitutive response of the fiber-reinforced material to depend on $\boldsymbol{\Gamma}$ via $\boldsymbol{\kappa}$; thus the strain energy is described by a (different) constitutive function $W(\boldsymbol{E},\boldsymbol{\kappa})$. For the sake of illustration we further assume the material to be uniform, and thus that W does not depend explicitly on $\boldsymbol{X}$.

To determine the associated response function $\boldsymbol{\mu}$ for use in (73), we use (cf. $(28)_1$)

$$\dot{\kappa}_i=\boldsymbol{d}_i\cdot\boldsymbol{a}',\quad\text{where}\quad\boldsymbol{a}=\mathrm{ax}(\dot{\boldsymbol{R}}\boldsymbol{R}^t), \tag{79}$$

in which the superposed dot again refers to the derivative with respect to the parameter in a one-parameter family of configurations. Accordingly, $\dot{\kappa}_i=\boldsymbol{D}_i\cdot\boldsymbol{R}^t\boldsymbol{a}'=-\boldsymbol{D}_i\cdot\boldsymbol{R}^t(\boldsymbol{R\omega})'$, yielding

$$\dot{\kappa}_i=(\boldsymbol{R}^t\boldsymbol{R}')\boldsymbol{D}_i\cdot\boldsymbol{\omega}-\omega_i',\quad\text{where}\quad\omega_i=\boldsymbol{\omega}\cdot\boldsymbol{D}_i. \tag{80}$$

Combining this with $\omega_i' = \omega_{i,A} D_A$ and fixing $\boldsymbol{E}$ ($\dot{\boldsymbol{E}} = \boldsymbol{0}$) we derive $\dot{W} = \boldsymbol{M} \cdot \dot{\boldsymbol{\kappa}}$, where

$$\boldsymbol{M} = M_i \boldsymbol{D}_i, \quad \text{with} \quad M_i = \partial W / \partial \kappa_i, \tag{81}$$

and

$$\dot{W} = \boldsymbol{\omega} \cdot [\mathrm{Div}(\boldsymbol{M} \otimes \boldsymbol{D}) + (\boldsymbol{R}^t \boldsymbol{R}')\boldsymbol{M}] - \mathrm{Div}[(\boldsymbol{M} \otimes \boldsymbol{D})^t \boldsymbol{\omega}]. \tag{82}$$

By equating this to the expression (A.10) for $\boldsymbol{\mu} \cdot \dot{\boldsymbol{\Gamma}}$ we conclude, on taking (A.9) into account, that

$$\boldsymbol{\mu} = \boldsymbol{M} \otimes \boldsymbol{D}, \tag{83}$$

and equation $(73)_2$ specializes, for uniform $\boldsymbol{D}(\boldsymbol{X})$, to

$$\boldsymbol{M}' + (\boldsymbol{R}^t \boldsymbol{R}')\boldsymbol{M} + \mathrm{ax}\{2\mathrm{Skw}[(\boldsymbol{\sigma} + \boldsymbol{\Lambda} \otimes \boldsymbol{D})\boldsymbol{E}^t]\} = \boldsymbol{0}, \tag{84}$$

where $\boldsymbol{M}' = (\nabla \boldsymbol{M})\boldsymbol{D}$, while the boundary condition $(74)_2$ becomes

$$\boldsymbol{c} = -(\boldsymbol{D} \cdot \boldsymbol{n})\boldsymbol{M}, \tag{85}$$

implying that $\boldsymbol{c}$ vanishes at points where the fibers lie parallel to the boundary.

The model may be recast in a form more easily recognizable from rod theory by introducing the field

$$\boldsymbol{m} = M_i \boldsymbol{d}_i = \boldsymbol{R}\boldsymbol{M}, \tag{86}$$

which may be used (Steigmann, 2012) to express (84) in the form

$$\boldsymbol{m}' + \boldsymbol{\chi}' \times \boldsymbol{\lambda} + \boldsymbol{R}\{\mathrm{ax}(\boldsymbol{\sigma}\boldsymbol{E}^t - \boldsymbol{E}\boldsymbol{\sigma}^t)\} = \boldsymbol{0}, \quad \text{where} \quad \boldsymbol{m}' = (\nabla \boldsymbol{m})\boldsymbol{D}, \tag{87}$$

whereas $(73)_1$ may be recast as

$$\boldsymbol{\lambda}' + \mathrm{Div}(\boldsymbol{R}\boldsymbol{\sigma} - p\boldsymbol{F}^*) = \boldsymbol{0}, \quad \text{where} \quad \boldsymbol{\lambda}' = (\nabla \boldsymbol{\lambda})\boldsymbol{D}, \tag{88}$$

in which the constraint of incompressibility has been incorporated. These may be regarded as the equilibrium equations for the reinforced solid.

Comparison with (35) furnishes the interpretation of $\boldsymbol{m}$ as a density of moment transmitted by a fiber, and confirms our earlier interpretation of $\boldsymbol{\lambda}$ as a density of force acting on a fiber. It also identifies the third term in (87) and the second term in (88), respectively, as a density of distributed couples exerted by the matrix on a fiber and a distributed force exerted on the fiber. Further, the contribution to the net moment from the embedded fibers reduces to

$$-\boldsymbol{R}\boldsymbol{c} = \boldsymbol{m}(\boldsymbol{D} \cdot \boldsymbol{n}). \tag{89}$$

The dependence of the strain-energy function on $\boldsymbol{\kappa}$ (or $\boldsymbol{\Gamma}$) introduces a natural length scale, l say, into the constitutive behavior which is on the order of that of the microstructure and hence of the diameter of a fiber cross section. Using this to define the dimensionless curvature-twist vector $\bar{\boldsymbol{\kappa}} = l\boldsymbol{\kappa}$, supposing the latter to be small in typical applications and assuming that the fibers carry no bending or twisting moments when straight and untwisted, we find that W is given to leading order in $\bar{\boldsymbol{\kappa}}$ by

$$W(\boldsymbol{E}, \boldsymbol{\kappa}) = W(\boldsymbol{E}, \boldsymbol{0}) + \tfrac{1}{2}\boldsymbol{\kappa} \cdot \boldsymbol{K}(\boldsymbol{E})\boldsymbol{\kappa}, \tag{90}$$

where $\boldsymbol{K}(\boldsymbol{E}) = W_{\boldsymbol{\kappa}\boldsymbol{\kappa}}(\boldsymbol{E}, \boldsymbol{0})$.

Open problems.

1. Develop the linear theory and the linearized theory of small deformations/rotations superposed on large.
2. Obtain an appropriate expression for the kinetic energy from microstructural considerations and develop the associated dynamical theory.
3. Investigate the present model as a limit of a conventional two-phase theory in which the fibers are much stiffer than the matrix and appropriately spaced on the micro-scale. What scalings, if any, yield the present model in the limit of a large number of closely spaced stiff fibers? The papers by Pideri and Seppecher (1997) and Berrehili and Marigo (2014) may offer some guidance in this regard.

4 Legendre–Hadamard Condition

We seek the analogue of the Legendre necessary condition (48) in the case of fiber-reinforced solids. This is referred to here as the Legendre–Hadamard condition, as in conventional elasticity theory. To this end we suppress the contribution of the boundary loading to the potential energy, which will prove to be irrelevant in any case, and consider

$$E = \int_{\kappa} W(\boldsymbol{E}, \boldsymbol{\kappa})dv. \tag{91}$$

The associated first and second variations are

$$\dot{E} = \int_{\kappa} (W_{\boldsymbol{E}} \cdot \dot{\boldsymbol{E}} + W_{\boldsymbol{\kappa}} \cdot \dot{\boldsymbol{\kappa}})dv \tag{92}$$

and

$$\ddot{E} = \int_\kappa (W_{\boldsymbol{E}} \cdot \ddot{\boldsymbol{E}} + W_{\boldsymbol{\kappa}} \cdot \ddot{\boldsymbol{\kappa}})dv + \int_\kappa \{\dot{\boldsymbol{E}} \cdot W_{\boldsymbol{EE}}[\dot{\boldsymbol{E}}] + (W_{\boldsymbol{E\kappa}})\dot{\boldsymbol{\kappa}} \cdot \dot{\boldsymbol{E}} + (W_{\boldsymbol{\kappa E}})\dot{\boldsymbol{E}} \cdot \dot{\boldsymbol{\kappa}} + \dot{\boldsymbol{\kappa}} \cdot (W_{\boldsymbol{\kappa\kappa}})\dot{\boldsymbol{\kappa}}\}dv. \tag{93}$$

From (A.1) and (A.2) of the Appendix, we have the first variations

$$\dot{\boldsymbol{E}} = \boldsymbol{\Omega}\boldsymbol{R}^t\boldsymbol{F} + \boldsymbol{R}^t\nabla\boldsymbol{u}, \quad \text{with} \quad \boldsymbol{u} = \dot{\boldsymbol{\chi}}, \quad \text{and} \quad \boldsymbol{\Omega} = \dot{\boldsymbol{R}}^t\boldsymbol{R}. \tag{94}$$

Then,

$$\dot{\boldsymbol{E}} = \boldsymbol{R}^t(\nabla\boldsymbol{u} - \boldsymbol{\alpha}\boldsymbol{F}), \quad \text{where} \quad \boldsymbol{\alpha} = \dot{\boldsymbol{R}}\boldsymbol{R}^t. \tag{95}$$

Further,

$$\dot{\boldsymbol{\kappa}} = \dot{\kappa}_i\boldsymbol{D}_i, \quad \text{where} \quad \dot{\kappa}_i = \boldsymbol{d}_i \cdot \boldsymbol{a}' \quad \text{and} \quad \boldsymbol{a} = \text{ax}\boldsymbol{\alpha}. \tag{96}$$

The latter result is the same as $(28)_1$ and yields

$$\dot{\boldsymbol{\kappa}} = \boldsymbol{R}^t\boldsymbol{a}'. \tag{97}$$

The second variations are

$$\ddot{\boldsymbol{\kappa}} = \boldsymbol{R}^t\boldsymbol{b}' + \dot{\boldsymbol{R}}^t\boldsymbol{a}' = \boldsymbol{R}^t\boldsymbol{b}' - \boldsymbol{R}^t\boldsymbol{\alpha}\boldsymbol{R}\dot{\boldsymbol{\kappa}}, \tag{98}$$

where $\boldsymbol{b} = \text{ax}\boldsymbol{\beta}$ with $\boldsymbol{\beta} = \dot{\boldsymbol{\alpha}}$, and

$$\ddot{\boldsymbol{E}} = \boldsymbol{R}^t(\nabla\boldsymbol{v} - \boldsymbol{\beta}\boldsymbol{F}) - \boldsymbol{R}^t\boldsymbol{\alpha}(\nabla\boldsymbol{u} + \boldsymbol{R}\dot{\boldsymbol{E}}), \tag{99}$$

where $\boldsymbol{v} = \ddot{\boldsymbol{\chi}}$.

Accordingly,

$$\int_\kappa (W_{\boldsymbol{E}} \cdot \ddot{\boldsymbol{E}} + W_{\boldsymbol{\kappa}} \cdot \ddot{\boldsymbol{\kappa}})dv = \int_\kappa [W_{\boldsymbol{E}} \cdot \boldsymbol{R}^t(\nabla\boldsymbol{v} - \boldsymbol{\beta}\boldsymbol{F}) + W_{\boldsymbol{\kappa}} \cdot \boldsymbol{R}^t\boldsymbol{b}']dv - \int_\kappa [W_{\boldsymbol{E}} \cdot \boldsymbol{R}^t\boldsymbol{\alpha}(\nabla\boldsymbol{u} + \boldsymbol{R}\dot{\boldsymbol{E}}) + W_{\boldsymbol{\kappa}} \cdot (\boldsymbol{R}^t\boldsymbol{\alpha}\boldsymbol{R})\dot{\boldsymbol{\kappa}}]dv. \tag{100}$$

At equilibrium we have

$$0 = \dot{E} = \int_\kappa [W_{\boldsymbol{E}} \cdot \boldsymbol{R}^t(\nabla\boldsymbol{u} - \boldsymbol{\alpha}\boldsymbol{F}) + W_{\boldsymbol{\kappa}} \cdot \boldsymbol{R}^t\boldsymbol{a}']dv \tag{101}$$

for all $\boldsymbol{u}$ and $\boldsymbol{a}$ such that

$$0 = \boldsymbol{D}_\alpha \cdot \dot{\boldsymbol{E}}\boldsymbol{D} = \boldsymbol{D}_\alpha \cdot \boldsymbol{R}^t(\nabla\boldsymbol{u} - \boldsymbol{\alpha}\boldsymbol{F})\boldsymbol{D} = \boldsymbol{d}_\alpha \cdot (\boldsymbol{u}' - \boldsymbol{a} \times \boldsymbol{\chi}'), \tag{102}$$

the latter following from differentiation of the constraint equation (56). This defines the class of variations that are admissible for the actual energy, as distinct from the extended energy used in Section 3. In the same way, admissible second variations satisfy

$$0 = \boldsymbol{D}_\alpha \cdot \ddot{\boldsymbol{E}}\boldsymbol{D} = \boldsymbol{D}_\alpha \cdot \boldsymbol{R}^t(\nabla \boldsymbol{v} - \boldsymbol{\beta}\boldsymbol{F})\boldsymbol{D} - \boldsymbol{D}_\alpha \cdot \boldsymbol{R}^t\boldsymbol{\alpha}(\nabla \boldsymbol{u} + \boldsymbol{R}\dot{\boldsymbol{E}})\boldsymbol{D}$$
$$= \boldsymbol{d}_\alpha \cdot (\boldsymbol{v}' - \boldsymbol{b} \times \boldsymbol{\chi}') - \boldsymbol{d}_\alpha \cdot \boldsymbol{a} \times (\boldsymbol{u}' + \boldsymbol{R}\dot{\boldsymbol{E}}\boldsymbol{D}). \tag{103}$$

Consider second variations satisfying

$$\boldsymbol{d}_\alpha \cdot (\boldsymbol{v}' - \boldsymbol{b} \times \boldsymbol{\chi}') = 0. \tag{104}$$

For these it follows from (101) and (102) that the first integral on the right-hand side of (100) vanishes, leaving

$$\ddot{E} = \int_\kappa \{\dot{\boldsymbol{E}} \cdot W_{\boldsymbol{EE}}[\dot{\boldsymbol{E}}] + (W_{\boldsymbol{E\kappa}})\dot{\boldsymbol{\kappa}} \cdot \dot{\boldsymbol{E}} + (W_{\boldsymbol{\kappa E}})\dot{\boldsymbol{E}} \cdot \dot{\boldsymbol{\kappa}} + \dot{\boldsymbol{\kappa}} \cdot (W_{\boldsymbol{\kappa\kappa}})\dot{\boldsymbol{\kappa}}\} dv$$
$$- \int_\kappa [W_{\boldsymbol{E}} \cdot \boldsymbol{R}^t\boldsymbol{\alpha}(\nabla \boldsymbol{u} + \boldsymbol{R}\dot{\boldsymbol{E}}) + W_{\boldsymbol{\kappa}} \cdot (\boldsymbol{R}^t\boldsymbol{\alpha}\boldsymbol{R})\dot{\boldsymbol{\kappa}}] dv, \tag{105}$$

subject to

$$\boldsymbol{d}_\alpha \cdot \boldsymbol{a} \times (\boldsymbol{u}' + \boldsymbol{R}\dot{\boldsymbol{E}}\boldsymbol{D}) = 0. \tag{106}$$

To solve the latter restriction we recall that

$$\boldsymbol{\chi}' = \boldsymbol{F}\boldsymbol{D} = \lambda \boldsymbol{d}, \tag{107}$$

where λ is the fiber stretch. The variational derivative yields

$$\boldsymbol{u}' = \dot{\lambda}\boldsymbol{d} + \boldsymbol{a} \times \boldsymbol{\chi}'. \tag{108}$$

We have

$$\dot{\boldsymbol{E}}\boldsymbol{D} = \boldsymbol{R}^t(\boldsymbol{u}' - \boldsymbol{a} \times \boldsymbol{\chi}'), \tag{109}$$

and so (106) reduces to

$$0 = \boldsymbol{d}_\alpha \cdot \boldsymbol{a} \times (2\dot{\lambda}\boldsymbol{d} + \boldsymbol{a} \times \boldsymbol{\chi}'). \tag{110}$$

This requires the existence of $c \in \mathbb{R}$ such that

$$2\dot{\lambda}\boldsymbol{a} \times \boldsymbol{d} + \lambda\boldsymbol{a} \times (\boldsymbol{a} \times \boldsymbol{d}) = c\boldsymbol{d}. \tag{111}$$

Taking the inner product with $\boldsymbol{a} \times \boldsymbol{d}$ yields $\dot{\lambda} = 0$, i.e.

$$\boldsymbol{u}' = \boldsymbol{a} \times \boldsymbol{\chi}', \tag{112}$$

and hence $\boldsymbol{a} \times (\boldsymbol{a} \times \boldsymbol{\chi}') = c\boldsymbol{d}$. Let $\boldsymbol{e} = \boldsymbol{a}/|\boldsymbol{a}|$. Then, using the identity $\boldsymbol{\chi}' = (\boldsymbol{e} \cdot \boldsymbol{\chi}')\boldsymbol{e} + \boldsymbol{e} \times (\boldsymbol{\chi}' \times \boldsymbol{e})$, we obtain

$$(c + \lambda |\boldsymbol{a}|^2)\boldsymbol{d} = \lambda |\boldsymbol{a}|^2 (\boldsymbol{e} \cdot \boldsymbol{d})\boldsymbol{e}. \tag{113}$$

The possibilities are: (i) $\boldsymbol{e} \cdot \boldsymbol{d} = 0$ and $c = -\lambda |\boldsymbol{a}|^2$, or (ii) $\boldsymbol{e} \cdot \boldsymbol{d} = \pm 1$ and $c = 0$. We conclude that (106) is satisfied at any $\boldsymbol{X} \in \kappa$ provided that $\boldsymbol{a} \perp \boldsymbol{d}$ or $\boldsymbol{a} \parallel \boldsymbol{d}$, i.e.

$$\boldsymbol{a} \in \text{Span}\{\boldsymbol{d}\} \quad \text{or} \quad \boldsymbol{a} \in \text{Span}\{\boldsymbol{d}_\alpha\}. \tag{114}$$

Let S measure arclength along the fiber passing through the point with position $\boldsymbol{X}$, and let Ω be an arbitrary simply-connected plane disc intersecting the fiber orthogonally. Then $\boldsymbol{X} = \boldsymbol{Z} + S\boldsymbol{D}$ with $\boldsymbol{Z} \in \Omega$. Consider the cylinder $\bar{\kappa} = \Omega \times \Delta$ with $\Delta = S_3 - S_1 (> 0)$, as in (42), and with $\operatorname{diam}\Omega$ and Δ small enough to ensure that $\bar{\kappa} \subset \kappa$ (Figure 2). Consider the variation $\boldsymbol{a}(\boldsymbol{X}) = f(S)\boldsymbol{g}(\boldsymbol{X})$ with $\boldsymbol{g}(\cdot)$ a smooth function compactly supported in $\bar{\kappa}$ and $f(\cdot)$ as in (40). This is continuous, and $f \to 0$ as $\Delta \to 0$. It follows from (97) that $|\dot{\boldsymbol{\kappa}}| \sim |\boldsymbol{a}'|$, where $\boldsymbol{a}' = f'\boldsymbol{g} + f\boldsymbol{g}' \to f'\boldsymbol{g}$, which remains finite as $\Delta \to 0$. From $(95)_1$ we also have that $|\boldsymbol{a}| \sim f$ and $\left|\dot{\boldsymbol{E}}\right| \sim \{|\nabla \boldsymbol{u}|, |\boldsymbol{a}|\} \sim \{f, \int^S f dx\}$, and conclude that $\boldsymbol{a}$ and $\dot{\boldsymbol{E}}$ vanish in the limit $\Delta \to 0$. We will show that

$$\int_{S_1}^{S_3} \dot{\boldsymbol{\kappa}} \cdot (W_{\boldsymbol{\kappa}\boldsymbol{\kappa}})\dot{\boldsymbol{\kappa}} dS = O(\Delta) \quad \text{as} \quad \Delta \to 0, \tag{115}$$

and hence that all other terms in the second variation $\ddot{E}$ are of order $o(\Delta)$.

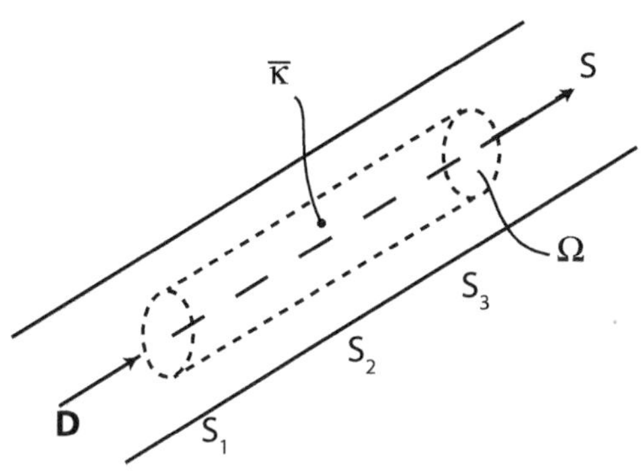

Figure 2. Cylindrical subdomain, on which the perturbation is supported.

Thus, assuming the equilibrium state in question to be an energy minimizer,

$$0 \leq \Delta^{-1}\ddot{E} = \int_{\Omega} \{\Delta^{-1} \int_{S_1}^{S_3} \dot{\boldsymbol{\kappa}} \cdot (W_{\boldsymbol{\kappa\kappa}})\dot{\boldsymbol{\kappa}} dS + \Delta^{-1} o(\Delta)\} da, \tag{116}$$

and passing to the limit yields

$$\int_{\Omega} \Delta^{-1} \int_{S_1}^{S_3} [\boldsymbol{a}' \cdot \boldsymbol{R}(W_{\boldsymbol{\kappa\kappa}})\boldsymbol{R}^t \boldsymbol{a}' dS] da \geq 0, \tag{117}$$

where we have used (97). Recalling that $\boldsymbol{a}' \rightarrow f'\boldsymbol{g}$ we have, from (40) and (41),

$$\begin{aligned} \int_{S_1}^{S_3} \boldsymbol{a}' \cdot \boldsymbol{R}(W_{\boldsymbol{\kappa\kappa}})\boldsymbol{R}^t \boldsymbol{a}' dS \ &\rightarrow \int_{S_1}^{S_2} \boldsymbol{g} \cdot \boldsymbol{R}(W_{\boldsymbol{\kappa\kappa}})\boldsymbol{R}^t \boldsymbol{g} dS \\ &+ \frac{\theta^2}{(1-\theta)^2} \int_{S_2}^{S_3} \boldsymbol{g} \cdot \boldsymbol{R}(W_{\boldsymbol{\kappa\kappa}})\boldsymbol{R}^t \boldsymbol{g} dS. \end{aligned} \tag{118}$$

Using (42) we then obtain

$$\begin{aligned} \Delta^{-1} \int_{S_1}^{S_3} \boldsymbol{a}' \cdot \boldsymbol{R}(W_{\boldsymbol{\kappa\kappa}})\boldsymbol{R}^t \boldsymbol{a}' dS \ &\rightarrow \theta \left[\boldsymbol{g} \cdot \boldsymbol{R}(W_{\boldsymbol{\kappa\kappa}})\boldsymbol{R}^t \boldsymbol{g}\right]_1 \\ &+ (1-\theta)\frac{\theta^2}{(1-\theta)^2} \left[\boldsymbol{g} \cdot \boldsymbol{R}(W_{\boldsymbol{\kappa\kappa}})\boldsymbol{R}^t \boldsymbol{g}\right]_2, \end{aligned} \tag{119}$$

where $[\cdot]_{1,2}$ are mean values in the intervals (S_1, S_2) and (S_2, S_3), respectively, and hence conclude that

$$\Delta^{-1} \int_{S_1}^{S_3} \boldsymbol{a}' \cdot \boldsymbol{R}(W_{\boldsymbol{\kappa\kappa}})\boldsymbol{R}^t \boldsymbol{a}' dS \rightarrow \frac{\theta}{1-\theta} \boldsymbol{g} \cdot \boldsymbol{R}(W_{\boldsymbol{\kappa\kappa}})\boldsymbol{R}^t \boldsymbol{g}, \tag{120}$$

which proves (115).

Finally, (117) furnishes

$$\frac{\theta}{1-\theta} \int_{\Omega} \boldsymbol{g} \cdot \boldsymbol{R}(W_{\boldsymbol{\kappa\kappa}})\boldsymbol{R}^t \boldsymbol{g} da \geq 0, \tag{121}$$

and the arbitrariness of Ω yields

$$\boldsymbol{g} \cdot \boldsymbol{R}(W_{\boldsymbol{\kappa\kappa}})\boldsymbol{R}^t \boldsymbol{g} \geq 0 \quad \text{at all} \quad \boldsymbol{X} \in \kappa, \tag{122}$$

where

$$\boldsymbol{R}(W_{\boldsymbol{\kappa\kappa}})\boldsymbol{R}^t = (\partial^2 W/\partial\kappa_i\partial\kappa_j)\boldsymbol{d}_i \otimes \boldsymbol{d}_j. \tag{123}$$

The constraints (114) require $\boldsymbol{g} \in \mathrm{Span}\{\boldsymbol{d}\}$ or $\boldsymbol{g} \in \mathrm{Span}\{\boldsymbol{d}_\alpha\}$. Accordingly, (122) implies that

$$\frac{\partial^2 W}{\partial\kappa_1^2} \geq 0 \quad \text{and} \quad \left(\frac{\partial^2 W}{\partial\kappa_\alpha\partial\kappa_\beta}\right) \quad \text{is positive definite.} \tag{124}$$

These are the Legendre–Hadamard necessary conditions for the present model.

Open problem. Establish the existence of energy minimizers, or solutions to the virtual-work statement, under reasonable hypotheses on W. We remark that the hypotheses of the existence theorem for Cosserat elasticity presented in Neff (2006) are not satisfied, although they subsume the Legendre–Hadamard conditions. Further, there are non-standard restrictions on admissible boundary data. For example, there is no solution to a problem in which non-trivial couples are assigned on a part of the boundary that lies parallel to the fibers (cf. (85)).

5 Material Symmetry and Transverse Isotropy

In this section we develop the theory of material symmetry for elastic Cosserat materials subject to the constraint (56). Our development borrows from that of Noll (1958) for conventional elasticity. We first describe the manner in which the constitutive function for the strain energy may be computed for any choice of reference when that pertaining to any particular choice is given.

5.1 Change of Reference Configuration

Let κ and μ be two references, and let $\boldsymbol{Y}(\boldsymbol{X})$ be the (invertible) map that takes points in κ to points in μ. The deformation gradients relative to κ and μ, denoted by $\boldsymbol{F}_\kappa$ and $\boldsymbol{F}_\mu$ respectively, are related by

$$\boldsymbol{F}_\kappa = \boldsymbol{F}_\mu \boldsymbol{H}, \quad \text{where} \quad \boldsymbol{H} = \nabla \boldsymbol{Y}. \tag{125}$$

We restrict attention to transformations $\boldsymbol{Y}$ with $\det \boldsymbol{H} = 1$, for reasons that are well known in conventional elasticity (Truesdell and Noll, 1965), and impose $\boldsymbol{Y}(\boldsymbol{X}_0) = \boldsymbol{X}_0$. The specification of such a *pivot* removes an inessential translational degree of freedom from the discussion of symmetry that follows.

We have seen that the presumed rigidity of the director triad leads to the existence of a rotation $\boldsymbol{R}$ such that (54) is satisfied; here we write $\boldsymbol{d}_i = \boldsymbol{R}_\kappa \boldsymbol{D}_i$. In the same way there is a rotation $\boldsymbol{R}_\mu$ such that $\boldsymbol{d}_i = \boldsymbol{R}_\mu \boldsymbol{G}_i$, where $\{\boldsymbol{G}_i(\boldsymbol{Y})\}$ is the positively-oriented orthonormal director field defined in μ. Thus,

$$\boldsymbol{R}_\kappa = \boldsymbol{R}_\mu \boldsymbol{L}, \tag{126}$$

where $\boldsymbol{L} = \boldsymbol{G}_i \otimes \boldsymbol{D}_i$ is the rotation field that maps the directors in κ to their images in μ. We have $\boldsymbol{d} = \boldsymbol{R}_\kappa \boldsymbol{D} = \boldsymbol{R}_\mu \boldsymbol{G}$, where $\boldsymbol{G}\,(= \boldsymbol{G}_1)$ is the unit-tangent field to fibers in μ, so that $\boldsymbol{G} = \boldsymbol{L}\boldsymbol{D}$. To ensure that $\boldsymbol{D}$ is a material vector (cf. (55)), it is thus necessary to impose

$$\boldsymbol{H}\boldsymbol{D} = |\boldsymbol{H}\boldsymbol{D}|\, \boldsymbol{L}\boldsymbol{D}. \tag{127}$$

Following Noll's characterization of solids (Truesdell and Noll, 1965) we assume the existence of an *undistorted* reference and suppose κ to be one of these. Thus we confine attention to proper-orthogonal $\boldsymbol{H}$. Further, we remove an inessential orientational degree of freedom in the local change of reference by requiring that it preserve the *pivotal axis* $\boldsymbol{D}$; thus,

$$\boldsymbol{D} = \boldsymbol{H}\boldsymbol{D} = \boldsymbol{L}\boldsymbol{D}, \tag{128}$$

in place of (127).

5.2 Application to the Present Model and Specialization to Transverse Isotropy

For the simple model discussed in Section 3, the strain energy depends on the rotation gradient via $\boldsymbol{\kappa} = \kappa_i \boldsymbol{D}_i$, where κ_i is given by (77) in which the prime refers to the fiber derivative in κ. In particular, for any function f we have $f' = (\nabla f)_\kappa \cdot \boldsymbol{D} = (\nabla f)_\mu \boldsymbol{H} \cdot \boldsymbol{D}$, where the subscripts κ and μ identify gradients with respect to $\boldsymbol{X} \in \kappa$ and $\boldsymbol{Y} \in \mu$, respectively. Thus, from (128), $f' = (\nabla f)_\mu \cdot \boldsymbol{H}^t \boldsymbol{D} = (\nabla f)_\mu \cdot \boldsymbol{D}$, implying that the fiber derivative is invariant under transformations of the reference configuration that preserve the fiber axis. Accordingly, the κ_i are also invariant. To see this we use (77) to derive

$$\begin{aligned}(\kappa_i)_\kappa &= \tfrac{1}{2} e_{ijk} \boldsymbol{D}_k \cdot \boldsymbol{R}^t_\kappa \boldsymbol{R}'_\kappa \boldsymbol{D}_j = \tfrac{1}{2} e_{ijk} \boldsymbol{L}^t \boldsymbol{G}_k \cdot \boldsymbol{L}^t \boldsymbol{R}^t_\mu \boldsymbol{R}'_\mu \boldsymbol{L}\boldsymbol{L}^t \boldsymbol{G}_j \\ &= \tfrac{1}{2} e_{ijk} \boldsymbol{G}_k \cdot \boldsymbol{R}^t_\mu \boldsymbol{R}'_\mu \boldsymbol{G}_j = (\kappa_i)_\mu,\end{aligned} \tag{129}$$

provided that $\boldsymbol{L}'$ vanishes. The curvature–twist vectors $\boldsymbol{\kappa}_\kappa$ and $\boldsymbol{\kappa}_\mu$ relative to the reference placements κ and μ are then related by

$$\boldsymbol{\kappa}_\kappa = \boldsymbol{L}^t \boldsymbol{\kappa}_\mu, \tag{130}$$

whereas in general

$$\boldsymbol{E}_\kappa = \boldsymbol{L}^t \boldsymbol{E}_\mu \boldsymbol{H}. \tag{131}$$

These hold whether or not the fibers are straight or untwisted in μ, i.e. whether or not $\boldsymbol{L}'$ vanishes, but then (77) must be adjusted to take the fiber derivatives of the $\boldsymbol{D}_i$ into account.

Because the choice of reference must not affect the state of the material, we require

$$W_\mu(\boldsymbol{E}_\mu, \boldsymbol{\kappa}_\mu) = W_\kappa(\boldsymbol{E}_\kappa, \boldsymbol{\kappa}_\kappa) = W_\kappa(\boldsymbol{L}^t \boldsymbol{E}_\mu \boldsymbol{H}, \boldsymbol{L}^t \boldsymbol{\kappa}_\mu). \tag{132}$$

If κ and μ are related by symmetry, then the associated strain-energy functions satisfy

$$W_\kappa(\boldsymbol{E}, \boldsymbol{\kappa}) = W_\mu(\boldsymbol{E}, \boldsymbol{\kappa}) \tag{133}$$

at the pivot point $\boldsymbol{X}_0$; combining with (130) and (131) yields the restriction

$$W_\kappa(\boldsymbol{E}, \boldsymbol{\kappa}) = W_\kappa(\boldsymbol{L}^t \boldsymbol{E} \boldsymbol{H}, \boldsymbol{L}^t \boldsymbol{\kappa}), \tag{134}$$

where the rotations $\boldsymbol{H}$ and $\boldsymbol{L}$ are connected by (128) but otherwise independent.

If the reinforced material is transversely isotropic, with the fibers perpendicular to the planes of isotropy, then (134) holds without further restrictions on $\boldsymbol{H}$ or $\boldsymbol{L}$; that is, for *all* rotations $\boldsymbol{H}, \boldsymbol{L} \in S$, where

$$S = \{\boldsymbol{Q} \in \mathrm{Orth}^+ \quad \text{with} \quad \boldsymbol{Q}\boldsymbol{D} = \boldsymbol{D}\}. \tag{135}$$

The micro-structural situation is depicted in Figure 3.

Remark. The study of material symmetry for Cosserat elasticity was taken up recently in Eremeyev and Pietraszkiewicz (2012). In that work it is assumed that the rotations $\boldsymbol{L}$ and $\boldsymbol{H}$ must coincide. This facilitates analysis using available representation theorems (Zheng, 1994). However, with reference to Figure 3 it may be concluded that $\boldsymbol{L}$ and $\boldsymbol{H}$ may be chosen independently, subject to the constraint that their axes coincide. For example, fixing $\boldsymbol{H} = \boldsymbol{I}$ in the matrix and choosing $\boldsymbol{L}$ in the fibers to be any rotation about $\boldsymbol{D}$ yields local configurations that are mechanically indistinguishable and hence related by material symmetry transformations.

Strain-energy functions of the type

$$W(\boldsymbol{E}, \boldsymbol{\kappa}) = W_1(\boldsymbol{E}) + W_2(\boldsymbol{E})(\boldsymbol{\kappa} \cdot \boldsymbol{D})^2 + W_3(\boldsymbol{E}) \left|\mathbf{1}\boldsymbol{\kappa}\right|^2, \tag{136}$$

with $\mathbf{1} = \boldsymbol{I} - \boldsymbol{D} \otimes \boldsymbol{D}$, furnish examples of (90). It is straightforward to verify that

$$\boldsymbol{\kappa} \cdot \boldsymbol{D} = \boldsymbol{L}^t \boldsymbol{\kappa} \cdot \boldsymbol{D} \quad \text{and} \quad |\mathbf{1}\boldsymbol{\kappa}| = \left|\mathbf{1}\boldsymbol{L}^t \boldsymbol{\kappa}\right| \tag{137}$$

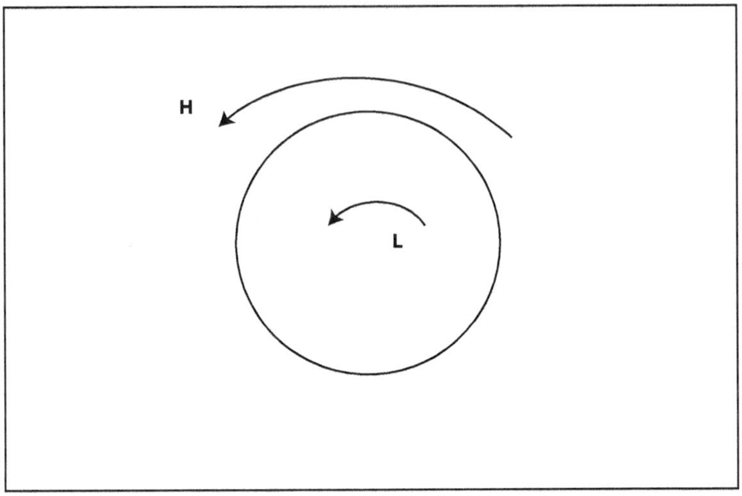

Figure 3. Symmetry transformations for an isotropic fiber embedded in an isotropic matrix.

for all $\boldsymbol{L} \in S$, and (134) is then satisfied for all $\boldsymbol{\kappa}$ provided that

$$W_i(\boldsymbol{E}) = W_i(\boldsymbol{L}^t \boldsymbol{E}\boldsymbol{H}), \quad i = 1, 2, 3. \tag{138}$$

To establish that this non-standard representation problem has a solution, we simply record a list of functionally independent scalars (Steigmann, 2012) that are easily shown to satisfy (138) individually, for all $\boldsymbol{L}, \boldsymbol{H} \in S$, namely

$$I = \{I_1, ..., I_9\}, \tag{139}$$

where

$$\begin{aligned} I_1 &= \operatorname{tr}(\boldsymbol{E}^t\boldsymbol{E}), \quad I_2 = \operatorname{tr}[(\boldsymbol{E}^t\boldsymbol{E})^2], \quad I_3 = \det \boldsymbol{E}, \\ I_4 &= \boldsymbol{D} \cdot \boldsymbol{E}\boldsymbol{D}, \quad I_5 = \boldsymbol{D} \cdot (\boldsymbol{E}^t\boldsymbol{E})\boldsymbol{D}, \quad I_6 = \boldsymbol{D} \cdot (\boldsymbol{E}\boldsymbol{E}^t)\boldsymbol{D}, \\ I_7 &= \boldsymbol{D} \cdot \boldsymbol{E}^*\boldsymbol{D}, \quad I_8 = \boldsymbol{D} \cdot (\boldsymbol{E}^t\boldsymbol{E})^2\boldsymbol{D}, \quad I_9 = \boldsymbol{D} \cdot (\boldsymbol{E}\boldsymbol{E}^t)^2\boldsymbol{D}, \end{aligned} \tag{140}$$

and $\boldsymbol{E}^* = (\det \boldsymbol{E})\boldsymbol{E}^{-t}$ is the cofactor of $\boldsymbol{E}$. Thus any function of the elements of I automatically satisfies (138), but of course we have not shown that I is a function basis for transverse isotropy.

We note in passing that $\det \boldsymbol{E} = \det \boldsymbol{F}$, $\boldsymbol{E}^t\boldsymbol{E} = \boldsymbol{C}$ and $\boldsymbol{E}\boldsymbol{E}^t = \boldsymbol{R}^t\boldsymbol{B}\boldsymbol{R}$, where $\boldsymbol{C} = \boldsymbol{F}^t\boldsymbol{F}$ and $\boldsymbol{B} = \boldsymbol{F}\boldsymbol{F}^t$ are the right and left Cauchy–Green deformation tensors respectively.

The response function $\boldsymbol{\sigma}$ derived from $(72)_3$ and (136) is given by

$$\boldsymbol{\sigma} = W_{\boldsymbol{E}} = (W_1)_{\boldsymbol{E}} + (\boldsymbol{\kappa}\cdot\boldsymbol{D})^2(W_2)_{\boldsymbol{E}} + |\mathbf{1}\boldsymbol{\kappa}|^2\,(W_3)_{\boldsymbol{E}}, \tag{141}$$

with

$$(W_i)_{\boldsymbol{E}} = \sum_j W_{ij}(I_j)_{\boldsymbol{E}}, \quad \text{where} \quad W_{ij} = \partial W_i/\partial I_j \tag{142}$$

and $(I_j)_{\boldsymbol{E}}$ are the gradients of the invariants with respect to $\boldsymbol{E}$. These are (Steigmann, 2012)

$$\begin{aligned}
(I_1)_{\boldsymbol{E}} &= 2\boldsymbol{E}, \quad (I_2)_{\boldsymbol{E}} = 4\boldsymbol{EC}, \quad (I_3)_{\boldsymbol{E}} = \boldsymbol{E}^*, \quad (I_4)_{\boldsymbol{E}} = \boldsymbol{D}\otimes\boldsymbol{D},\\
(I_5)_{\boldsymbol{E}} &= 2\boldsymbol{E}(\boldsymbol{D}\otimes\boldsymbol{D}), \quad (I_6)_{\boldsymbol{E}} = 2(\boldsymbol{D}\otimes\boldsymbol{D})\boldsymbol{E},\\
(I_7)_{\boldsymbol{E}} &= I_7\boldsymbol{E}^{-t} - I_3\boldsymbol{E}^{-t}(\boldsymbol{D}\otimes\boldsymbol{D})\boldsymbol{E}^{-t},\\
(I_8)_{\boldsymbol{E}} &= 2\boldsymbol{E}[(\boldsymbol{D}\otimes\boldsymbol{D})\boldsymbol{C} + \boldsymbol{C}(\boldsymbol{D}\otimes\boldsymbol{D})],\\
(I_9)_{\boldsymbol{E}} &= 2[(\boldsymbol{D}\otimes\boldsymbol{D})\boldsymbol{EC} + \boldsymbol{EE}^t(\boldsymbol{D}\otimes\boldsymbol{D})\boldsymbol{E}].
\end{aligned} \tag{143}$$

To obtain the response function $\boldsymbol{M}$ (cf. (81)) we require the gradients

$$(\boldsymbol{\kappa}\cdot\boldsymbol{D})_{\boldsymbol{\kappa}} = \boldsymbol{D} \quad \text{and} \quad (|\mathbf{1}\boldsymbol{\kappa}|^2)_{\boldsymbol{\kappa}} = 2\mathbf{1}\boldsymbol{\kappa}. \tag{144}$$

Equations (81) and (86) then deliver

$$\boldsymbol{M} = W_{\boldsymbol{\kappa}} = 2W_2(\boldsymbol{E})(\boldsymbol{\kappa}\cdot\boldsymbol{D})\boldsymbol{D} + 2W_3(\boldsymbol{E})\mathbf{1}\boldsymbol{\kappa} \tag{145}$$

and

$$\boldsymbol{m} = 2W_2(\boldsymbol{E})(\boldsymbol{\kappa}\cdot\boldsymbol{D})\boldsymbol{d} + 2W_3(\boldsymbol{E})\kappa_\alpha\boldsymbol{d}_\alpha, \tag{146}$$

where $\kappa_\alpha\boldsymbol{d}_\alpha = \boldsymbol{d}\times\boldsymbol{d}', \boldsymbol{d}' = (\nabla\boldsymbol{d})\boldsymbol{D}$. We impose $W_{2,3} > 0$ in accordance with (124), which in turn imply that the tensor $\boldsymbol{K}(\boldsymbol{E})$ of (90) is positive definite.

Open problem. Derive general representations for functions satisfying (134), or the simpler restriction (138), for independent rotations $\boldsymbol{L}$ and $\boldsymbol{H}$ having a common axis.

6 Examples

To illustrate the theory we consider simple examples in which

$$2W_1(\boldsymbol{E}) = \mu(I_1 - 3), \quad 2W_2(\boldsymbol{E}) = T \quad \text{and} \quad 2W_3(\boldsymbol{E}) = F, \tag{147}$$

where μ, T and F are positive constants. These deliver the response functions

$$\boldsymbol{\sigma} = \mu\boldsymbol{E} \quad \text{and} \quad \boldsymbol{m} = T\kappa\boldsymbol{d} + F\boldsymbol{d}\times\boldsymbol{d}', \quad \text{where} \quad \kappa = \boldsymbol{\kappa}\cdot\boldsymbol{D}. \tag{148}$$

This models neo-Hookean materials with shear modulus μ reinforced by fibers having torsional and flexural moduli T and F, respectively. We observe that in this case $\boldsymbol{\sigma}\boldsymbol{E}^t \in \mathrm{Sym}$, implying (cf. (87)) that the matrix transmits no distributed couples to the fibers. Evidently distributed couples may exist if the strain energy depends on the invariants I_4, I_6, I_7 or I_9. We emphasize the fact that the neo-Hookean energy is chosen merely for the sake of illustration and may yield predictions deviating significantly from relevant data on fiber-reinforced solids.

Using $\boldsymbol{R}\boldsymbol{\sigma} = \mu\boldsymbol{F}$ in the present circumstances, together with the rule $J\mathrm{div}\,\boldsymbol{A}_1 = \mathrm{Div}\,\boldsymbol{A}_2$, with $\boldsymbol{A}_2 = \boldsymbol{A}_1\boldsymbol{F}^*$ and $J = \det\boldsymbol{F}$, we find that the balance equations (87) and (88) reduce to

$$\boldsymbol{m}' + \boldsymbol{\chi}' \times \boldsymbol{\lambda} = \boldsymbol{0} \quad \text{and} \quad \boldsymbol{\lambda}' + \mu\,\mathrm{div}\,\boldsymbol{B} = \mathrm{grad}\,p, \tag{149}$$

respectively.

We consider deformations belonging to the *controllable* class introduced by Ericksen (1954). These are the deformations for which an equilibrium pressure field exists for *every* isotropic, uniform, incompressible *conventional* elastic solid in the absence of body force. They furnish solutions for neo-Hookean solids in particular; that is, they yield a pressure field satisfying $\mu\,\mathrm{div}\,\boldsymbol{B} = \mathrm{grad}\,p$, and thus provide solutions for fiber-reinforced neo-Hookean solids provided that

$$\boldsymbol{\lambda}' = \boldsymbol{0} \quad \text{and} \quad \boldsymbol{m}' + \boldsymbol{\chi}' \times \boldsymbol{\lambda} = \boldsymbol{0}. \tag{150}$$

These of course are the equations of equilibrium for an isolated fiber in the absence of distributed forces or couples (cf. (34)). Accordingly, in the examples studied there is no local mechanical interaction between the fibers and matrix.

6.1 Torsion

Torsion is of course a well known controllable deformation (Beatty, 1987). In the present context, the problem of finite torsion of a right circular cylinder was solved in Steigmann (2012). We reproduce the analysis here for the reader's convenience.

The reference placement κ of the body is the region defined by $0 \le r \le a$, $0 \le \theta < 2\pi$, $0 \le z \le L$ in a cylindrical polar coordinate system (r, θ, z). Position of a material point in this region is given by

$$\boldsymbol{X} = r\boldsymbol{e}_r(\theta) + z\boldsymbol{k}, \tag{151}$$

where $\boldsymbol{e}_r$ is the radial unit vector at azimuth θ, directed away from the cylinder axis, $\boldsymbol{k}$ is the fixed unit vector along the axis and $\boldsymbol{e}_\theta = \boldsymbol{k} \times \boldsymbol{e}_r$. We pursue a standard semi-inverse strategy and seek a deformation of the form

$$\chi(\boldsymbol{X}) = r\boldsymbol{e}_r(\phi) + z\boldsymbol{k}, \quad \text{where} \quad \phi = \theta + \tau z, \tag{152}$$

in which τ – the twist per unit length – is constant. The associated deformation gradient is (Ogden, 1997)

$$\boldsymbol{F} = \boldsymbol{Q}[\boldsymbol{I} + r\tau\boldsymbol{e}_\theta(\theta) \otimes \boldsymbol{k}], \tag{153}$$

where $\boldsymbol{Q} = \boldsymbol{e}_r(\phi) \otimes \boldsymbol{e}_r(\theta) + \boldsymbol{e}_\theta(\phi) \otimes \boldsymbol{e}_r(\theta) + \boldsymbol{k} \otimes \boldsymbol{k}$ is a rotation. This deformation is isochoric and hence kinematically admissible in an incompressible material. Accordingly, we suppose the incompressibility constraint to be in effect.

The fibers are assumed to be everywhere aligned with the axis of the cylinder in the reference placement; thus, $\boldsymbol{D} = \boldsymbol{k}$, the fiber derivative is $(\cdot)' = \partial(\cdot)/\partial z$, and the unit tangent $\boldsymbol{d}$ to a deformed fiber and the fiber stretch λ are given respectively by

$$\lambda\boldsymbol{d} = \boldsymbol{F}\boldsymbol{k} = \boldsymbol{k} + r\tau\boldsymbol{e}_\theta(\phi), \quad \lambda = \sqrt{1 + r^2\tau^2}. \tag{154}$$

This is sufficient to determine the action of the Cosserat rotation tensor on $\boldsymbol{k}$, i.e. $\boldsymbol{R}\boldsymbol{k} = \boldsymbol{d}$; we do not require the complete expression for $\boldsymbol{R}$. The trajectory of a fiber piercing a cross section at the point with coordinates (r, θ) is obtained by fixing the latter in the expression (154), yielding a circular helix of constant pitch. Such configurations are known to furnish equilibria for rods that are isolated in the sense that the distributed forces and moments exerted on them vanish identically (Chouaieb and Maddocks, 2004).

We have

$$\boldsymbol{B} = \boldsymbol{F}\boldsymbol{F}^t = \boldsymbol{I} + r\tau[\boldsymbol{e}_\theta(\phi) \otimes \boldsymbol{k} + \boldsymbol{k} \otimes \boldsymbol{e}_\theta(\phi)] + r^2\tau^2\boldsymbol{e}_\theta(\phi) \otimes \boldsymbol{e}_\theta(\phi), \tag{155}$$

and

$$\operatorname{div}\boldsymbol{B} = -r\tau^2\boldsymbol{e}_r(\phi). \tag{156}$$

With reference to (149) and (150), the constraint pressure p reduces to a function of r only with $dp/dr = -\mu r\tau^2$, yielding

$$p(r) = p_0 - \tfrac{1}{2}\mu\tau^2 r^2, \tag{157}$$

where p_0 is a constant.

If $(150)_2$ is scalar-multiplied by $\boldsymbol{d}$ we find, using $(148)_2$, that the fiber twist $\kappa \, (= \boldsymbol{k} \cdot \boldsymbol{\kappa})$ is such that $\kappa' = 0$, i.e. κ is independent of z. Using the expression (154) for $\boldsymbol{d}$, together with

$$\boldsymbol{d} \times \boldsymbol{d}' = \lambda^{-2} r\tau^2[r\tau\boldsymbol{k} - \boldsymbol{e}_\theta(\phi)], \tag{158}$$

we then derive

$$\boldsymbol{m}' = \lambda^{-1} r\tau^2(\lambda^{-1}F\tau - T\kappa)\boldsymbol{e}_r(\phi). \tag{159}$$

This is sufficient to determine the fiber force $\boldsymbol{\lambda}$. For, $\boldsymbol{\lambda}\cdot\boldsymbol{d}$ vanishes identically (cf. $(72)_{1,2}$), whereas $(150)_1$ and (159) yield $\boldsymbol{\lambda}\times\boldsymbol{d}$. We obtain

$$\boldsymbol{\lambda} = (\boldsymbol{\lambda}\cdot\boldsymbol{d})\boldsymbol{d} + \boldsymbol{d}\times(\boldsymbol{\lambda}\times\boldsymbol{d}) = -\lambda^{-3} r\tau^2(\lambda^{-1}F\tau - T\kappa)[r\tau\boldsymbol{k} - \boldsymbol{e}_\theta(\phi)]. \tag{160}$$

From $(150)_1$ we have $\boldsymbol{\lambda}' = \boldsymbol{0}$. Because $\boldsymbol{e}_\theta' = -\tau\boldsymbol{e}_r(\phi)$ is non-zero, this in turn requires that

$$\kappa = \lambda^{-1}(F/T)\tau, \tag{161}$$

yielding the fiber twist as a function of r (cf. $(154)_2$), which is maximized on the axis of the cylinder. With this result we find that $\boldsymbol{\lambda}$ and $\boldsymbol{m}'$ vanish separately, so that $(150)_2$ is identically satisfied. After some algebra we also find, from $(148)_2$, that

$$\boldsymbol{m} = F\tau\boldsymbol{k}, \tag{162}$$

implying that every fiber transmits the same moment. This result is interesting in light of the fact that the individual terms in $(148)_2$ associated with fiber twisting and bending are non-uniform.

To complete the solution we impose the traction condition (cf. $(74)_1$ with $\boldsymbol{D} = \boldsymbol{k}$)

$$(\boldsymbol{R\sigma} - p\boldsymbol{F}^*)\boldsymbol{e}_r(\theta) = \boldsymbol{0} \quad \text{at} \quad r = a. \tag{163}$$

This is equivalent to $(\boldsymbol{R\sigma})\boldsymbol{F}^t\boldsymbol{e}_r(\phi) = p\boldsymbol{e}_r(\phi)$ and thus, in the present circumstances, to

$$\mu\boldsymbol{B}\boldsymbol{e}_r(\phi) = p\boldsymbol{e}_r(\phi) \quad \text{at} \quad r = a, \tag{164}$$

furnishing $p(a) = \mu$ and hence $p(r) = \frac{1}{2}\mu\tau^2(a^2 - r^2) + \mu$, finally yielding

$$(\boldsymbol{R\sigma})\boldsymbol{F}^t - p\boldsymbol{I} = \mu[\tfrac{1}{2}\tau^2(r^2 - a^2) - 1]\boldsymbol{I} + \mu\boldsymbol{B}. \tag{165}$$

We observe that $\boldsymbol{R}$ and $\boldsymbol{\sigma}$ never occur separately in the equilibrium equations or boundary conditions. In fact, neither is determined by the analysis.

The overall response of the cylinder may be determined by computing the net force on a cross section and the net torque required to effect the torsion. These in turn require the traction

$$\boldsymbol{t} = [(\boldsymbol{R\sigma})\boldsymbol{F}^t - p\boldsymbol{I}]\boldsymbol{k} = \tfrac{1}{2}\mu\tau^2(r^2 - a^2)\boldsymbol{k} + \mu r\tau\boldsymbol{e}_\theta(\phi) \tag{166}$$

acting on a cross section. This is the same as the traction appearing in $(74)_1$ because there is no change in cross-sectional area in the course of the deformation. The resultant force is

$$\boldsymbol{f} = \int_0^{2\pi}\int_0^a \boldsymbol{t}\, r\,dr\,d\phi = f(\tau)\boldsymbol{k}, \tag{167}$$

where

$$f(\tau) = -\tfrac{1}{4}\pi a^4 \mu \tau^2, \tag{168}$$

and is a manifestation of the well-known normal-stress effect in nonlinear elasticity.

Finally, the torque is (cf. (89) with $\boldsymbol{D} = \boldsymbol{k} = \boldsymbol{n}$)

$$\boldsymbol{\rho} = \int_0^{2\pi} \int_0^a (\boldsymbol{\chi} \times \boldsymbol{t} + \boldsymbol{m}) r dr d\phi = \rho(\tau)\boldsymbol{k}, \tag{169}$$

where

$$\rho(\tau) = \pi a^2 \tau (F + \tfrac{1}{2}\mu a^2). \tag{170}$$

6.2 Flexure

To describe flexure of a rectangular block we use Cartesian coordinates in the reference placement and polars in the current placement (Ogden, 1997). Specifically,

$$\boldsymbol{X} = x\boldsymbol{i} + y\boldsymbol{j} + z\boldsymbol{k} \quad \text{and} \quad \boldsymbol{\chi}(\boldsymbol{X}) = r\boldsymbol{e}_r(\theta) + z\boldsymbol{k}, \tag{171}$$

where $r = f(x)$ and $\theta = g(y)$, for some functions f and g, to be determined. The deformation gradient is

$$\boldsymbol{F} = f'\boldsymbol{e}_r \otimes \boldsymbol{i} + fg'\boldsymbol{e}_\theta \otimes \boldsymbol{j} + \boldsymbol{k} \otimes \boldsymbol{k}, \tag{172}$$

yielding

$$1 = J = f(x)f'(x)g'(y) \tag{173}$$

in the case of incompressibility. This implies that $g' = C$, a constant. We assume g to be an odd function and conclude that

$$g = Cy \quad \text{and} \quad f = \sqrt{C^{-1}x + B}, \tag{174}$$

where B is constant and we take $C > 0$ without loss of generality. We consider two cases

Case (a): $\boldsymbol{D} = \boldsymbol{i}$. The fibers are initially horizontal and mapped by the deformation to rays through the origin. Indeed, from (55),

$$\lambda \boldsymbol{d} = \boldsymbol{F}\boldsymbol{i} = f'(x)\boldsymbol{e}_r, \tag{175}$$

yielding

$$\boldsymbol{d} = \boldsymbol{e}_r(\theta) \quad \text{and} \quad \lambda = f'(x). \tag{176}$$

Then, $\boldsymbol{d}' = \boldsymbol{d}_{,x} = \boldsymbol{e}_\theta \theta_{,x}$, and this vanishes because θ is a function of y alone. Accordingly, the fibers remain straight in the course of the deformation, and the constitutive equation $(148)_2$ furnishes

$$\boldsymbol{m} = T\kappa \boldsymbol{e}_r. \tag{177}$$

Equation $(150)_2$ implies that

$$0 = \boldsymbol{d} \cdot \boldsymbol{m}' = \boldsymbol{e}_r \cdot \boldsymbol{m}_{,x} = T\kappa_{,x} \tag{178}$$

and hence that $\kappa' (= \kappa_{,x})$ vanishes. This in turn implies that $\boldsymbol{m}'$ vanishes and $(150)_2$ reduces to $\boldsymbol{d} \times \boldsymbol{\lambda} = \boldsymbol{0}$. Because $\boldsymbol{d} \cdot \boldsymbol{\lambda} = 0$ we then have $\boldsymbol{\lambda} = \boldsymbol{0}$. Further, if there are no twisting couples at the vertical boundaries $x =$ constant, then κ and $\boldsymbol{m}$ vanish everywhere; the fibers are effectively inactive. Of course there exists an equilibrating pressure field because this deformation is controllable. We leave the remaining details to the interested reader.

Case (b): $\boldsymbol{D} = \boldsymbol{j}$. The fibers are initially vertical and hence mapped by the deformation to concentric circles. In place of (175) we have

$$\lambda \boldsymbol{d} = \boldsymbol{F}\boldsymbol{j} = fg' \boldsymbol{e}_\theta, \tag{179}$$

and hence

$$\boldsymbol{d} = \boldsymbol{e}_\theta \quad \text{and} \quad \lambda = fg' = C\sqrt{C^{-1}x + B}. \tag{180}$$

Then, $\boldsymbol{d}' = \boldsymbol{d}_{,y} = \boldsymbol{e}_\theta' g'(y) = -C\boldsymbol{e}_r$ and, from $(148)_2$,

$$\boldsymbol{m} = T\kappa \boldsymbol{e}_\theta + FC\boldsymbol{k}. \tag{181}$$

Equation $(150)_2$ yields

$$0 = \boldsymbol{d} \cdot \boldsymbol{m}' = \boldsymbol{e}_\theta \cdot \boldsymbol{m}_{,y} = T\kappa_{,y}, \tag{182}$$

implying that $\kappa' (= \kappa_{,y})$ vanishes. If no twisting couples are applied at the horizontal boundaries $y =$ constant, then the fiber twist vanishes everywhere and

$$\boldsymbol{m} = FC\boldsymbol{k}. \tag{183}$$

This yields $\boldsymbol{m}' = \boldsymbol{0}$ and hence $\boldsymbol{\lambda} = \boldsymbol{0}$, as before. The moment $\boldsymbol{m}$ combines with the overall bending moment generated by the matrix material (Ogden, 1997).

6.3 Bending, Stretching and Shearing of a Block (Beatty, 1987)

This is another controllable deformation, obtained by composing transverse shear with flexure. First we deform the block by flexure to the configuration defined by $\boldsymbol{x}_1 = \boldsymbol{\chi}_1(\boldsymbol{X})$, where $\boldsymbol{\chi}_1$ is given by $(171)_2$. Then the block is sheared to the configuration

$$\boldsymbol{x}_2 = \boldsymbol{\chi}_2(\boldsymbol{x}_1) = r\boldsymbol{e}_r(\theta) + \varsigma\boldsymbol{k}, \quad \text{where} \quad \varsigma = z + \beta\theta, \tag{184}$$

with β a positive constant. This maps a plane $z =$ constant to a helicoidal surface. We obtain $\boldsymbol{F} = \boldsymbol{F}_2\boldsymbol{F}_1$ with $\boldsymbol{F}_1$ given by (172) and

$$\boldsymbol{F}_2 = \boldsymbol{I} + \beta r^{-1}\boldsymbol{k} \otimes \boldsymbol{e}_\theta. \tag{185}$$

For the case $\boldsymbol{D} = \boldsymbol{j}$ we compute

$$\lambda\boldsymbol{d} = \boldsymbol{F}\boldsymbol{j} = C(r\boldsymbol{e}_\theta + \beta\boldsymbol{k}), \tag{186}$$

yielding

$$\lambda = C\sqrt{r^2 + \beta^2}, \tag{187}$$

a function of x alone. The fibers $r =$ constant are circular helices of constant pitch. From these results we find

$$\boldsymbol{d}' = \boldsymbol{d}_{,y} = \frac{-Cr}{\sqrt{r^2+\beta^2}}\boldsymbol{e}_r(\theta) \quad \text{and} \quad \boldsymbol{d} \times \boldsymbol{d}' = \frac{Cr^2}{r^2+\beta^2}(\boldsymbol{k} - \beta r^{-1}\boldsymbol{e}_\theta). \tag{188}$$

Using $(148)_2$ and $(150)_2$ we again obtain $\kappa' \; (= \kappa_{,y}) = 0$ and conclude that the fiber twist is a function of x alone. To determine it we substitute into $(150)_2$, obtaining

$$T\kappa\boldsymbol{d}' + F\boldsymbol{d} \times \boldsymbol{d}'' = \lambda\boldsymbol{\lambda} \times \boldsymbol{d}, \tag{189}$$

and hence

$$\boldsymbol{\lambda} \times \boldsymbol{d} = \lambda^{-1}\left(F\frac{C^2\beta r}{r^2+\beta^2} - T\frac{\kappa C r}{\sqrt{r^2+\beta^2}}\right)\boldsymbol{e}_r. \tag{190}$$

This yields $\boldsymbol{\lambda} = \boldsymbol{d} \times (\boldsymbol{\lambda} \times \boldsymbol{d})$ in terms of $\kappa(x)$.

A force-free solution ($\boldsymbol{\lambda} = \boldsymbol{0}$) is available, with fiber twist

$$\kappa(x) = \frac{F}{T}\frac{C\beta}{\sqrt{r^2+\beta^2}}, \quad \text{where} \quad r = f(x). \tag{191}$$

Then $\boldsymbol{m}$ is fully determined by the constitutive equation $(148)_2$, and includes both bending and twisting components. We have satisfied (150), and so $(149)_2$ yields the existence of an equilibrating pressure field by virtue of the controllability of the deformation.

Open problem. Develop analytical and numerical methods to treat general boundary-value problems.

Appendix: Variational Formulae

Consider the one-parameter families $\boldsymbol{F}(\boldsymbol{X};\epsilon)$ and $\boldsymbol{R}(\boldsymbol{X};\epsilon)$ of deformation and rotation fields, and let superposed dots stand for derivatives with respect to the parameter, evaluated at $\epsilon = 0$. Then,

$$\boldsymbol{\Omega} = \dot{\boldsymbol{R}}^t \boldsymbol{R} \tag{A.1}$$

is skew, and

$$\dot{\boldsymbol{E}} = \boldsymbol{\Omega}\boldsymbol{R}^t\boldsymbol{F} + \boldsymbol{R}^t\nabla\dot{\boldsymbol{\chi}}, \quad \dot{E}_{AB} = \Omega_{AC}E_{CB} + R_{iA}\dot{\chi}_{i,B}. \tag{A.2}$$

Further,

$$\dot{\boldsymbol{\Gamma}} = \dot{\Gamma}_{AB}\boldsymbol{E}_A \otimes \boldsymbol{E}_B, \quad \dot{\Gamma}_{AB} = \tfrac{1}{2}e_{CEA}(\dot{R}_{iC,B}R_{iD} + R_{iC,B}\dot{R}_{iD}), \tag{A.3}$$

in which

$$\dot{R}_{iA} = R_{iB}\Omega_{AB}. \tag{A.4}$$

The induced variation of the energy density is

$$\dot{W} = \sigma_{AB}\dot{E}_{AB} + \mu_{AB}\dot{\Gamma}_{AB}, \tag{A.5}$$

where

$$\sigma_{AB} = \partial W/\partial E_{AB} \quad \text{and} \quad \mu_{AB} = \partial W/\partial \Gamma_{AB}.$$

We seek an expression for this in terms of $\dot{\chi}_i$ and Ω_{AB}. To this end we write

$$\begin{aligned}\mu_{AB}\dot{\Gamma}_{AB} &= \tfrac{1}{2}e_{BAE}\mu_{EC}(R_{iB,C}\dot{R}_{iA} + R_{iA}\dot{R}_{iB,C})\\ &= \tfrac{1}{2}e_{BAE}[\mu_{EC}R_{iB,C}\dot{R}_{iA} - (\mu_{EC}R_{iA})_{,C}\dot{R}_{iB} + (\mu_{EC}R_{iA}\dot{R}_{iB})_{,C}]\\ &= \tfrac{1}{2}e_{BAE}[\mu_{EC}(R_{iB,C}\dot{R}_{iA} - R_{iA,C}\dot{R}_{iB})] - \tfrac{1}{2}e_{BAE}\mu_{EC,C}R_{iA}\dot{R}_{iB}\\ &+ \tfrac{1}{2}e_{BAE}(\mu_{EC}R_{iA}\dot{R}_{iB})_{,C}.\end{aligned} \tag{A.6}$$

Because the term in square brackets in the last equality is skew in the subscripts B, A, we may simplify the expression to

$$\begin{aligned}\mu_{AB}\dot{\Gamma}_{AB} = e_{BAE}\mu_{EC}R_{iB,C}R_{iD}\Omega_{AD} - \tfrac{1}{2}e_{BAE}\mu_{EC,C}\Omega_{BA}\\ +\tfrac{1}{2}e_{BAE}(\mu_{EC}\Omega_{BA})_{,C},\end{aligned} \tag{A.7}$$

where (A.4) has been used with $R_{iA}R_{iD} = \delta_{AD}$.

Let $\boldsymbol{\omega} = \text{ax}\boldsymbol{\Omega}$ be the axial vector of $\boldsymbol{\Omega}$. Then, $e_{BAE}\Omega_{BA} = -2\omega_E$, $\Omega_{AD} = e_{DAF}\omega_F$ and $e_{BAE}e_{DAF}\omega_F = \omega_E\delta_{BD} - \omega_B\delta_{ED}$, the last of these following from one of the $e-\delta$ identities. We substitute into (A.7) and use $R_{iB,C}R_{iB} = (R_{iB}R_{iB})_{,C} - R_{iB}R_{iB,C}$ with $R_{iB}R_{iB} = 3$, obtaining

$$\mu_{AB}\dot{\Gamma}_{AB} = \omega_E(\mu_{EC,C} - \mu_{BC}R_{iE,C}R_{iB}) - (\omega_E\mu_{EC})_{,C}. \tag{A.8}$$

Using the inverse of $(60)_2$ in the form

$$R_{iB}R_{iE,C} = e_{EBD}\Gamma_{DC}, \tag{A.9}$$

we finally arrive at

$$\mu_{AB}\dot{\Gamma}_{AB} = \omega_E(\mu_{EC,C} + e_{EDB}\mu_{BC}\Gamma_{DC}) - (\omega_E\mu_{EC})_{,C}. \tag{A.10}$$

To reduce the first expression in $(\text{A.5})_1$ we use (A.2) to obtain

$$\begin{aligned}\sigma_{AB}\dot{E}_{AB} &= R_{iA}\sigma_{AB}\dot{\chi}_{i,B} + e_{CAD}\sigma_{AB}E_{CB}\omega_D \\ &= (R_{iA}\sigma_{AB}\dot{\chi}_i)_{,B} - \dot{\chi}_i(R_{iA}\sigma_{AB})_{,B} + e_{CAD}\sigma_{AB}E_{CB}\omega_D.\end{aligned} \tag{A.11}$$

The variation of the expression in (70) involving the constraint is

$$(\Lambda_\alpha \boldsymbol{D}_\alpha \cdot \boldsymbol{E}\boldsymbol{D})^{\cdot} = \dot{\Lambda}_\alpha \boldsymbol{D}_\alpha \cdot \boldsymbol{E}\boldsymbol{D} + \boldsymbol{\Lambda} \cdot \dot{\boldsymbol{E}}\boldsymbol{D}, \tag{A.12}$$

where $\boldsymbol{\Lambda} = \Lambda_\alpha \boldsymbol{D}_\alpha$. Using (A.2) we reduce the second term to

$$\boldsymbol{\Lambda} \cdot \dot{\boldsymbol{E}}\boldsymbol{D} = \boldsymbol{\lambda} \otimes \boldsymbol{D} \cdot \nabla\dot{\boldsymbol{\chi}} + \text{ax}(\boldsymbol{\Lambda} \otimes \boldsymbol{E}\boldsymbol{D} - \boldsymbol{E}\boldsymbol{D} \otimes \boldsymbol{\Lambda}) \cdot \boldsymbol{\omega}, \tag{A.13}$$

where $\boldsymbol{\lambda} = \boldsymbol{R}\boldsymbol{\Lambda} = \boldsymbol{\Lambda}_\alpha \boldsymbol{d}_\alpha$. The variation of the term in (75) involving the constraint of incompressibility is

$$\overline{[p(\det \boldsymbol{E} - 1)]}^{\cdot} = \dot{p}(\det \boldsymbol{E} - 1) + p\boldsymbol{E}^* \cdot \dot{\boldsymbol{E}}, \tag{A.14}$$

where

$$\boldsymbol{E}^* \cdot \dot{\boldsymbol{E}} = \boldsymbol{E}^*\boldsymbol{E}^t \cdot \boldsymbol{\Omega} + \boldsymbol{R}\boldsymbol{E}^* \cdot \nabla\dot{\boldsymbol{\chi}} \tag{A.15}$$

in which the first term on the right-hand side vanishes identically and $\boldsymbol{R}\boldsymbol{E}^* = \boldsymbol{R}\boldsymbol{R}^t\boldsymbol{F}^* = \boldsymbol{F}^*$, yielding $\boldsymbol{E}^* \cdot \dot{\boldsymbol{E}} = \boldsymbol{F}^* \cdot \nabla\dot{\boldsymbol{\chi}}$.

With all constraints incorporated we then have

$$\begin{aligned}\dot{\tilde{W}} &= \boldsymbol{\omega} \cdot \{\text{Div}\,\boldsymbol{\mu} + 2\text{ax}[\text{Skw}(\boldsymbol{\sigma}\boldsymbol{E}^t + \boldsymbol{\mu}\boldsymbol{\Gamma}^t + \boldsymbol{\Lambda} \otimes \boldsymbol{E}\boldsymbol{D})]\} \\ &\quad - \dot{\boldsymbol{\chi}} \cdot \text{Div}(\boldsymbol{R}\boldsymbol{\sigma} - p\boldsymbol{F}^* + \boldsymbol{\lambda} \otimes \boldsymbol{D}) \\ &\quad + \text{Div}[(\boldsymbol{R}\boldsymbol{\sigma} - p\boldsymbol{F}^* + \boldsymbol{\lambda} \otimes \boldsymbol{D})^t\dot{\boldsymbol{\chi}}] - \text{Div}(\boldsymbol{\mu}^t\boldsymbol{\omega}) + \dot{\Lambda}_\alpha(\boldsymbol{D}_\alpha \cdot \boldsymbol{E}\boldsymbol{D}) \\ &\quad - \dot{p}(\det \boldsymbol{E} - 1).\end{aligned} \tag{A.16}$$

Acknowledgement.

I thank P. Rangamani for help with the manuscript and for designing the figures.

Bibliography

S. S. Antman. *Nonlinear Problems of Elasticity.* Springer, Berlin, 2005.

M. F. Beatty. Topics in finite elasticity: hyperelasticity of rubber, elastomers, and biological materials - with examples. *Appl. Mech. Rev.*, 40:1699–1734, 1987.

Y. Berrehili and J.-J. Marigo. The homogenized behavior of unidirectional fiber-reinforced composite materials in the case of debonded fibers. *Math. Mech. Complex Syst.*, 2:181–207, 2014.

N. Chouaieb and J. H. Maddocks. Kirchhoff's problem of helical equilibria of uniform rods. *J. Elasticity*, 77:221–247, 2004.

G. deBotton and I. Hariton. Out-of-plane shear deformation of a neo-Hookean fiber composite. *Phys. Lett. A*, 354: 156-160, 2006.

E. H. Dill. Kirchhoff's theory of rods. *Arch. Hist. Exact Scis.*, 44:1–23, 1992.

V. A. Eremeyev and W. Pietraszkiewicz. Material symmetry group of the non-linear polar-elastic continuum. *Int. J. Solids Structures*, 49:1993–2005, 2012.

J. L. Ericksen. Deformations possible in every isotropic, incompressible, perfectly elastic body. *Z. Angew. Math. Phys.*, 5:466–489, 1954.

W. T. Koiter. Couple-stresses in the theory of elasticity. *Nederl. Akad. Wetensch. Proc. B*, 67:17–44, 1964.

L. D. Landau and E. M. Lifshitz. *Theory of Elasticity*, 3rd edition. Pergamon Press, Oxford, 1986.

R. D. Mindlin and H. F. Tiersten. Effects of couple-stresses in linear elasticity. *Arch. Ration. Mech. Anal.*, 11:415–448, 1962.

P. Neff. Existence of minimizers for a finite-strain micromorphic elastic solid. *Proc. Roy. Soc. Edinburgh A*, 136:997–1012, 2006.

W. Noll. A mathematical theory of the mechanical behavior of continuous media. *Arch. Ration. Mech. Anal.*, 2:198–226,1958.

R. W. Ogden. *Non-linear Elastic Deformations.* Dover, New York, 1997.

W. Pietraszkiewicz and V. A. Eremeyev. On natural strain measures of the nonlinear micropolar continuum. *Int. J. Solids Structures*, 46:774–787, 2009.

C. Pideri and P. Seppecher. A second gradient material resulting from the homogenization of an heterogeneous linear elastic medium. *Contin. Mech. Thermodyn.*, 9:241–257, 1997.

A. C. Pipkin. Stress analysis for fiber-reinforced materials. *Adv. Appl. Mech.*, 19:1–51, 1979.

E. Reissner. A further note on finite-strain force and moment stress elasticity. *Z. Angew. Math. Phys.*, 38:665–673, 1987.

A. J. M. Spencer. *Deformations of Fibre-Reinforced Materials.* Oxford University Press, 1972.

A. J. M. Spencer and K. P. Soldatos. Finite deformations of fibre-reinforced elastic solids with fibre bending stiffness. *Int. J. Non-Linear Mech.*, 42:355–368, 2007.

D. J. Steigmann. The variational structure of a nonlinear theory for spatial lattices. *Meccanica*, 31:441–455, 1996.

D. J. Steigmann. Theory of elastic solids reinforced with fibers resistant to extension, flexure and twist. *Int. J. Non-Linear Mech.*, 47:734–742, 2012.

R. A. Toupin. Theories of elasticity with couple stress. *Arch. Ration. Mech. Anal.*, 17:85–112, 1964.

C. Truesdell and W. Noll. The Non-Linear Field Theories of Mechanics. In S. Flügge, editor, *Handbuch der Physik*, Vol. III/3. Springer, Berlin, 1965.

Q.-S. Zheng. Theory of representations for tensor functions—a unified invariant approach to constitutive equations. *Appl. Mech. Rev.*, 47:545–587, 1994.

Kolibri 360 GmbH
Ferdinand-Jühlke-Straße 7
99095 Erfurt, Deutschland
produktsicherheit@kolibri360.de